Springer Monographs in Mathematics

T0122418

For other titles published in this series, go to
www.springer.com/series/3733

Peter Butkovič

Max-linear Systems:
Theory and Algorithms

 Springer

Peter Butkovič
School of Mathematics
University of Birmingham
Birmingham, UK

ISSN 1439-7382
ISBN 978-1-4471-2583-9   ISBN 978-1-84996-299-5 (eBook)
DOI 10.1007/978-1-84996-299-5
Springer London Dordrecht Heidelberg New York

British Library Cataloguing in Publication Data
A catalogue record for this book is available from the British Library

Mathematics Subject Classification (2000): 15A80

Cover design: deblik

Printed on acid-free paper

Springer is part of Springer Science+Business Media (www.springer.com)

To Eva, Evička and Alenka

Preface

Max-algebra provides mathematical theory and techniques for solving nonlinear problems that can be given the form of linear problems, when arithmetical addition is replaced by the operation of maximum and arithmetical multiplication is replaced by addition. Problems of this kind are sometimes of a managerial nature, arising in areas such as manufacturing, transportation, allocation of resources and information processing technology.

The aim of this book is to present max-algebra as a modern modelling and solution tool. The first five chapters provide the fundamentals of max-algebra, focusing on one-sided max-linear systems, the eigenvalue-eigenvector problem and maxpolynomials. The theory is self-contained and covers both irreducible and reducible matrices. Advanced material is presented from Chap. 6 onwards.

The book is intended for a wide-ranging readership, from undergraduate and postgraduate students to researchers and mathematicians working in industry, commerce or management. No prior knowledge of max-algebra is assumed. We concentrate on linear-algebraic aspects, presenting both classical and new results. Most of the theory is illustrated by numerical examples and complemented by exercises at the end of every chapter.

Chapter 1 presents essential definitions, examples and basic results used throughout the book. It also introduces key max-algebraic tools: the maximum cycle mean, transitive closures, conjugation and the assignment problem, and presents their basic properties and corresponding algorithms. Section 1.3 introduces applications which were the main motivation for this book and towards which it is aimed: feasibility and reachability in multi-machine interactive processes. Many results in Chaps. 6–10 find their use in solving feasibility and reachability problems.

Chapter 2 has a specific aim: to explain two special features of max-algebra particularly useful for its applications. The first is the possibility of efficiently describing the set of *all* solutions to a problem which may otherwise be awkward or even impossible to do. This methodology may be used to find solutions satisfying further requirements. The second feature is the ability of max-algebra to describe a class of problems in combinatorics or combinatorial optimization in algebraic terms. This chapter may be skipped without loss of continuity whilst reading the book.

Most of Chap. 3 contains material on one-sided systems and the geometry of subspaces. It is presented here in full generality with all the proofs. The main results are: a straightforward way of solving one-sided systems of equations and inequalities both algebraically and combinatorially, characterization of bases of max-algebraic subspaces and a proof that finitely generated max-algebraic subspaces have an essentially unique basis. Linear independence is a rather tricky concept in max-algebra and presented dimensional anomalies illustrate the difficulties. Advanced material on linear independence can be found in Chap. 6.

Chapter 4 presents the max-algebraic eigenproblem. It contains probably the first book publication of the complete solution to this problem, that is, characterization and efficient methods for finding all eigenvalues and describing all eigenvectors for any square matrix over $\mathbb{R} \cup \{-\infty\}$ with all the necessary proofs.

The question of factorization of max-algebraic polynomials (briefly, maxpolynomials) is easier than in conventional linear algebra, and it is studied in Chap. 5. A related topic is that of characteristic maxpolynomials, which are linked to the job rotation problem. A classical proof is presented showing that similarly to conventional linear algebra the greatest corner is equal to the principal eigenvalue. The complexity of finding all coefficients of a characteristic maxpolynomial still seems to be an unresolved problem but a polynomial algorithm is presented for finding all essential coefficients.

Chapter 6 provides a unifying overview of the results published in various research papers on linear independence and simple image sets. It is proved that three types of regularity of matrices can be checked in $O(n^3)$ time. Two of them, strong regularity and Gondran–Minoux regularity, are substantially linked to the assignment problem. The chapter includes an application of Gondran–Minoux regularity to the minimal-dimensional realization problem for discrete-event dynamic systems.

Unlike in conventional linear algebra, two-sided max-linear systems are substantially harder to solve than their one-sided counterparts. An account of the existing methodology for solving two-sided systems (homogenous, nonhomogenous, or with separated variables) is given in Chap. 7. The core ideas are those of the Alternating Method and symmetrized semirings. This chapter is concluded by the proof of a result of fundamental theoretical importance, namely that the solution set to a two-sided system is finitely generated.

Following the complete resolution of the eigenproblem, Chap. 8 deals with the problem of reachability of eigenspaces by matrix orbits. First it is shown how matrix scaling can be useful in visualizing spectral properties of matrices. This is followed by presenting the classical theory of the periodic behavior of matrices in max-algebra and then it is shown how the reachability question for irreducible matrices can be answered in polynomial time. Matrices whose orbit from every starting vector reaches an eigenvector are called robust. An efficient characterization of robustness for both irreducible and reducible matrices is presented.

The generalized eigenproblem is a relatively new and hard area of research. Existing methodology is restricted to a few solvability conditions, a number of solvable special cases and an algorithm for narrowing the search for generalized eigenvalues. An account of these results can be found in Chap. 9. Almost all of Sect. 9.3 is original research never published before.

Chapter 10 presents theory and algorithms for solving max-linear programs subject to one or two-sided max-linear constraints (both minimization and maximization). The emphasis is on the two-sided case. We present criteria for the objective function to be bounded and we prove that the bounds are always attained, if they exist. Finally, bisection methods for localizing the optimal value with a given precision are presented. For programs with integer entries these methods turn out to be exact, of pseudopolynomial computational complexity.

The last chapter contains a brief summary of the book and a list of open problems.

In a text of this size, it would be impossible to give a fully comprehensive account of max-algebra. In particular this book does not cover (or does so only marginally) control, discrete-event systems, stochastic systems or case studies; material related to these topics may be found in e.g. [8, 102] and [112]. On the other hand, max-algebra as presented in this book provides the linear-algebraic background to the rapidly developing field of tropical mathematics.

This book is the result of many years of my work in max-algebra. Throughout the years I worked with many colleagues but I would like to highlight my collaboration with Ray Cuninghame-Green, with whom I was privileged to work for almost a quarter of a century and whose mathematical style and elegance I will always admire. Without Ray's encouragement, for which I am extremely grateful, this book would never exist. I am also indebted to Hans Schneider, with whom I worked in recent years, for his advice which played an important role in the preparation of this book. His vast knowledge of linear algebra made it possible to solve a number of problems in max-algebra.

I would like to express gratitude to my teachers, in particular to Ernest Jucovič for his vision and leadership, and to Karel Zimmermann, who in 1974 introduced me to max-algebra and to Miroslav Fiedler who introduced me to numerical linear algebra.

Sections 8.3–8.5 of this book have been prepared in collaboration with my research fellow Sergeĭ Sergeev, whose enthusiasm for max-algebra and achievement of several groundbreaking results in a short span of time make him one of the most promising researchers of his generation. His comments on various parts of the book have helped me to improve the presentation.

Numerical examples and exercises have been checked by my students Abdulhadi Aminu, Kin Po Tam and Vikram Dokka. I am of course taking full responsibility for any outstanding errors or omissions.

I wish to thank the Engineering and Physical Sciences Research Council for their support expressed by the award of three research grants without which many parts of this book would not exist.

I am grateful to my parents, to my wife Eva and daughters Evička and Alenka for their tremendous support and love, and for their patience and willingness to sacrifice many evenings and weekends when I was conducting my research.

Birmingham   Peter Butkovič

Contents

List of Symbols

Chapter 1
Introduction

In this chapter we introduce max-algebra, give the essential definitions and study the concepts that play a key role in max-algebra: the maximum cycle mean, transitive closures, conjugation and the assignment problem. In Sect. 1.3 we briefly introduce two types of problems that are of particular interest in this book: feasibility and reachability.

1.1 Notation, Definitions and Basic Properties

Throughout this book[1] we use the following notation:

$$\overline{\mathbb{R}} = \mathbb{R} \cup \{-\infty\},$$
$$\overline{\overline{\mathbb{R}}} = \overline{\mathbb{R}} \cup \{+\infty\},$$
$$\overline{\mathbb{Z}} = \mathbb{Z} \cup \{-\infty\},$$
$$a \oplus b = \max(a, b)$$

and

$$a \otimes b = a + b$$

for $a, b \in \overline{\overline{\mathbb{R}}}$. Note that by definition

$$(-\infty) + (+\infty) = -\infty = (+\infty) + (-\infty).$$

By max-algebra we understand the analogue of linear algebra developed for the pair of operations $(\oplus, \otimes)$, after extending these to matrices and vectors. This notation is of key importance since it enables us to formulate and in many cases also solve

---

[1]Except Sect. 1.4 and in the proof of Theorem 8.1.4.

P. Butkovič, *Max-linear Systems: Theory and Algorithms*,
Springer Monographs in Mathematics 151,
DOI 10.1007/978-1-84996-299-5_1, © Springer-Verlag London Limited 2010

certain nonlinear problems in a way similar to that in linear algebra. Note that we could alternatively define

$$a \oplus b = \min(a, b)$$

for $a, b \in \overline{\overline{\mathbb{R}}}$. The corresponding theory would then be called min-algebra or also "tropical algebra" [104, 141]. However, in this book, $\oplus$ will always denote the max operator.

Some authors use the expression "max-plus algebra", to highlight the difference from "max-times algebra" (see Sect. 1.4). We use the shorter version "max-algebra", since the structures are isomorphic and we can easily form the adjective "max-algebraic". Other names used in the past include "path algebra" [45] and "schedule algebra" [95].

Max-algebra has been studied in research papers and books from the early 1960's. Perhaps the first paper was that of R.A. Cuninghame-Green [57] in 1960, followed by [58, 60, 63, 65] and numerous other articles. Independently, a number of pioneering articles were published, e.g. by B. Giffler [95, 96], N.N. Vorobyov [144, 145], M. Gondran and M. Minoux [97–100], B.A. Carré [45], G.M. Engel and H. Schneider [80, 81, 129] and L. Elsner [77]. Intensive development of max-algebra has followed since 1985 in the works of M. Akian, R. Bapat, R.E. Burkard, G. Cohen, B. De Schutter, P. van den Driessche, S. Gaubert, M. Gavalec, R. Goverde, J. Gunawardena, B. Heidergott, M. Joswig, R. Katz, G. Litvinov, J.-J. Loiseau, W. McEneaney, G.-J. Olsder, J. Plávka, J.-P. Quadrat, I. Singer, S. Sergeev, E. Wagneur, K. Zimmermann, U. Zimmermann and many others. Note that idempotency of addition makes max-algebra part of idempotent mathematics [101, 108, 110].

Our aim is to develop a theory of max-algebra over $\overline{\mathbb{R}}$; $+\infty$ appears as a necessary element only when using certain techniques, such as dual operations and conjugation (see Sect. 1.6.3). We do not attempt to develop a concise max-algebraic theory over $\overline{\overline{\mathbb{R}}}$.

In max-algebra the pair of operations $(\oplus, \otimes)$ is extended to matrices and vectors similarly as in linear algebra. That is if $A = (a_{ij})$, $B = (b_{ij})$ and $C = (c_{ij})$ are matrices with elements from $\overline{\mathbb{R}}$ of compatible sizes, we write $C = A \oplus B$ if $c_{ij} = a_{ij} \oplus b_{ij}$ for all i, j, $C = A \otimes B$ if $c_{ij} = \sum_k^{\oplus} a_{ik} \otimes b_{kj} = \max_k(a_{ik} + b_{kj})$ for all i, j and $\alpha \otimes A = A \otimes \alpha = (\alpha \otimes a_{ij})$ for $\alpha \in \overline{\mathbb{R}}$. The symbol A^T stands for the *transpose* of the matrix A. The standard order $\leq$ of real numbers is extended to matrices (including vectors) componentwise, that is, if $A = (a_{ij})$ and $B = (b_{ij})$ are of the same size then $A \leq B$ means that $a_{ij} \leq b_{ij}$ for all i, j.

Throughout the book we denote $-\infty$ by ε and for convenience we also denote by the same symbol any vector or matrix whose every component is ε. If $a \in \overline{\overline{\mathbb{R}}}$ then the symbol a^{-1} stands for $-a$.

So $2 \oplus 3 = 3, 2 \otimes 3 = 5, 4^{-1} = -4$,

$$(5, 9) \otimes \begin{pmatrix} -3 \\ \varepsilon \end{pmatrix} = 2$$

and the system

$$\begin{pmatrix} 1 & -3 \\ 5 & 2 \end{pmatrix} \otimes \begin{pmatrix} x_1 \\ x_2 \end{pmatrix} = \begin{pmatrix} 3 \\ 7 \end{pmatrix}$$

in conventional notation reads

$$\max(1 + x_1, -3 + x_2) = 3,$$
$$\max(5 + x_1, 2 + x_2) = 7.$$

The possibility of working in a formally linear way is based on the fact that the following statements hold for $a, b, c \in \overline{\mathbb{R}}$ (their proofs are either trivial or straightforward from the definitions):

$$a \oplus b = b \oplus a$$
$$(a \oplus b) \oplus c = a \oplus (b \oplus c)$$
$$a \oplus \varepsilon = a = \varepsilon \oplus a$$
$$a \oplus b = a \text{ or } b$$
$$a \oplus b \geq a$$
$$a \oplus b = a \iff a \geq b$$
$$a \otimes b = b \otimes a$$
$$(a \otimes b) \otimes c = a \otimes (b \otimes c)$$
$$a \otimes 0 = a = 0 \otimes a$$
$$a \otimes \varepsilon = \varepsilon = \varepsilon \otimes a$$
$$a \otimes a^{-1} = 0 = a^{-1} \otimes a \quad \text{for } a \in \mathbb{R}$$
$$(a \oplus b) \otimes c = a \otimes c \oplus b \otimes c$$
$$a \geq b \implies a \oplus c \geq b \oplus c$$
$$a \geq b \implies a \otimes c \geq b \otimes c$$
$$a \otimes c \geq b \otimes c, \quad c \in \mathbb{R} \implies a \geq b.$$

Let us denote by I any square matrix, called the *unit matrix*, whose diagonal entries are 0 and off-diagonal ones are ε. For matrices (including vectors) A, B, C and I of compatible sizes over $\overline{\mathbb{R}}$ and $a \in \overline{\mathbb{R}}$ we have:

$$A \oplus B = B \oplus A$$
$$(A \oplus B) \oplus C = A \oplus (B \oplus C)$$
$$A \oplus \varepsilon = A = \varepsilon \oplus A$$
$$A \oplus B \geq A$$

$$A \oplus B = A \quad \Longleftrightarrow \quad A \geq B$$

$$(A \otimes B) \otimes C = A \otimes (B \otimes C)$$

$$A \otimes I = A = I \otimes A$$

$$A \otimes \varepsilon = \varepsilon = \varepsilon \otimes A$$

$$(A \oplus B) \otimes C = A \otimes C \oplus B \otimes C$$

$$A \otimes (B \oplus C) = A \otimes B \oplus A \otimes C$$

$$a \otimes (B \oplus C) = a \otimes B \oplus a \otimes C$$

$$a \otimes (B \otimes C) = B \otimes (a \otimes C).$$

It follows that $(\overline{\mathbb{R}}, \oplus, \otimes)$ is a commutative idempotent semiring and $(\overline{\mathbb{R}}^n, \oplus)$ is a semimodule (for definitions and further properties see [8, 146, 147]). Hence many of the tools known from linear algebra are available in max-algebra as well. The neutral elements are of course different: ε is neutral for $\oplus$ and 0 for $\otimes$. In the case of matrices the neutral elements are the matrix (of appropriate dimensions) with all entries ε (for $\oplus$) and I for $\otimes$.

On the other hand, in contrast to linear algebra, the operation $\oplus$ is not invertible. However, $\oplus$ is idempotent and this provides the possibility of constructing alternative tools, such as transitive closures of matrices or conjugation (see Sect. 1.6), for solving problems such as the eigenvalue-eigenvector problem and systems of linear equations or inequalities.

One of the most frequently used elementary property is *isotonicity* of both $\oplus$ and $\otimes$ which we formulate in the following lemma for ease of reference.

Lemma 1.1.1 *If A, B, C are matrices over $\overline{\overline{\mathbb{R}}}$ of compatible sizes and $c \in \overline{\overline{\mathbb{R}}}$ then*

$$A \geq B \quad \Longrightarrow \quad A \oplus C \geq B \oplus C,$$

$$A \geq B \quad \Longrightarrow \quad A \otimes C \geq B \otimes C,$$

$$A \geq B \quad \Longrightarrow \quad C \otimes A \geq C \otimes B,$$

$$A \geq B \quad \Longrightarrow \quad c \otimes A \geq c \otimes B.$$

Proof The first and last statements follow from the scalar versions immediately since max-algebraic addition and multiplication by scalars are defined component-wise. For the second implication assume $A \geq B$, then $A \oplus B = A$ and $(A \oplus B) \otimes C = A \otimes C$. Hence $A \otimes C \oplus B \otimes C = A \otimes C$, yielding finally $A \otimes C \geq B \otimes C$. The third implication is proved in a similar way. □

Corollary 1.1.2 *If $A, B \in \overline{\overline{\mathbb{R}}}^{m \times n}$ and $x, y \in \overline{\overline{\mathbb{R}}}^n$ then the following hold:*

$$A \geq B \quad \Longrightarrow \quad A \otimes x \geq B \otimes x,$$

$$x \geq y \quad \Longrightarrow \quad A \otimes x \geq A \otimes y.$$

Throughout the book, unless stated otherwise, we will assume that m and n are given integers, $m, n \geq 1$, and M and N will denote the sets $\{1, \ldots, m\}$ and $\{1, \ldots, n\}$, respectively.

An $n \times n$ matrix is called *diagonal*, notation $\text{diag}(d_1, \ldots, d_n)$, or just $\text{diag}(d)$, if its diagonal entries are $d_1, \ldots, d_n \in \mathbb{R}$ and off-diagonal entries are ε. Thus $I = \text{diag}(0, \ldots, 0)$. Any matrix which can be obtained from the unit (diagonal) matrix by permuting the rows and/or columns will be called a *permutation matrix(generalized permutation matrix)*. Obviously, for any generalized permutation matrix $A = (a_{ij}) \in \overline{\mathbb{R}}^{n \times n}$ there is a permutation π of the set N such that for all $i, j \in N$ we have:

$$a_{ij} \in \mathbb{R} \quad \Longleftrightarrow \quad j = \pi(i). \tag{1.1}$$

The position of generalized permutation matrices in max-algebra is slightly more special than in conventional linear algebra as they are the only matrices having an *inverse*:

Theorem 1.1.3 [60] *Let $A = (a_{ij}) \in \overline{\mathbb{R}}^{n \times n}$. Then a matrix $B = (b_{ij})$ such that*

$$A \otimes B = I = B \otimes A \tag{1.2}$$

exists if and only if A is a generalized permutation matrix.

Proof Suppose that A is a permutation matrix and π a permutation satisfying (1.1). Define $B = (b_{ij}) \in \overline{\mathbb{R}}^{n \times n}$ so that

$$b_{\pi(i),i} = (a_{i,\pi(i)})^{-1}$$

and

$$b_{ji} = \varepsilon \quad \text{if } j \neq \pi(i).$$

It is easily seen then that $A \otimes B = I = B \otimes A$.

Suppose now that (1.2) is satisfied, that is,

$$\sum_{k \in N}^{\oplus} a_{ik} \otimes b_{kj} = \sum_{k \in N}^{\oplus} b_{ik} \otimes a_{kj} = \begin{cases} 0 & \text{if } i = j, \\ \varepsilon & \text{if } i \neq j. \end{cases}$$

Hence for every $i \in N$ there is an $r \in N$ such that $a_{ir} \otimes b_{ri} = 0$, thus $a_{ir}, b_{ri} \in \mathbb{R}$. If there was an $a_{il} \in \mathbb{R}$ for an $l \neq r$ then $b_{ri} \otimes a_{il} \in \mathbb{R}$ which would imply

$$\sum_{k \in N}^{\oplus} b_{rk} \otimes a_{kl} > \varepsilon,$$

a contradiction. Therefore every row of A contains a unique finite entry. It is proved in a similar way that the same holds about every column of A. Hence A is a generalized permutation matrix. $\qquad\square$

Clearly, if an inverse matrix to A exists then it is unique and we may therefore denote it by A^{-1}. We will often need to work with the inverse of a diagonal matrix. If $X = \mathrm{diag}(x_1, \ldots, x_n)$, $x_1, \ldots, x_n \in \mathbb{R}$ then

$$X^{-1} = \mathrm{diag}(x_1^{-1}, \ldots, x_n^{-1}).$$

As usual a matrix A is called *blockdiagonal* if it consists of blocks and all off-diagonal blocks are ε.

If A is a square matrix then the iterated product $A \otimes A \otimes \cdots \otimes A$, in which the letter A stands k-times, will be denoted as A^k. By definition $A^0 = I$ for any square matrix A.

The symbol a^k applies similarly to scalars, thus a^k is simply ka and $a^0 = 0$. This definition immediately extends to $a^x = xa$ for any real x (but not for matrices).

The (i, j) entry of A^k will usually be denoted by $a_{ij}^{(k)}$ and should not be confused with a_{ij}^k, which is the kth power of a_{ij}. The symbol $a_{ij}^{[k]}$ will be used to denote the (i, j) entry of the kth matrix in a sequence $A^{[1]}, A^{[2]}, \ldots$.

Idempotency of $\oplus$ enables us to deduce the following formula, specific for max-algebra:

Lemma 1.1.4 *The following holds for every $A \in \overline{\mathbb{R}}^{n \times n}$ and nonnegative integer k:*

$$(I \oplus A)^k = I \oplus A \oplus A^2 \oplus \cdots \oplus A^k. \tag{1.3}$$

Proof By induction, straightforwardly from definitions.   □

We finish this section with some more terminology and notation used throughout the book, unless stated otherwise. As an analogue to "stochastic", $A = (a_{ij}) \in \overline{\mathbb{R}}^{m \times n}$ will be called *column* (*row*) $\mathbb{R}$-*astic* [60] if $\sum_{i \in M}^{\oplus} a_{ij} \in \mathbb{R}$ for every $j \in N$ (if $\sum_{j \in N}^{\oplus} a_{ij} \in \mathbb{R}$ for every $i \in M$), that is, when A has no ε column (no ε row). The matrix A will be called *doubly* $\mathbb{R}$-*astic* if it is both row and column $\mathbb{R}$-astic. Also, we will call A *finite* if none of its entries is $-\infty$. Similarly for vectors and scalars.

If

$$1 \leq i_1 < i_2 < \cdots < i_k \leq m,$$

$$1 \leq j_1 < j_2 < \cdots < j_l \leq n,$$

$$K = \{i_1, \ldots, i_k\}, \qquad L = \{j_1, \ldots, j_l\},$$

then $A[K, L]$ denotes the submatrix

$$\begin{pmatrix} a_{i_1 j_1} & \cdots & a_{i_1 j_l} \\ \cdots & \cdots & \cdots \\ a_{i_k j_1} & \cdots & a_{i_k j_l} \end{pmatrix}$$

of the matrix $A = (a_{ij}) \in \overline{\mathbb{R}}^{m \times n}$ and $x[L]$ denotes the subvector $(x_{j_1}, \ldots, x_{j_l})^T$ of the vector $x = (x_1, \ldots, x_n)^T$. If $K = L$ then, as usual, we say that $A[K, L]$ is a *principal submatrix* of A; $A[K, K]$ will be abbreviated to $A[K]$.

If X is a set then $|X|$ stands for the size of X. By convention, $\max \emptyset = \varepsilon$.

1.2 Examples

We present a few simple examples illustrating how a nonlinear formulation is converted to a linear one in max-algebra (we briefly say, "max-linear"). This indicates the key strength of max-algebra, namely converting a nonlinear problem into another one, which is linear with respect to the pair of operators $(\oplus, \otimes)$. These examples are introductory; more substantial applications of max-algebra are presented in Sect. 1.3 and in Chap. 2. The first two examples are related to the role of max-algebra as a "schedule algebra", see [95, 96].

Example 1.2.1 Suppose two trains leave two different stations but arrive at the same station from which a third train, connecting to the first two, departs. Let us denote the departure times of the trains as x_1 and x_2, respectively and the duration of the journeys of the first two trains (including the necessary times for changing the trains) by a_1 and a_2, respectively (Fig. 1.1). Let x_3 be the earliest departure time of the third train. Then

$$x_3 = \max(x_1 + a_1, x_2 + a_2)$$

which in the max-algebraic notation reads

$$x_3 = x_1 \otimes a_1 \oplus x_2 \otimes a_2.$$

Thus x_3 is a max-algebraic scalar product of the vectors (x_1, x_2) and (a_1, a_2). If the departure times of the first two trains is given, then the earliest possible departure time of the third train is calculated as a max-algebraic scalar product of two vectors.

Example 1.2.2 Consider two flights from airports A and B, arriving at a major airport C from which two other connecting flights depart. The major airport has many gates and transfer time between them is nontrivial. Departure times from C (and therefore also gate closing times) are given and cannot be changed: for the above mentioned flights they are b_1 and b_2. The transfer times between the two arrival and two departure gates are given in the matrix

$$A = \begin{pmatrix} a_{11} & a_{12} \\ a_{21} & a_{22} \end{pmatrix}.$$

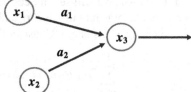

Fig. 1.1 Connecting train

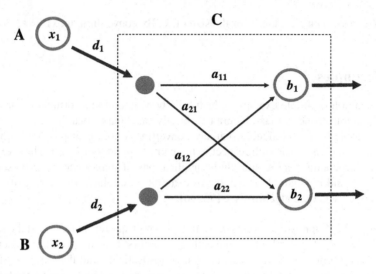

Fig. 1.2 Transfer between connecting flights

Durations of the flights from A to C and B to C are d_1 and d_2, respectively. The task is to determine the departure times x_1 and x_2 from A and B, respectively, so that all passengers arrive at the departure gates on time, but as close as possible to the closing times (Fig. 1.2).

We can express the gate closing times in terms of departure times from airports A and B:

$$b_1 = \max(x_1 + d_1 + a_{11}, x_2 + d_2 + a_{12})$$
$$b_2 = \max(x_1 + d_1 + a_{21}, x_2 + d_2 + a_{22}).$$

In max-algebraic notation this system gets a more formidable form, of a system of linear equations:

$$b = A \otimes x.$$

We will see in Sects. 3.1 and 3.2 how to solve such systems. For those that have no solution, Sect. 3.5 provides a simple max-algebraic technique for finding the "tightest" solution to $A \otimes x \leq b$.

Example 1.2.3 One of the most common operational tasks is to find the shortest distances between all pairs of places in a network for which a direct-distances matrix, say $A = (a_{ij})$, is known. We will see in Sect. 1.4 that there is no substantial difference between max-algebra and min-algebra and for continuity we will consider the task of finding the longest distances. Consider the matrix $A^2 = A \otimes A$: its elements are

$$\sum_{k \in N}^{\oplus} a_{ik} \otimes a_{kj} = \max_{k \in N}(a_{ik} + a_{kj}),$$

that is, the weights of longest $i - j$ paths of length 2 (if any) for all $i, j \in N$. Similarly the elements of A^k $(k = 1, 2, \ldots.)$ are the weights of longest paths of length k for all pairs of places. Therefore the matrix

$$A \oplus A^2 \oplus \cdots \tag{1.4}$$

represents the weights of longest paths of all lengths. In particular, its diagonal entries are the weights of longest cycles in the network. It is known that the longest-distances matrix exists if and only if there is no cycle of positive weight in the network (Lemma 1.5.4). Assuming this, and under the natural assumption $a_{ii} = 0$ for all $i \in N$, we will prove later in this chapter that the infinite series (1.4) converges and is equal to A^{n-1}, where n is the number of places in the network. Thus the longest- (and shortest-) distances matrix can max-algebraically be described simply as a power of the direct-distances matrix.

1.3 Feasibility and Reachability

Throughout the years (since the 1960's) max-algebra has found a considerable number of practical interpretations [8, 51, 60, 91]. Note that [102] is devoted to applications of max-algebra in the Dutch railway system.

One of the aims of this book is to study problems in max-algebra that are motivated by feasibility or reachability problems. In this section we briefly introduce these type of problems.

1.3.1 Multi-machine Interactive Production Process: A Managerial Application

The first model is of special significance as it is used as a basis for subsequent models. It is called the *multi-machine interactive production process* [58] (MMIPP) and is formulated as follows.

Products $P_1, \ldots, P_m$ are prepared using n machines (or processors), every machine contributing to the completion of each product by producing a partial product. It is assumed that every machine can work for all products simultaneously and that all these actions on a machine start as soon as the machine starts to work. Let a_{ij} be the duration of the work of the jth machine needed to complete the partial product for P_i $(i = 1, \ldots, m; j = 1, \ldots, n)$. If this interaction is not required for some i and j then a_{ij} is set to $-\infty$. Let us denote by x_j the starting time of the jth machine $(j = 1, \ldots, n)$. Then all partial products for P_i $(i = 1, \ldots, m)$ will be ready at time

$$\max(x_1 + a_{i1}, \ldots, x_n + a_{in}).$$

Hence if $b_1, \ldots, b_m$ are given completion times then the starting times have to satisfy the system of equations:

$$\max(x_1 + a_{i1}, \ldots, x_n + a_{in}) = b_i \quad \text{for all } i \in M.$$

Using max-algebra this system can be written in a compact form as a system of linear equations:

$$A \otimes x = b. \tag{1.5}$$

The matrix A is called the *production matrix*. The problem of solving (1.5) is a feasibility problem. A system of the form (1.5) is called a *one-sided system of max-linear equations* (or briefly a *one-sided max-linear system* or just a *max-linear system*). Such systems are studied in Chap. 3.

1.3.2 MMIPP: Synchronization and Optimization

Now suppose that independently, as part of a wider MMIPP, k other machines prepare partial products for products $Q_1, \ldots, Q_m$ and the duration and starting times are b_{ij} and y_j, respectively. Then the *synchronization problem* is to find starting times of all $n + k$ machines so that each pair (P_i, Q_i) $(i = 1, \ldots, m)$ is completed at the same time. This task is equivalent to solving the system of equations

$$\max(x_1 + a_{i1}, \ldots, x_n + a_{in}) = \max(y_1 + b_{i1}, \ldots, y_k + b_{ik}) \quad (i \in M). \tag{1.6}$$

It may also be given that P_i is not completed before a particular time c_i and similarly Q_i not before time d_i. Then the equations are

$$\max(x_1 + a_{i1}, \ldots, x_n + a_{in}, c_i) = \max(y_1 + b_{i1}, \ldots, y_k + b_{ik}, d_i) \quad (i \in M). \tag{1.7}$$

Again, using max-algebra and denoting $K = \{1, \ldots, k\}$ we can write this system as a system of linear equations:

$$\sum_{j \in N}^{\oplus} a_{ij} \otimes x_j \oplus c_i = \sum_{j \in K}^{\oplus} b_{ij} \otimes y_j \oplus d_i \quad (i \in M). \tag{1.8}$$

To distinguish such systems from those of the form (1.5), the system (1.7) (and also (1.8)) is called a *two-sided system of max-linear equations* (or briefly a *two-sided max-linear system*). Such systems are studied in Chap. 7. It is shown there that we may assume without loss of generality that (1.8) has the same variables on both sides, that is, in the matrix-vector notation it has the form

$$A \otimes x \oplus c = B \otimes x \oplus d.$$

This is another feasibility problem; Chap. 7 provides solution methods for this generalization.

Another variant of (1.6) is the task when $n = k$ and the starting times are linked, for instance it is required that there be a fixed interval between the starting times of the first and second system, that is, the starting times x_j, y_j of each pair of machines differ by the same value. If we denote this (unknown) value by λ then the equations read

$$\max(x_1 + a_{i1}, \ldots, x_n + a_{in}) = \max(\lambda + x_1 + b_{i1}, \ldots, \lambda + x_n + b_{in}) \qquad (1.9)$$

for $i = 1, \ldots, m$. In max-algebraic notation this system gets the form

$$\sum_{j \in N}^{\oplus} a_{ij} \otimes x_j = \lambda \otimes \sum_{j \in N}^{\oplus} b_{ij} \otimes x_j \quad (i \in M) \qquad (1.10)$$

which in a compact form is a "generalized eigenproblem":

$$A \otimes x = \lambda \otimes B \otimes x.$$

This is another feasibility problem and is studied in Chap. 9.

In applications it may be required that the starting times be optimized with respect to a given criterion. In Chap. 10 we consider the case when the objective function is *max-linear*, that is,

$$f(x) = f^T \otimes x = \max(f_1 + x_1, \ldots, f_n + x_n)$$

and $f(x)$ has to be either minimized or maximized. Thus the studied *max-linear programs (MLP)* are of the form

$$f^T \otimes x \longrightarrow \text{min or max}$$

subject to

$$A \otimes x \oplus c = B \otimes x \oplus d.$$

This is an example of a reachability problem.

1.3.3 Steady Regime and Its Reachability

Other reachability problems are obtained when the MMIPP is considered as a multi-stage rather than a one-off process.

Suppose that in the MMIPP the machines work in stages. In each stage all machines simultaneously produce components necessary for the next stage of some or all other machines. Let $x_i(r)$ denote the starting time of the rth stage on machine i $(i = 1, \ldots, n)$ and let a_{ij} denote the duration of the operation at which the jth machine prepares a component necessary for the ith machine in the $(r + 1)$st stage $(i, j = 1, \ldots, n)$. Then

$$x_i(r + 1) = \max(x_1(r) + a_{i1}, \ldots, x_n(r) + a_{in}) \quad (i = 1, \ldots, n; r = 0, 1, \ldots)$$

or, in max-algebraic notation

$$x(r + 1) = A \otimes x(r) \quad (r = 0, 1, \ldots)$$

where $A = (a_{ij})$ is, as before, the *production matrix*. We say that the system reaches a *steady regime* [58] if it eventually moves forward in regular steps, that is, if for some λ and r_0 we have $x(r + 1) = \lambda \otimes x(r)$ for all $r \geq r_0$. This implies $A \otimes x(r) = \lambda \otimes x(r)$ for all $r \geq r_0$. Therefore a steady regime is reached if and only if for some λ and r, $x(r)$ is a solution to

$$A \otimes x = \lambda \otimes x.$$

Systems of this form describe the max-algebraic eigenvalue-eigenvector problem and can be considered as two-sided max-linear systems with a parameter. Obviously, a steady regime is reached immediately if $x(0)$ is a (max-algebraic) eigenvector of A corresponding to a (max-algebraic) eigenvalue λ (these concepts are defined and studied in Chap. 4). However, if the choice of a start-time vector is restricted, we may need to find out for which vectors a steady regime will be reached. The set of such vectors will be called the attraction space. The problem of finding the attraction space for a given matrix is a reachability problem (see Sects. 8.4 and 8.5).

Another reachability problem is to characterize production matrices for which a steady regime is reached with any start-time vector, that is, the attraction space is the whole space (except ε). In accordance with the terminology in control theory such matrices are called *robust* and it is the primary objective of Sect. 8.6 to provide a characterization of such matrices.

Note that a different type of reachability has been studied in [88].

1.4 About the Ground Set

The semiring $(\overline{\mathbb{R}}, \oplus, \otimes)$ could be introduced in more general terms as follows: Let $\mathcal{G}$ be a linearly ordered commutative group (LOCG). Let us denote the group operation by $\otimes$ and the linear order by $\leq$. Thus $\mathcal{G} = (G, \otimes, \leq)$, where G is a set. We can then denote $\overline{G} = G \cup \{\varepsilon\}$, where ε is an adjoined element such that $\varepsilon < a$ for all $a \in G$, and define $a \oplus b = \max(a, b)$ for $a, b \in \overline{G}$ and extend $\otimes$ to $\overline{G}$ by setting $a \otimes \varepsilon = \varepsilon = \varepsilon \otimes a$. It is easily seen that $(\overline{G}, \oplus, \otimes)$ is an idempotent commutative semiring (see p. 3). Max-algebra as defined in Sect. 1.1 corresponds to the case when $\mathcal{G}$ is the additive group of reals, that is, $\mathcal{G} = (\mathbb{R}, +, \leq)$ where $\leq$ is the natural ordering of real numbers. This LOCG will be denoted by $\mathcal{G}_0$ and called the *principal interpretation* [60].

Let us list a few other linearly ordered commutative groups which will be useful later in the book (here $\mathbb{R}^+$ ($\mathbb{Q}^+, \mathbb{Z}^+$) are the sets of positive reals (rationals, integers), $\mathbb{Z}_2$ is the set of even integers):

$$\mathcal{G}_1 = (\mathbb{R}, +, \geq),$$

$$\mathcal{G}_2 = (\mathbb{R}^+, \cdot, \leq),$$

$$\mathcal{G}_3 = (\mathbb{Z}, +, \leq),$$

$$\mathcal{G}_4 = (\mathbb{Z}_2, +, \leq),$$

$$\mathcal{G}_5 = (\mathbb{Q}^+, \cdot, \leq),$$

$$\mathcal{G}_6 = (\mathbb{Z}^+, +, \geq).$$

Obviously both $\mathcal{G}_1$ and $\mathcal{G}_2$ are isomorphic with $\mathcal{G}_0$ (the isomorphism in the first case is $f(x) = -x$, in the second case it is $f(x) = \log(x)$). This book presents results for max-algebra over the principal interpretation but due to the isomorphism these results usually immediately extend to max-algebra over $\mathcal{G}_1$ and $\mathcal{G}_2$. A rare exception is strict visualization (Theorem 8.1.4), where the proof has to be done in $\mathcal{G}_2$ and then transformed to $\mathcal{G}_0$. Many (but not all) of the results in this book are applicable to general LOCG. In a few cases we will present results for groups other than $\mathcal{G}_0$, $\mathcal{G}_1$ and $\mathcal{G}_2$. The theory corresponding to $\mathcal{G}_1$ is usually called *min-algebra*, or *tropical algebra*.

A linearly ordered group $\mathcal{G} = (G, \otimes, \leq)$ is called *dense* if for any $a, b \in G, a < b$, there is a $c \in G$ satisfying $a < c < b$; it is called *sparse* if it is not dense. A group $(G, \otimes)$ is called *radicable* if for any $a \in G$ and positive integer k there is a $b \in G$ satisfying $b^k = a$. Observe that in a radicable group

$$a < \sqrt{a \otimes b} < b$$

if $a < b$ and so every radicable group is dense.

Thus $\mathcal{G}_0, \mathcal{G}_1, \mathcal{G}_2$ and $\mathcal{G}_5$ are dense, $\mathcal{G}_3, \mathcal{G}_4, \mathcal{G}_6$ are sparse, $\mathcal{G}_0, \mathcal{G}_1$ and $\mathcal{G}_2$ are radicable and $\mathcal{G}_3, \mathcal{G}_4, \mathcal{G}_5, \mathcal{G}_6$ are not. It will turn out in Sect. 6.2 that the density of groups is important for strong regularity of matrices and in Sect. 4.2 that radicability is crucial for the existence of eigenvalues.

1.5 Digraphs and Matrices

We will often use the language of directed graphs (briefly digraphs). A *digraph* is an ordered pair $D = (V, E)$ where V is a nonempty finite set (of *nodes*) and $E \subseteq V \times V$ (the set of *arcs*). A *subdigraph* of D is any digraph $D' = (V', E')$ such that $V' \subseteq V$ and $E' \subseteq E$. If $e = (u, v) \in E$ for some $u, v \in V$ then we say that e is *leaving* u and *entering* v. Any arc of the form (u, u) is called a *loop*.

Let $D = (V, E)$ be a given digraph. A sequence $\pi = (v_1, \ldots, v_p)$ of nodes in D is called a *path* (in D) if $p = 1$ or $p > 1$ and $(v_i, v_{i+1}) \in E$ for all $i = 1, \ldots, p - 1$. The node v_1 is called the *starting node* and v_p the *endnode* of π, respectively. The number $p - 1$ is called the *length* of π and will be denoted by $l(\pi)$. If u is the starting node and v is the endnode of π then we say that π is a $u - v$ path. If there is a $u - v$ path in D then v is said to be *reachable* from u, notation $u \to v$. Thus $u \to u$ for any $u \in V$. If π is a $u - v$ path and π' is a $v - w$ path in D, then $\pi \circ \pi'$ stands for the concatenation of these two paths.

A path $(v_1, \ldots, v_p)$ is called a *cycle* if $v_1 = v_p$ and $p > 1$ and it is called an *elementary cycle* if, moreover, $v_i \neq v_j$ for $i, j = 1, \ldots, p - 1, i \neq j$. If there is no cycle in D then D is called *acyclic*. Note that the word "cycle" will also be used to refer to cyclic permutations, see Sect. 1.6.4, as no confusion should arise from the use of the same word in completely different circumstances.

A digraph D is called *strongly connected* if $u \to v$ for all nodes u, v in D. A subdigraph D' of D is called a *strongly connected component* of D if it is a maximal strongly connected subdigraph of D, that is, D' is a strongly connected subdigraph of D and if D' is a subdigraph of a strongly connected subdigraph D'' of D then $D' = D''$. All strongly connected components of a given digraph $D = (V, E)$ can be identified in $O(|V| + |E|)$ time [142]. Note that a digraph consisting of one node and no arc is strongly connected and acyclic; however, if a strongly connected digraph has at least two nodes then it obviously cannot be acyclic. Because of this singularity we will have to assume in some statements that $|V| > 1$.

If $A = (a_{ij}) \in \overline{\mathbb{R}}^{n \times n}$ then the symbol F_A (Z_A) will denote the digraph with the node set N and arc set $E = \{(i, j); a_{ij} > \varepsilon\}$ $(E = \{(i, j); a_{ij} = 0\})$. Z_A will be called the *zero digraph* of the matrix A. If F_A is strongly connected then A is called *irreducible* and *reducible* otherwise.

Lemma 1.5.1 *If $A \in \overline{\mathbb{R}}^{n \times n}$ is irreducible and $n > 1$ then A is doubly $\mathbb{R}$-astic.*

Proof It follows from irreducibility that an arc leaving and an arc entering a node exist for every node in F_A. Hence every row and column of A has a finite entry. $\square$

Note that a matrix may be reducible even if it is doubly $\mathbb{R}$-astic (e.g. I).

Lemma 1.5.2 *If $A \in \overline{\mathbb{R}}^{n \times n}$ is column $\mathbb{R}$-astic and $x \neq \varepsilon$ then $A^k \otimes x \neq \varepsilon$ for every nonnegative integer k. Hence if $A \in \overline{\mathbb{R}}^{n \times n}$ is column $\mathbb{R}$-astic then A^k is column $\mathbb{R}$-astic for every such k. This is true in particular when A is irreducible and $n > 1$.*

Proof If $x_j \neq \varepsilon$ and $a_{ij} \neq \varepsilon$ then the ith component of $A \otimes x$ is finite and the first statement follows by repeating this argument; the second one by setting x to be any column of A. The third one follows from Lemma 1.5.1. $\square$

Lemma 1.5.3 *If $A \in \overline{\mathbb{R}}^{n \times n}$ is row or column $\mathbb{R}$-astic then F_A contains a cycle.*

Proof Without loss of generality suppose that $A = (a_{ij})$ is row $\mathbb{R}$-astic and let $i_1 \in N$ be any node. Then $a_{i_1 i_2} > \varepsilon$ for some $i_2 \in N$. Similarly $a_{i_2 i_3} > \varepsilon$ for some $i_3 \in N$ and so on. Hence F_A has arcs $(i_1, i_2), (i_2, i_3), \ldots$. By finiteness of N in the sequence $i_1, i_2, \ldots$, some i_r will eventually recur; this proves the existence of a cycle in F_A. $\square$

A *weighted digraph* is $D = (V, E, w)$ where (V, E) is a digraph and w is a real function on E. All definitions for digraphs are naturally extended to weighted digraphs. If $\pi = (v_1, \ldots, v_p)$ is a path in (V, E, w) then the *weight* of π is

$w(\pi) = w(v_1, v_2) + w(v_2, v_3) + \cdots + w(v_{p-1}, v_p)$ if $p > 1$ and ε if $p = 1$. A path π is called *positive* if $w(\pi) > 0$. In contrast, a cycle $\sigma = (u_1, \ldots, u_p)$ is called a *zero cycle* if $w(u_k, u_{k+1}) = 0$ for all $k = 1, \ldots, p - 1$. Since w stands for "weight" rather than "length", from now on we will use the word "heaviest path/cycle" instead of "longest path/cycle".

The following is a basic combinatorial optimization property.

Lemma 1.5.4 *If $D = (V, E, w)$ is a weighted digraph with no positive cycles then for every $u, v \in V$ a heaviest $u - v$ path exists if at least one $u - v$ path exists. In this case at least one heaviest $u - v$ path has length $|V|$ or less.*

Proof If π is a $u - v$ path of length greater than $|V|$ then it contains a cycle as a subpath. By successive deletions of all such subpaths (necessarily of nonpositive weight) we obtain a $u - v$ path π' of length not exceeding $|V|$ such that $w(\pi') \geq w(\pi)$. A heaviest $u - v$ path of length $|V|$ or less exists since the set of such paths is finite, and the statement follows.   □

Given $A = (a_{ij}) \in \overline{\mathbb{R}}^{n \times n}$ the symbol D_A will denote the weighted digraph (N, E, w) where $F_A = (N, E)$ and $w(i, j) = a_{ij}$ for all $(i, j) \in E$. If $\pi = (i_1, \ldots, i_p)$ is a path in D_A then we denote $w(\pi, A) = w(\pi)$ and it now follows from the definitions that $w(\pi, A) = a_{i_1 i_2} + a_{i_2 i_3} + \cdots + a_{i_{p-1} i_p}$ if $p > 1$ and ε if $p = 1$.

If $D = (N, E, w)$ is an arc-weighted digraph with the weight function $w : E \to \mathbb{R}$ then A_D will denote the matrix $(a_{ij}) \in \overline{\mathbb{R}}^{n \times n}$ defined by

$$a_{ij} = \begin{cases} w(i, j), & \text{if } (i, j) \in E, \\ \varepsilon, & \text{else,} \end{cases} \qquad \text{for all } i, j \in N.$$

A_D will be called the *direct-distances matrix* of the digraph D.

If $D = (N, E)$ is a digraph and $K \subseteq N$ then $D[K]$ denotes the *induced subdigraph* of D, that is

$$D[K] = (K, E \cap (K \times K)).$$

It follows from the definitions that $D_{A[K]} = D[K]$.

Various types of transformations between matrices will be used in this book. We say that matrices A and B are

- *equivalent* (notation $A \equiv B$) if $B = P^{-1} \otimes A \otimes P$ for some permutation matrix P, that is, B can be obtained by a simultaneous permutation of the rows and columns of A;
- *directly similar* (notation $A \sim B$) if $B = C \otimes A \otimes D$ for some diagonal matrices C and D, that is, B can be obtained by adding finite constants to the rows and/or columns of A;
- *similar* (notation $A \approx B$) if $B = P \otimes A \otimes Q$ for some generalized permutation matrices P and Q, that is, B can be obtained by permuting the rows and/or columns and by adding finite constants to the rows and columns of A.

We also say that B is obtained from A by *diagonal similarity scaling* (briefly, *matrix scaling*, or just *scaling*) if

$$B = X^{-1} \otimes A \otimes X$$

for some diagonal matrix X. Clearly all these four relations are relations of equivalence.

Observe that A and B are similar if they are either directly similar or equivalent. Scaling is a special case of direct similarity.

If $A \sim B$ then $F_A = F_B$; if $A \approx B$ then F_A can be obtained from F_B by a renumbering of the nodes and finally, if $A \equiv B$ then D_A can be obtained from D_B by a renumbering of the nodes.

Matrix scaling preserves crucial spectral properties of matrices and we conclude this section by a simple but important statement that is behind this fact (more properties of this type can be found in Lemma 8.1.1):

Lemma 1.5.5 *Let* $A = (a_{ij})$, $B = (b_{ij}) \in \overline{\mathbb{R}}^{n \times n}$ *and* $B = X^{-1} \otimes A \otimes X$ *where* $X = \mathrm{diag}(x_1, \ldots, x_n)$, $x_1, \ldots, x_n \in \mathbb{R}$. *Then* $w(\sigma, A) = w(\sigma, B)$ *for every cycle* σ *in* $F_A (= F_B)$.

Proof $B = X^{-1} \otimes A \otimes X$ implies

$$b_{ij} = -x_i + a_{ij} + x_j$$

for all $i, j \in N$, hence for $\sigma = (i_1, \ldots, i_{p-1}, i_p = i_1)$ we have

$$
\begin{aligned}
w(\sigma, B) &= b_{i_1 i_2} + b_{i_2 i_3} + \cdots + b_{i_{p-1} i_1} \\
&= -x_{i_1} + a_{i_1 i_2} + x_{i_2} - \cdots - x_{i_{p-1}} + a_{i_{p-1} i_1} + x_{i_1} \\
&= a_{i_1 i_2} + a_{i_2 i_3} + \cdots + a_{i_{p-1} i_1} = w(\sigma, A). \quad \square
\end{aligned}
$$

1.6 The Key Players

Since the operation $\oplus$ in max-algebra is not invertible, inverse matrices are almost non-existent (Theorem 1.1.3) and thus some tools used in linear algebra are unavailable. It was therefore necessary to develop an alternative methodology that helps to solve basic problems such as systems of inequalities and equations, the eigenvalue-eigenvector problem, linear dependence and so on.

In this section we introduce and prove basic properties of the maximum cycle mean and transitive closures. We also discuss conjugation and the assignment problem. All these four concepts will play a key role in solving problems in max-algebra.

1.6.1 Maximum Cycle Mean

Everywhere in this book, given $A \in \overline{\mathbb{R}}^{n \times n}$, the symbol $\lambda(A)$ will stand for the *maximum cycle mean* of A, that is:

$$\lambda(A) = \max_{\sigma} \mu(\sigma, A), \tag{1.11}$$

where the maximization is taken over all elementary cycles in D_A, and

$$\mu(\sigma, A) = \frac{w(\sigma, A)}{l(\sigma)} \tag{1.12}$$

denotes the *mean* of a cycle σ. Clearly, $\lambda(A)$ always exists since the number of elementary cycles is finite. It follows from this definition that D_A is acyclic if and only if $\lambda(A) = \varepsilon$.

Example 1.6.1 If

$$A = \begin{pmatrix} -2 & 1 & -3 \\ 3 & 0 & 3 \\ 5 & 2 & 1 \end{pmatrix}$$

then the means of elementary cycles of length 1 are $-2, 0, 1$, of length 2 are $2, 1, 5/2$, of length 3 are 3 and $2/3$. Hence $\lambda(A) = 3$.

Lemma 1.6.2 $\lambda(A)$ *remains unchanged if the maximization in* (1.11) *is taken over all cycles.*

Proof We only need to prove that $\mu(\sigma, A) \leq \lambda(A)$ for any cycle σ in D_A.

Let σ be a cycle. Then σ can be partitioned into elementary cycles $\sigma_1, \ldots, \sigma_t$ $(t \geq 1)$. Hence

$$\mu(\sigma, A) = \frac{w(\sigma, A)}{l(\sigma)} = \frac{\sum_{i=1}^{t} w(\sigma_i, A)}{\sum_{i=1}^{t} l(\sigma_i)}$$

$$\leq \frac{\sum_{i=1}^{t} l(\sigma_i) \lambda(A)}{\sum_{i=1}^{t} l(\sigma_i)} = \lambda(A). \qquad \square$$

The maximum cycle mean of a matrix is of fundamental importance in max-algebra because for any square matrix A it is the greatest (max-algebraic) eigenvalue of A, and every eigenvalue of A is the maximum cycle mean of some principal submatrix of A (see Sects. 1.6.2, 2.2.2 and Chap. 4 for details).

In this subsection we first prove a few basic properties of $\lambda(A)$ that will be useful later on and then we show how it can be calculated.

Lemma 1.6.3 *If* $A = (a_{ij}) \in \overline{\mathbb{R}}^{n \times n}$ *is row or column* $\mathbb{R}$*-astic then* $\lambda(A) > \varepsilon$*. This is true in particular when A is irreducible and $n > 1$.*

Proof The statement follows from Lemmas 1.5.1 and 1.5.3.   □

Lemma 1.6.4 *Let* $A \in \overline{\mathbb{R}}^{n \times n}$. *Then for every* $\alpha \in \mathbb{R}$ *the sets of arcs (and therefore also the sets of cycles) in* D_A *and* $D_{\alpha \otimes A}$ *are equal and* $\mu(\sigma, \alpha \otimes A) = \alpha \otimes \mu(\sigma, A)$ *for every cycle* σ *in* D_A.

Proof For any $A = (a_{ij}) \in \overline{\mathbb{R}}^{n \times n}$, cycle $\sigma = (i_1, \ldots, i_k, i_1)$ and $\alpha \in \mathbb{R}$ we have

$$\mu(\sigma, \alpha \otimes A) = \frac{\alpha + a_{i_1 i_2} + \alpha + a_{i_2 i_3} + \cdots + \alpha + a_{i_{k-1} i_k} + \alpha + a_{i_k i_1}}{k}$$

$$= \alpha + \frac{a_{i_1 i_2} + a_{i_2 i_3} + \cdots + a_{i_{k-1} i_k} + a_{i_k i_1}}{k}$$

$$= \alpha \otimes \mu(\sigma, A).$$ □

A matrix A is called *definite* if $\lambda(A) = 0$ [45, 60]. Thus a matrix is definite if and only if all cycles in D_A are nonpositive and at least one has weight zero.

Theorem 1.6.5 *Let* $A \in \overline{\mathbb{R}}^{n \times n}$ *and* $\alpha \in \mathbb{R}$. *Then* $\lambda(\alpha \otimes A) = \alpha \otimes \lambda(A)$ *for any* $\alpha \in \mathbb{R}$. *Hence* $(\lambda(A))^{-1} \otimes A$ *is definite whenever* $\lambda(A) > \varepsilon$.

Proof For any $A \in \overline{\mathbb{R}}^{n \times n}$ and $\alpha \in \mathbb{R}$ we have by Lemma 1.6.4:

$$\lambda(\alpha \otimes A) = \max_{\sigma} \mu(\sigma, \alpha \otimes A) = \max_{\sigma} \alpha \otimes \mu(\sigma, A)$$

$$= \alpha \otimes \max_{\sigma} \mu(\sigma, A) = \alpha \otimes \lambda(A).$$

Also, $\lambda((\lambda(A))^{-1} \otimes A) = \lambda(A)^{-1} \otimes \lambda(A) = 0$.   □

The matrix $(\lambda(A))^{-1} \otimes A$ will be denoted in this book by A_λ.
For $A \in \overline{\mathbb{R}}^{n \times n}$ we denote

$$N_c(A) = \{i \in N; \exists \sigma = (i = i_1, \ldots, i_k, i_1) \text{ in } D_A : \mu(\sigma, A) = \lambda(A)\}.$$

The elements of $N_c(A)$ are called *critical nodes* or *eigennodes* of A since they play an essential role in solving the eigenproblem (Lemma 4.2.3). And a cycle σ is called *critical* (in D_A) if $\mu(\sigma, A) = \lambda(A)$. Hence $N_c(A)$ is the set of the nodes of all critical cycles in D_A. If $i, j \in N_c(A)$ belong to the same critical cycle then i and j are called *equivalent* and we write $i \sim j$; otherwise they are called *nonequivalent* and we write $i \nsim j$. Clearly, $\sim$ constitutes a relation of equivalence on $N_c(A)$.

Lemma 1.6.6 *Let* $A \in \overline{\mathbb{R}}^{n \times n}$. *Then for every* $\alpha \in \mathbb{R}$ *we have* $N_c(\alpha \otimes A) = N_c(A)$.

Proof By Lemma 1.6.4 we have

$$\mu(\sigma, \alpha \otimes A) = \alpha \otimes \mu(\sigma, A)$$

for any $A \in \overline{\mathbb{R}}^{n \times n}$ and $\alpha \in \mathbb{R}$. Hence the critical cycles in D_A and $D_{\alpha \otimes A}$ are the same.   □

The *critical digraph* of A is the digraph $C(A)$ with the set of nodes N; the set of arcs, notation $E_c(A)$, is the set of arcs of all critical cycles. A strongly connected component of $C(A)$ is called *trivial* if it consists of a single node without a loop, *nontrivial* otherwise. Nontrivial strongly connected components of $C(A)$ will be called *critical components*.

Remark 1.6.7 [8, 102] It is not difficult to prove from the definitions that all cycles in a critical digraph are critical. We will see this as Corollary 8.1.7.

Computation of the maximum cycle mean from the definition is difficult except for small matrices since the number of elementary cycles in a digraph may be prohibitively large in general. The task of finding the maximum cycle mean of a matrix was studied also in combinatorial optimization, independently of max-algebra. Publications presenting a method are e.g. [60, 72, 106, 109, 144]. One of the first methods was Vorobyov's $O(n^4)$ formula, following directly from Lemma 1.6.2 and the longest path interpretation of matrix powers, see Example 1.2.3:

$$\lambda(A) = \max_{k \in N} \max_{i \in N} \frac{a_{ii}^{(k)}}{k}$$

where $A^k = (a_{ij}^{(k)}), k \in N$.

Example 1.6.8 For the matrix A of Example 1.6.1 we get

$$A^2 = \begin{pmatrix} 4 & 1 & 4 \\ 8 & 5 & 4 \\ 6 & 6 & 5 \end{pmatrix},$$

$$A^3 = \begin{pmatrix} 9 & 6 & 5 \\ 9 & 9 & 8 \\ 10 & 7 & 9 \end{pmatrix},$$

hence $\lambda(A) = \max(1, 5/2, 9/3) = 3$.

A linear programming method has been designed in [60], see Remark 1.6.30. Another one is Lawler's [109] of computational complexity $O(n^3 \log n)$ based on Theorem 1.6.5 and existing $O(n^3)$ methods for checking the existence of a positive cycle. It uses a bivalent search for a value of α such that $\lambda(\alpha \otimes A) = 0$.

We present *Karp's algorithm* [106] which finds the maximum cycle mean of an $n \times n$ matrix A in $O(n|E|)$ time where E is the set of arcs of D_A. Note that for the computation of the maximum cycle mean of a matrix we may assume without loss of generality that A is irreducible since any cycle is wholly contained in one strongly connected component and, as already mentioned, all strongly connected components can be recognized in $O(|V| + |E|)$ time [142]. Let $A = (a_{ij}) \in \overline{\mathbb{R}}^{n \times n}$

and $s \in N$ be an arbitrary fixed node of $D_A = (N, E, (a_{ij}))$. For every $j \in N$, and every positive integer k we define $F_k(j)$ as the maximum weight of an $s - j$ path of length k; if no such path exists then $F_k(j) = \varepsilon$.

Theorem 1.6.9 (Karp) *If $A = (a_{ij}) \in \overline{\mathbb{R}}^{n \times n}$ is irreducible then*

$$\lambda(A) = \max_{j \in N} \min_{k \in N} \frac{F_{n+1}(j) - F_k(j)}{n + 1 - k}. \tag{1.13}$$

Proof The statement holds for $n = 1$. If $n > 1$ then $\lambda(A) > \varepsilon$. By subtracting $\lambda(A)$ from the weight of every arc of D_A the value of $F_k(j)$ decreases by $k\lambda(A)$ and thus the right-hand side in (1.13) decreases by $\lambda(A)$. Hence it is sufficient to prove that

$$\max_{j \in N} \min_{k \in N} \frac{F_{n+1}(j) - F_k(j)}{n + 1 - k} = 0 \tag{1.14}$$

if A is definite. If A is definite then there are no positive cycles in D_A and by Lemma 1.5.4 a heaviest $s - j$ path of length n or less exists for every $j \in N$ (since at least one such path exists by strong connectivity of D_A). Let us denote this maximum weight by $w(j)$. Then

$$F_{n+1}(j) \leq w(j) = \max_{k \in N} F_k(j),$$

hence

$$\min_{k \in N}(F_{n+1}(j) - F_k(j)) = F_{n+1}(j) - \max_{k \in N} F_k(j)$$

$$= F_{n+1}(j) - w(j) \leq 0$$

holds for every $j \in N$. It remains to show that equality holds for at least one j. Let σ be a cycle of weight zero and i be any node in σ. Let π be any $s - i$ path of maximum weight $w(i)$. Then π extended by any number of repetitions of σ is also an $s - i$ path of weight $w(i)$ and therefore any subpath of such an extension starting at s is also a heaviest path from s to its endnode. By using a sufficient number of repetitions of σ we may assume that the extension of π is of length $n + 1$ or more. Let us denote one such extension by π'. A subpath of π' starting at s of length $n + 1$ exists. Its endnode is the sought j. $\qquad\square$

The quantities $F_k(j)$ can be computed by the recurrence

$$F_k(j) = \max_{(i,j) \in E}(F_{k-1}(i) + a_{ij}) \quad (k = 2, \ldots, n + 1) \tag{1.15}$$

with the initial conditions $F_1(j) = a_{sj}$ for all $j \in N$. The computation of $F_k(j)$ from (1.15) for a fixed k and for all j requires $O(|E|)$ operations as every arc will be used once. Hence the number of operations needed for the computation of all quantities $F_k(j)$ ($j \in N, k = 1, \ldots, n + 1$) is $O(n|E|)$. The application of (1.13) is obviously $O(n^2)$. By connectivity we have $n \leq |E|$ and the overall complexity bound $O(n|E|)$ now follows.

Specially designed algorithms find the maximum cycle mean for some types of matrices with computational complexity lower than $O(n^3)$ [33, 94, 122]. See also [46, 121].

There are also other, fast methods for finding the maximum cycle mean for general matrices whose performance bound is not known. See for instance Howard's algorithm or the power method [8, 17, 49, 77, 78, 84, 102].

1.6.2 Transitive Closures

1.6.2.1 Transitive Closures, Eigenvectors and Subeigenvectors

Given $A \in \overline{\mathbb{R}}^{n \times n}$ we define the following infinite series

$$\Gamma(A) = A \oplus A^2 \oplus A^3 \oplus \cdots \qquad (1.16)$$

and

$$\Delta(A) = I \oplus \Gamma(A) = I \oplus A \oplus A^2 \oplus A^3 \oplus \cdots . \qquad (1.17)$$

If these series converge to matrices that do not contain $+\infty$, then the matrix $\Gamma(A)$ is called the *weak transitive closure* of A and $\Delta(A)$ is the *strong transitive closure* of A. These names are motivated by the digraph representation if A is a $\{0, -1\}$ matrix since the existence of arcs (i, j) and (j, k) in $Z_{\Gamma(A)}$ implies that also the arc (i, k) exists.

The matrices $\Gamma(A)$ and $\Delta(A)$ are of fundamental importance in max-algebra. This follows from the fact that they enable us to efficiently describe all solutions (called *eigenvectors,* if different from ε) to

$$A \otimes x = \lambda \otimes x, \quad \lambda \in \overline{\mathbb{R}} \qquad (1.18)$$

in the case of $\Gamma(A)$, and all finite solutions to

$$A \otimes x \leq \lambda \otimes x, \quad \lambda \in \overline{\mathbb{R}} \qquad (1.19)$$

in the case of $\Delta(A)$. Solutions to (1.19) different from ε are called *subeigenvectors.* The possibility of finding all (finite) solutions is an important feature of max-algebra and we illustrate the benefits of this on an application in Sect. 2.1.

If $A \in \overline{\mathbb{R}}^{n \times n}$ and $\lambda \in \mathbb{R}$, we will denote the set of finite subeigenvectors by $V^*(A, \lambda)$, that is

$$V^*(A, \lambda) = \{x \in \mathbb{R}^n; A \otimes x \leq \lambda \otimes x\},$$

and for convenience also

$$V^*(A) = V^*(A, \lambda(A)),$$

$$V_0^*(A) = V^*(A, 0).$$

We will first show how $\Gamma(A)$ and $\Delta(A)$ can be used for finding one solution to (1.18) and (1.19), respectively. Then we describe all finite solutions to (1.19) using $\Delta(A)$. The description of all solutions to (1.18) will follow from the theory presented in Chap. 4.

It has been observed in Example 1.2.3 that the entries of $A^2 = A \otimes A$ are the weights of heaviest paths of length 2 for all pairs of nodes in D_A. Similarly the elements of A^k ($k = 1, 2, \ldots$) are the weights of heaviest paths of length k for all pairs of nodes. Therefore the matrix $\Gamma(A)$ (if the infinite series converges) represents the weights of heaviest paths of any length for all pairs of nodes. Motivated by this fact $\Gamma(A)$ is also called the *metric matrix* corresponding to the matrix A [60]. Note that $\Delta(A)$ is often called the *Kleene star* [3].

1.6.2.2 Weak Transitive Closure

If $\lambda(A) \leq 0$ then all cycles in D_A have nonpositive weights and so by Lemma 1.5.4 we have:

$$A^k \leq A \oplus A^2 \oplus \cdots \oplus A^n \qquad (1.20)$$

for every $k \geq 1$, and therefore $\Gamma(A)$ for any matrix with $\lambda(A) \leq 0$, and in particular for definite matrices, exists and is equal to $A \oplus A^2 \oplus \cdots \oplus A^n$. On the other hand if $\lambda(A) > 0$ then a positive cycle in D_A exists, thus the value of at least one position in A^k is unbounded as $k \longrightarrow \infty$ and, consequently, at least one entry of $\Gamma(A)$ is $+\infty$. Also, $\Gamma(A)$ is finite if A is irreducible since $\Gamma(A)$ is the matrix of the weights of heaviest paths in D_A and in a strongly connected digraph there is a path between any pair of nodes. We have proved:

Proposition 1.6.10 *Let $A \in \overline{\mathbb{R}}^{n \times n}$. Then (1.16) converges to a matrix with no $+\infty$ if and only if $\lambda(A) \leq 0$. If $\lambda(A) \leq 0$ then*

$$\Gamma(A) = A \oplus A^2 \oplus \cdots \oplus A^k$$

for every $k \geq n$. If A is also irreducible and $n > 1$ then $\Gamma(A)$ is finite.

A matrix $A = (a_{ij}) \in \overline{\mathbb{R}}^{n \times n}$ is called *increasing* if $a_{ii} \geq 0$ for all $i \in N$. Obviously, $A = I \oplus A$ when A is increasing and so then there is no difference between $\Gamma(A)$ and $\Delta(A)$.

Lemma 1.6.11 *If $A = (a_{ij}) \in \overline{\mathbb{R}}^{n \times n}$ is increasing then $x \leq A \otimes x$ for every $x \in \overline{\mathbb{R}}^n$. Hence*

$$A \leq A^2 \leq A^3 \leq \cdots . \qquad (1.21)$$

Proof If A is increasing then $I \leq A$ and thus $x = I \otimes x \leq A \otimes x$ for any $x \in \overline{\mathbb{R}}^n$ by Corollary 1.1.2. The rest follows by taking the individual columns of A for x and repeating the argument. $\qquad \square$

A matrix $A = (a_{ij}) \in \overline{\mathbb{R}}^{n \times n}$ is called *strongly definite* if it is definite and increasing. Since the diagonal entries of A are the weights of cycles (loops) we have that $a_{ii} = 0$ for all $i \in N$ if A is strongly definite.

Proposition 1.6.12 If $A \in \overline{\mathbb{R}}^{n \times n}$ is strongly definite then

$$\Delta(A) = \Gamma(A) = A^{n-1} = A^n = A^{n+1} = \cdots.$$

Proof Since $A \leq A^2 \leq A^3 \leq \cdots$ we have $\Gamma(A) = A \oplus A^2 \oplus \cdots \oplus A^k = A^k$ for any $k \geq n$ straightforwardly by Proposition 1.6.10. Also, we deduce that all diagonal entries of all powers are nonnegative; they are all actually zero as a positive diagonal entry would indicate a positive cycle. To prove the case $k = n - 1$ consider $a_{ij}^{(n-1)}$ and $a_{ij}^{(n)}$, that is, the (i, j) entries in A^{n-1} and A^n for some $i, j \in N$, respectively. If

$$a_{ij}^{(n-1)} < a_{ij}^{(n)} \tag{1.22}$$

then $i \neq j$ (since all diagonal entries in all powers are zero) and the greatest weight of an $i - j$ path, say π, of length n is greater than the greatest weight of an $i - j$ path of length $n - 1$. However π contains a cycle, say σ, as a subpath. Since $w(\sigma, A) \leq 0$ by removing σ from π we obtain an $i - j$ path, say π', $l(\pi') < n$, $w(\pi', A) \geq w(\pi, A)$ which contradicts (1.22). Hence $a_{ij}^{(n-1)} = a_{ij}^{(n)}$ for all $i, j \in N$. $\square$

Remark 1.6.13 As a by-product of Proposition 1.6.12 we may compile a simple and fast power method [65] for finding $\Gamma(A)$ if A is strongly definite, since we only need to find a sufficiently high power of A. We calculate $A^2, A^4 = (A^2)^2, A^8 = (A^4)^2, \ldots, A^{2^k}, \ldots$ and we stop as soon as $2^k \geq n - 1$, that is, when $k \geq \log_2(n - 1)$, yielding an $O(n^3 \log n)$ method.

Another useful property of strongly definite matrices immediately follows from Lemma 1.6.11:

Lemma 1.6.14 If $A \in \overline{\mathbb{R}}^{n \times n}$ is strongly definite and $x \in \overline{\mathbb{R}}^n$ then $A \otimes x = x$ if and only if $A \otimes x \leq x$.

1.6.2.3 Strong Transitive Closure (Kleene Star)

The matrix $\Delta(A)$ also has some remarkable properties. A key to understanding these is Lemma 1.1.4 which immediately implies another formula:

$$\Delta(A) = \Gamma(I \oplus A). \tag{1.23}$$

Proposition 1.6.15 If $A \in \overline{\mathbb{R}}^{n \times n}$ and $\lambda(A) \leq 0$ then

$$\Delta(A) = I \oplus A \oplus \cdots \oplus A^{n-1}, \tag{1.24}$$

$$(\Delta(A))^k = \Delta(A) \tag{1.25}$$

for every $k \geq 1$ and

$$A \otimes \Delta(A) = \Gamma(A). \tag{1.26}$$

Proof If $\lambda(A) \leq 0$ then $I \oplus A$ is both definite and increasing, hence by (1.23), Lemma 1.1.4 and Proposition 1.6.12 we have

$$\Delta(A) = \Gamma(I \oplus A) = (I \oplus A)^{n-1} = I \oplus A \oplus \cdots \oplus A^{n-1}.$$

The other two formulae straightforwardly follow from the first. $\qquad\square$

Corollary 1.6.16 $A = (a_{ij}) \in \overline{\mathbb{R}}^{n \times n}$ *is a Kleene star if and only if $A^2 = A$ and $a_{ii} = 0$ for all $i \in N$.*

Suppose $\lambda(A) \leq 0$, then by (1.20)

$$A \otimes \Gamma(A) = A^2 \oplus \cdots \oplus A^{n+1} \leq A \oplus A^2 \oplus \cdots \oplus A^{n+1} = \Gamma(A)$$

and similarly by (1.24)

$$A \otimes \Delta(A) = A \oplus \cdots \oplus A^n = \Gamma(A) \leq \Delta(A).$$

Hence every column of $\Delta(A)$ or $\Gamma(A)$ is a solution to $A \otimes x \leq x$ if $\lambda(A) \leq 0$. If, moreover, A is also increasing then

$$\Delta(A) = \Gamma(A) = A^{n-1} = A^n = A^{n+1} = \cdots$$

and so $A \otimes \Delta(A) = \Delta(A)$ and $A \otimes \Gamma(A) = \Gamma(A)$. We readily deduce:

Proposition 1.6.17 *If $A \in \overline{\mathbb{R}}^{n \times n}$ is strongly definite then every column of $\Delta(A)(= \Gamma(A))$ is a solution to $A \otimes x = x$.*

We will show in Chap. 4 how to use $\Gamma(A)$ for finding all solutions to $A \otimes x = x$ for definite matrices A. Consequently, this will enable us to describe all solutions and all finite solutions to $A \otimes x = \lambda \otimes x$.

Now we use the strong transitive closure to provide a description of all finite solutions to $A \otimes x \leq \lambda \otimes x$ for any $\lambda \in \overline{\mathbb{R}}$ and all solutions for $\lambda \geq \lambda(A)$ and $\lambda > \varepsilon$. Note that $A \otimes x \leq \lambda \otimes x$ may have a solution $x \in \overline{\mathbb{R}}^n, x \neq \varepsilon$ even if $\lambda < \lambda(A)$, see Theorem 4.5.14.

Observe that if $A = \varepsilon$ then every $x \in \overline{\mathbb{R}}^n$ is a solution to $A \otimes x \leq \lambda \otimes x$.

Theorem 1.6.18 [40, 59, 80, 128] *Let $A = (a_{ij}) \in \overline{\mathbb{R}}^{n \times n}, A \neq \varepsilon$. Then the following statements hold:*

(a) $A \otimes x \leq \lambda \otimes x$ *has a finite solution if and only if $\lambda \geq \lambda(A)$ and $\lambda > \varepsilon$.*

(b) *If $\lambda \geq \lambda(A)$ and $\lambda > \varepsilon$ then*

$$V^*(A, \lambda) = \{\Delta(\lambda^{-1} \otimes A) \otimes u; u \in \mathbb{R}^n\}.$$

(c) *If $\lambda \geq \lambda(A)$ and $\lambda > \varepsilon$ then*

$$A \otimes x \leq \lambda \otimes x, \quad x \in \overline{\mathbb{R}}^n$$

if and only if

$$x = \Delta(\lambda^{-1} \otimes A) \otimes u, \quad u \in \overline{\mathbb{R}}^n.$$

Proof (a) Suppose $A \otimes x \leq \lambda \otimes x, x \in \mathbb{R}^n$. Since $A \neq \varepsilon$ we have $\lambda > \varepsilon$. If $\lambda(A) = \varepsilon$ then also $\lambda > \lambda(A)$. Suppose now that $\lambda(A) > \varepsilon$, thus D_A contains a cycle. Let $\sigma = (i_1, \ldots, i_k, i_{k+1} = i_1)$ be any cycle in D_A. Then we have

$$a_{i_1 i_2} + x_{i_2} \leq \lambda + x_{i_1}$$
$$a_{i_2 i_3} + x_{i_3} \leq \lambda + x_{i_2}$$
$$\cdots$$
$$a_{i_k i_1} + x_{i_1} \leq \lambda + x_{i_k}.$$

If we add up these inequalities and simplify, we get

$$\lambda \geq \frac{a_{i_1 i_2} + a_{i_2 i_3} + \cdots + a_{i_{k-1} i_k} + a_{i_k i_1}}{k} = \mu(\sigma, A).$$

It follows that $\lambda \geq \max_\sigma \mu(\sigma, A) = \lambda(A)$.

For the converse suppose $\lambda \geq \lambda(A)$ and $\lambda > \varepsilon$, thus $\lambda(\lambda^{-1} \otimes A) \leq 0$ and take $u \in \mathbb{R}^n$. We show that

$$A \otimes x \leq \lambda \otimes x, \quad x \in \mathbb{R}^n$$

is satisfied by $x = \Delta(\lambda^{-1} \otimes A) \otimes u$. Since $\Delta(\lambda^{-1} \otimes A) \geq I$ we have that $x \geq u$ and thus $x \in \mathbb{R}^n$. Also,

$$\Delta(\lambda^{-1} \otimes A) \otimes x = (\Delta(\lambda^{-1} \otimes A))^2 \otimes u = \Delta(\lambda^{-1} \otimes A) \otimes u = x$$

by (1.25). Hence we have

$$(\lambda^{-1} \otimes A) \otimes x \leq \Delta(\lambda^{-1} \otimes A) \otimes x = x$$

and the statement follows.

(b) Suppose $\lambda \geq \lambda(A)$, $\lambda > \varepsilon$ and $A \otimes x \leq \lambda \otimes x, x \in \mathbb{R}^n$, thus

$$(\lambda^{-1} \otimes A) \otimes x \leq x$$

and

$$x \oplus (\lambda^{-1} \otimes A) \otimes x = x.$$

Hence

$$(I \oplus \lambda^{-1} \otimes A) \otimes x = x,$$

and by (1.3) and (1.24) we have

$$\Delta(\lambda^{-1} \otimes A) \otimes x = (I \oplus \lambda^{-1} \otimes A)^{n-1} \otimes x = x.$$

The proof of sufficiency follows the second part of the proof of (a).

(c) The proof is the same as that of part (b) except the reasoning that $x \in \mathbb{R}^n$. $\square$

1.6.2.4 Two properties of subeigenvectors

The following two statements provide information that will be helpful later on.

Lemma 1.6.19 *Let* $A \in \overline{\mathbb{R}}^{n \times n}$ *and* $\lambda(A) > \varepsilon$. *If* $x \in V^*(A)$ *and* $(i, j) \in E_c(A)$ *then*

$$a_{ij} \otimes x_j = \lambda(A) \otimes x_i.$$

Proof The inequality $a_{ij} \otimes x_j \leq \lambda(A) \otimes x_i$ for all i, j follows from Theorem 1.6.18. Suppose it is strict for some $(i, j) \in E_c(A)$. Since (i, j) belongs to a critical cycle, say $\sigma = (j_1 = i, j_2 = j, j_3, \ldots, j_k, j_{k+1} = j_1)$, we have

$$a_{j_r j_{r+1}} \otimes x_{j_{r+1}} \leq \lambda(A) \otimes x_{j_r}$$

for all $r = 1, \ldots, k$. Since the first of these inequalities is strict, by multiplying them out using $\otimes$ and cancellations of all x_j we get the strict inequality

$$a_{j_1 j_2} \otimes \cdots \otimes a_{j_k j_1} < (\lambda(A))^k,$$

which is a contradiction with the assumption that σ is critical. $\square$

Lemma 1.6.20 *The set* $V^*(A, \lambda)$ *is convex for any* $A \in \overline{\mathbb{R}}^{n \times n}$ *and* $\lambda \in \overline{\mathbb{R}}$.

Proof If $\lambda = \varepsilon$ then $V^*(A, \lambda)$ is either empty (if $A \neq \varepsilon$) or $\mathbb{R}^n$ (if $A = \varepsilon$).

If $\lambda > \varepsilon$ then $A \otimes x \leq \lambda \otimes x$ is in conventional notation equivalent to

$$a_{ij} + x_j \leq \lambda + x_i$$

for all $i, j \in N$ such that $a_{ij} > \varepsilon$; which is a system of conventional linear inequalities, hence the solution set is convex. $\square$

1.6.2.5 Computation of Transitive Closures

We finish this section with computational observations. The product of two $n \times n$ matrices from the definition uses $O(n^3)$ operations of $\oplus$ and $\otimes$ and unlike in conventional linear algebra a faster way of finding this product does not seem to be known (see Chap. 11 for a list of open problems). This implies that the computation of $\Gamma(A)$ (and therefore also $\Delta(A)$) for a matrix A with $\lambda(A) \leq 0$ from the definition needs $O(n^4)$ operations. However, a classical method can do better:

Algorithm 1.6.21 FLOYD−WARSHALL
Input: $A = (a_{ij}) \in \overline{\mathbb{R}}^{n \times n}$.
Output: $\Gamma(A) = (\gamma_{ij})$ or an indication that there is a positive cycle in D_A (and hence $\Gamma(A)$ contains $+\infty$).
$\quad \gamma_{ij} := a_{ij}$ for all $i, j \in N$
$\quad$ for all $p = 1, \ldots, n$ do
$\quad\quad$ for all $i = 1, \ldots, n, i \neq p$ do
$\quad\quad\quad$ for all $j = 1, \ldots, n, i \neq p$ do
$\quad\quad\quad$ begin
$\quad\quad\quad$ if $\gamma_{ij} < \gamma_{ip} + \gamma_{pj}$ then $\gamma_{ij} := \gamma_{ip} + \gamma_{pj}$
$\quad\quad\quad$ if $i = j$ and $\gamma_{ij} > 0$ then stop (Positive cycle exists)
$\quad\quad\quad$ end

Theorem 1.6.22 [120] *The algorithm FLOYD−WARSHALL is correct and terminates after $O(n^3)$ operations.*

Proof Correctness: Let

$$G^{[p]} = \left(\gamma_{ij}^{[p]}\right)$$

be the matrix obtained at the end of the $(p-1)$st run of the main (outer) loop of the algorithm, $p = 1, 2, \ldots, n+1$. Hence the algorithm starts with the matrix $G^{[1]} = A$ and constructs a sequence of matrices

$$G^{[2]}, \ldots, G^{[n+1]}.$$

The formula used in the algorithm is

$$\gamma_{ij}^{[p+1]} := \max\left(\gamma_{ij}^{[p]}, \gamma_{ip}^{[p]} + \gamma_{pj}^{[p]}\right) \quad (i, j \in N; i, j \neq p). \qquad (1.27)$$

It is sufficient to prove that each $\gamma_{ij}^{[p]}$ ($i, j \in N, p = 1, \ldots, n+1$) calculated in this way is the greatest weight of an $i - j$ path not containing nodes $p, p+1, \ldots, n$ as intermediate nodes because then $G^{[n+1]}$ is the matrix of weights of heaviest paths (without any restriction) for all pairs of nodes, that is, $\Gamma(A)$. We show this by induction on p.

The statement is true for $p = 1$ because $G^{[1]} = A$ is the direct-distances matrix (in which no intermediate nodes are allowed).

For the second induction step realize that a heaviest $i - j$ path, say π, not containing nodes $p + 1, \ldots, n$ as intermediate nodes either does or does not contain node p. In the first case it consists of two subpaths, without loss of generality both elementary, one being an $i - p$ path, the other a $p - j$ path; neither of them contains node p as an intermediate node. By optimality both are heaviest paths and therefore the weight of π is $\gamma_{ip}^{[p]} + \gamma_{pj}^{[p]}$. In the second case π is a heaviest $i - j$ path not containing p, thus its weight is $\gamma_{ij}^{[p]}$. The correctness of the transition formula (1.27) now follows.

Complexity bound: Two inner nested loops, each of length $n - 1$, contain two lines which require a constant number of operations. The outer loop has length n, thus the complexity bound is $O(n(n - 1)^2) = O(n^3)$.   □

Example 1.6.23 For the matrix A of Example 1.6.1 we have $\lambda(A) = 3$, hence by subtracting 3 from every entry of A we obtain the definite matrix A_λ:

$$\begin{pmatrix} -5 & -2 & -6 \\ 0 & -3 & 0 \\ 2 & -1 & -2 \end{pmatrix}.$$

We may calculate $\Gamma(A_\lambda)$ from the definition as $A_\lambda \oplus A_\lambda^2 \oplus A_\lambda^3$. Since

$$A_\lambda^2 = \begin{pmatrix} -2 & -5 & -2 \\ 2 & -1 & -2 \\ 0 & 0 & -1 \end{pmatrix}, \qquad A_\lambda^3 = \begin{pmatrix} 0 & -3 & -4 \\ 0 & 0 & -1 \\ 1 & -2 & 0 \end{pmatrix}$$

we see that

$$\Gamma(A_\lambda) = \begin{pmatrix} 0 & -2 & -2 \\ 2 & 0 & 0 \\ 2 & 0 & 0 \end{pmatrix}.$$

Alternatively we may use the algorithm FLOYD−WARSHALL:

$$A_\lambda \quad = \quad \begin{pmatrix} -5 & -2 & -6 \\ 0 & -3 & 0 \\ 2 & -1 & -2 \end{pmatrix} \xrightarrow{p=1} \begin{pmatrix} -5 & -2 & -6 \\ 0 & -2 & 0 \\ 2 & 0 & -2 \end{pmatrix}$$

$$\xrightarrow{p=2} \begin{pmatrix} -2 & -2 & -2 \\ 0 & -2 & 0 \\ 2 & 0 & 0 \end{pmatrix} \xrightarrow{p=3} \begin{pmatrix} 0 & -2 & -2 \\ 2 & 0 & 0 \\ 2 & 0 & 0 \end{pmatrix}.$$

Remark 1.6.24 The transitive closure of Boolean matrices A (in conventional linear algebra) can be calculated in $O(n^2 + m^\alpha log(m))$ time [115], where m is the number of strongly connected components of D_A and α is the *matrix multiplication constant* (currently $\alpha = 2.376$ [56]). This immediately yields an $O(n^2 + m^\alpha \log(m))$ algorithm for finding the weak and strong transitive closures of matrices over $\{0, -\infty\}$ in max-algebra. Note that the transitive closure of every irreducible matrix over $\{0, -\infty\}$ is the zero matrix.

1.6.3 Dual Operators and Conjugation

Other tools that help to overcome the difficulties caused by the absence of subtraction and matrix inversion are the *dual pair of operations* $(\oplus', \otimes')$ and the *matrix conjugation* respectively [59, 60]. These are defined as follows. For $a, b \in \overline{\mathbb{R}}$ set

$$a \oplus' b = \min(a, b),$$

$$a \otimes' b = a + b \quad \text{if } \{a, b\} \neq \{-\infty, +\infty\}$$

and

$$(-\infty) \otimes' (+\infty) = +\infty = (+\infty) \otimes' (-\infty).$$

The pair of operations $(\oplus', \otimes')$ is extended to matrices (including vectors) in the same way as $(\oplus, \otimes)$ and it is easily verified that all properties described in Sect. 1.1 hold dually if $\oplus$ is replaced by $\oplus'$, $\otimes$ by $\otimes'$ and by reverting the inequality signs.

The *conjugate* of $A = (a_{ij}) \in \overline{\overline{\mathbb{R}}}^{m \times n}$ is $A^* = -A^T \in \overline{\overline{\mathbb{R}}}^{n \times m}$. The significance of the dual operators and conjugation is indicated by the following statement which will be proved in Sect. 3.2, where we also show more of their properties.

Theorem 1.6.25 [59] *If* $A \in \overline{\overline{\mathbb{R}}}^{m \times n}$, $b \in \overline{\overline{\mathbb{R}}}^{m}$ *and* $x \in \overline{\overline{\mathbb{R}}}^{n}$ *then*

$$A \otimes x \leq b \text{ if and only if } x \leq A^* \otimes' b.$$

Corollary 1.6.26 *If* $A \in \overline{\overline{\mathbb{R}}}^{m \times n}$ *and* $v \in \overline{\overline{\mathbb{R}}}^{m}$ *then* $A \otimes (A^* \otimes' v) \leq v$.

Corollary 1.6.27 *If* $A \in \overline{\overline{\mathbb{R}}}^{m \times n}$ *and* $B \in \overline{\overline{\mathbb{R}}}^{m \times k}$ *then*

$$A \otimes (A^* \otimes' B) \leq B.$$

Conjugation can also be used to conveniently express the maximum cycle mean of A in terms of its finite subeigenvectors:

Lemma 1.6.28 *Let* $A \in \overline{\mathbb{R}}^{n \times n}$ *and* $\lambda(A) > \varepsilon$. *If* $z \in V^*(A)$ *then*

$$\lambda(A) = z^* \otimes A \otimes z = \min_{x \in \mathbb{R}^n} x^* \otimes A \otimes x.$$

Proof It follows from the definition of $V^*(A)$ that $z^* \otimes A \otimes z \leq \lambda(A)$. At the same time

$$z^* \otimes A \otimes z = \max_{i, j \in N} (-z_i + a_{ij} + z_j) \geq \lambda(A)$$

by Lemma 1.6.19.

On the other hand, if $x^* \otimes A \otimes x = \lambda$ for $x \in \mathbb{R}^n$ then $A \otimes x \leq \lambda \otimes x$ and $\lambda \geq \lambda(A)$ by Theorem 1.6.18. $\qquad\square$

We conclude this subsection by an observation that was proved many years ago and inspired a linear programming method for finding $\lambda(A)$ [60, 80, 128]. See also [40].

Theorem 1.6.29 *If* $A = (a_{ij}) \in \overline{\mathbb{R}}^{n \times n}$ *then*

$$\lambda(A) = \inf\{\lambda;\; A \otimes x \le \lambda \otimes x, x \in \mathbb{R}^n\}. \tag{1.28}$$

If $\lambda(A) > \varepsilon$ *or* $A = \varepsilon$ *then the infimum in* (1.28) *is attained.*

Proof The statement follows from Theorem 1.6.18 and Lemma 1.6.28. □

Note that using the spectral theory of Sect. 4.5 we will be able to prove a more general result, Theorem 4.5.14.

Remark 1.6.30 If $\lambda(A) > \varepsilon$ then formula (1.28) suggests that $\lambda(A)$ is the optimal value of the linear program

$$\lambda \longrightarrow \min$$

s.t.

$$\lambda + x_i - x_j \ge a_{ij}, \quad (i, j) \in F_A.$$

This idea was used in [60] to design a linear programming method for finding the maximum cycle mean of a matrix.

1.6.4 The Assignment Problem and Its Variants

By P_n we denote in this book the set of all permutations of the set N. The symbol id will stand for the identity permutation. As usual, *cyclic permutations* (or, briefly, *cycles* if no confusion arises) are of the form $\sigma : i_1 \longrightarrow i_2 \longrightarrow \cdots \longrightarrow i_k \longrightarrow i_1$. We will also write $\sigma = (i_1 i_2 \cdots i_k)$. Every permutation of the set N can be written as a product of cyclic permutations of subsets of N, called *constituent cycles*. For instance, if $n = 5$ then the permutation

$$\pi = \begin{pmatrix} 1 & 2 & 3 & 4 & 5 \\ 4 & 5 & 1 & 3 & 2 \end{pmatrix}$$

is the product of cyclic permutations $1 \longrightarrow 4 \longrightarrow 3 \longrightarrow 1$ and $2 \longrightarrow 5 \longrightarrow 2$, that is, $\pi = (143)(25)$.

Let $A = (a_{ij}) \in \overline{\mathbb{R}}^{n \times n}$. The *max-algebraic permanent* (or briefly *permanent*) of A is

$$\text{maper}(A) = \sum_{\pi \in P_n}^{\oplus} \prod_{i \in N}^{\otimes} a_{i,\pi(i)},$$

which in conventional notation reads

$$\text{maper}(A) = \max_{\pi \in P_n} \sum_{i \in N} a_{i,\pi(i)}.$$

For $\pi \in P_n$ the value

$$w(\pi, A) = \prod_{i \in N}^{\otimes} a_{i,\pi(i)} = \sum_{i \in N} a_{i,\pi(i)}$$

is called the *weight* of the permutation π (with respect to A). The problem of finding a permutation $\pi \in P_n$ of maximum weight (called *optimal permutation* or *optimal solution*) is the *assignment problem* for the matrix A solvable in $O(n^3)$ time using e.g. the *Hungarian method* (see for instance [21, 22, 120] or textbooks on combinatorial optimization). Hence the max-algebraic permanent of A is the optimal value to the assignment problem for A and, in contrast to the linear-algebraic permanent, it can be found efficiently. To mark this link we denote the set of optimal solutions to the assignment problem by ap(A), that is,

$$\text{ap}(A) = \{\pi \in P_n; \, w(\pi, A) = \text{maper}(A)\}.$$

The permanent plays a key role in a number of max-algebraic problems because of the absence of the determinant due to the lack of subtraction. It turns out that the structure of the set of optimal solutions is related to some max-algebraic properties, in particular to questions such as the regularity of matrices.

Example 1.6.31 If

$$A = \begin{pmatrix} 3 & 7 & 2 \\ 4 & 1 & 5 \\ 2 & 6 & 3 \end{pmatrix}$$

then maper(A) = 14, ap(A) = {(123), (1)(23), (12)(3)}.

A very simple property, on which the Hungarian method is based, is that the set of optimal solutions to the assignment problem for A does not change by adding a constant to a row or column of A. We can express this fact conveniently in max-algebraic terms: adding the constants $c_1, \ldots, c_n$ to the rows and $d_1, \ldots, d_n$ to the columns of A means to multiply $C \otimes A \otimes D$, where $C = \text{diag}(c_1, \ldots, c_n)$ and $D = (d_1, \ldots, d_n)$.

Lemma 1.6.32 *If $A \sim B$ then* ap(A) = ap(B).

Proof Let $\pi \in P_n$ and $B = C \otimes A \otimes D$. Then

$$w(\pi, B) = \prod_{i \in N}^{\otimes} b_{i,\pi(i)} = \prod_{i \in N}^{\otimes} c_i \otimes a_{i,\pi(i)} \otimes d_{\pi(i)}$$

$$= \prod_{i \in N}^{\otimes} c_i \otimes \prod_{i \in N}^{\otimes} a_{i,\pi(i)} \otimes \prod_{i \in N}^{\otimes} d_{\pi(i)} = c \otimes w(\pi, A) \otimes d,$$

where $c = \prod_{i \in N}^{\otimes} c_i$ and $d = \prod_{i \in N}^{\otimes} d_i$. Hence optimal permutations for B are exactly the same as for A. □

The Hungarian method applied to a matrix A assumes without loss of generality that $w(\pi, A)$ is finite for at least one $\pi \in P_n$ or, equivalently, maper$(A) > \varepsilon$ (otherwise ap$(A) = P_n$). Any such matrix is transformed by adding suitable constants to the rows and columns to produce a nonpositive matrix B with $w(\pi, B) = 0$ for at least one $\pi \in P_n$ and thus maper$(B) = 0$. By Lemma 1.6.32 we have ap$(A) = $ ap(B). Because of the special form of B we then have that optimal permutations for B (and A) are exactly those that select only zeros from B that is

$$\text{ap}(A) = \text{ap}(B) = \{\pi \in P_n; b_{i,\pi(i)} = 0\}.$$

Example 1.6.33 The Hungarian method transforms the matrix A of Example 1.6.31 using

$$C = \text{diag}(-4, -5, -3), \qquad D = \text{diag}(1, -3, 0)$$

to

$$\begin{pmatrix} 0 & 0 & -2 \\ 0 & -7 & 0 \\ 0 & 0 & 0 \end{pmatrix},$$

from which we can readily identify ap(A).

We may immediately deduce from the Hungarian method the following, otherwise rather nontrivial statement:

Theorem 1.6.34 *Let $A \in \overline{\mathbb{R}}^{n \times n}$ and suppose that $w(\pi, A)$ is finite for at least one $\pi \in P_n$. Then diagonal matrices C, D such that*

$$\text{maper}(C \otimes A \otimes D) = 0$$

and

$$C \otimes A \otimes D \leq 0$$

exist and can be found in $O(n^3)$ time.

The assignment problem plays a prominent role in various max-algebraic problems, see Chaps. 5, 6, 7 and 9. Therefore we will now discuss some computational aspects of the assignment problem relevant to max-algebra. First we mention that the diagonal entries in C and D in Theorem 1.6.34 are components of a dual optimal solution when the assignment problem is considered as a linear program and therefore using the duality of linear programming it is possible to improve the complexity bound in that theorem if an optimal solution is known [22, 120]:

Theorem 1.6.35 *Let* $A \in \overline{\mathbb{R}}^{n \times n}$ *and suppose that a* $\pi \in \mathrm{ap}(A)$ *is known. Then diagonal matrices* C, D *such that*

$$\mathrm{maper}(C \otimes A \otimes D) = 0$$

and

$$C \otimes A \otimes D \leq 0$$

can be found in $O(n)$ *time.*

It will be essential in Chap. 6 to decide whether an optimal permutation to the assignment problem is unique, that is, whether $|\mathrm{ap}(A)| = 1$. If this is the case then we say that A has *strong permanent*. For answering this question (see Theorem 1.6.39 below) it will be useful to transform a given matrix by permuting the rows and/or columns to a form where the diagonal entries of the matrix form an optimal solution, that is, where $\mathrm{id} \in \mathrm{ap}(A)$. We say that $A \in \overline{\mathbb{R}}^{n \times n}$ is *diagonally dominant* if $\mathrm{id} \in \mathrm{ap}(A)$. We therefore first make some observations on diagonally dominant matrices.

It is a straightforward matter to transform any square matrix to a diagonally dominant by suitably permuting the rows and/or columns once an optimal permutation has been found for this matrix. This transformation clearly does not change the size of the set of optimal permutations and can be described as a multiplication of the matrix by permutation matrices, that is, a transformation of the matrix to a similar one. Using Lemma 1.6.32 we readily get:

Lemma 1.6.36 *If* $A \approx B$ *then* $|\mathrm{ap}(A)| = |\mathrm{ap}(B)|$.

An example of a class of diagonally dominant matrices is the set of strongly definite matrices, since the weight of every permutation is the sum of the weights of constituent cycles, which are all nonpositive and the weight of id is 0.

A nonpositive matrix with zero diagonal is called *normal* (thus every normal matrix is strongly definite but not conversely). A normal matrix whose all off-diagonal elements are negative is called *strictly normal*. Obviously, a strictly normal matrix has strong permanent. We have

strictly normal $\Longrightarrow$ normal $\Longrightarrow$ strongly definite $\Longrightarrow$ diagonally dominant.
(1.29)

As a consequence of Theorem 1.6.34 we have:

Theorem 1.6.37 *Every square matrix A with finite* $\mathrm{maper}(A)$ *is similar to a normal matrix, that is, there exist generalized permutation matrices P and Q such that* $P \otimes A \otimes Q$ *is normal.*

A normal matrix similar to a matrix A may not be unique. Any such matrix will be called a *normal form* of A.

Corollary 1.6.38 *A normal form of any square matrix $A \in \overline{\mathbb{R}}^{n \times n}$ with finite* maper(A) *can be found using the Hungarian method in* $O(n^3)$ *time.*

Not every square matrix is similar to a strictly normal (for instance a constant matrix). This question is related to strong regularity of matrices in max-algebra and will be revisited in Chap. 6.

We are now ready to present a method for checking whether a matrix has strong permanent. Let $A = (a_{ij}) \in \overline{\mathbb{R}}^{n \times n}$. If maper$(A) = \varepsilon$ then A does not have strong permanent. Suppose now that maper$(A) > \varepsilon$. Due to the Hungarian method we can find a normal matrix B similar to A. By Lemma 1.6.36 A has strong permanent if and only if B has the same property. Every permutation is a product of elementary cycles, therefore if $w(\pi, B) = 0$ for some $\pi \neq$ id then at least one of the constituent cycles of π is of length two or more or, equivalently, there is a cycle of length two or more in the digraph Z_B. Conversely, every such cycle can be extended using the complementary diagonal zeros in B to a permutation of zero weight with respect to B, different from id. Thus we have:

Theorem 1.6.39 [24] *A square matrix has strong permanent if and only if the zero digraph of any (and thus of all) of its normal forms contains no cycles other than the loops (that is, it becomes acyclic once all loops are removed).*

Checking that a digraph is acyclic can be done using standard techniques [120] in linear time expressed in terms of the number of arcs.

Note that an early paper [82] on matrix scaling contains results which are closely related to Theorem 1.6.39.

Another aspect of the assignment problem that will be useful is the following simple transformation: Once an optimal solution to the assignment problem for a matrix A is known, it is trivial to permute the columns of A so that id $\in$ ap(A). By subtracting the diagonal entries from their columns we readily get a matrix that is not only diagonally dominant but also has all diagonal entries equal to 0. Hence this matrix is strongly definite. We summarize:

Proposition 1.6.40 *If $A \in \overline{\mathbb{R}}^{n \times n}$ has finite* maper(A) *then there is a generalized permutation matrix Q such that $A \otimes Q$ is strongly definite. The matrix Q can be found using $O(n^3)$ operations.*

Finally we discuss the question of parity of optimal permutations for the assignment problem, which will be useful in Chap. 6.

As usual [111], we define the *sign* of a cyclic permutation (cycle) $\sigma = (i_1 i_2 \cdots i_k)$ as sgn$(\sigma) = (-1)^{k-1}$. The integer k is called the *length* of the cycle σ. If $\pi_1, \ldots, \pi_r$ are the constituent cycles of a permutation $\pi \in P_n$ then the *sign* of π is

$$\text{sgn}(\pi) = \text{sgn}(\pi_1) \cdots \text{sgn}(\pi_k).$$

A permutation π is *odd* if $\mathrm{sgn}(\pi) = -1$ and *even* otherwise. We denote the set of odd (even) permutations of N by P_n^- (P_n^+). Straightforwardly from the definitions we get:

Lemma 1.6.41 *If π is an odd permutation then at least one of the constituent cycles of π has an even length.*

In Chap. 6 it will important to decide whether all permutations in $ap(A)$ are of the same parity. We therefore denote

$$\mathrm{ap}^+(A) = \mathrm{ap}(A) \cap P_n^+,$$
$$\mathrm{ap}^-(A) = \mathrm{ap}(A) \cap P_n^-$$

and

$$\mathrm{maper}^+(A) = \max_{\pi \in P_n^+} \sum_{i \in N} a_{i,\pi(i)},$$
$$\mathrm{maper}^-(A) = \max_{\pi \in P_n^-} \sum_{i \in N} a_{i,\pi(i)}.$$

Example 1.6.42 For the matrix A of Example 1.6.31 we have

$$\mathrm{ap}^+(A) = \{(123)\},$$
$$\mathrm{ap}^-(A) = \{(1)(23), (12)(3)\}$$

and

$$\mathrm{maper}^+(A) = \mathrm{maper}(A) = \mathrm{maper}^-(A).$$

It is obvious that the following three statements are equivalent:

$$\mathrm{ap}^+(A) \neq \mathrm{ap}(A) \neq \mathrm{ap}^-(A),$$
$$\mathrm{maper}^+(A) = \mathrm{maper}^-(A),$$
$$\mathrm{ap}^+(A) \neq \emptyset \quad \text{and} \quad \mathrm{ap}^-(A) \neq \emptyset.$$

Adding a constant to a row or column affects neither $\mathrm{ap}^+(A)$ nor $\mathrm{ap}^-(A)$. On the other hand a permutation of the rows or columns either swaps these two sets or leaves them unchanged. Hence we deduce:

Lemma 1.6.43 *If $A \approx B$ then either $\mathrm{ap}^+(A) = \mathrm{ap}^+(B)$ and $\mathrm{ap}^-(A) = \mathrm{ap}^-(B)$ or $\mathrm{ap}^+(A) = \mathrm{ap}^-(B)$ and $\mathrm{ap}^-(A) = \mathrm{ap}^+(B)$.*

Due to Lemma 1.6.43 and Theorem 1.6.37 we may assume that A is normal, thus $\mathrm{id} \in \mathrm{ap}(A)$ and therefore the question whether all optimal permutations are of the same parity reduces to deciding whether $\mathrm{ap}^-(A) \neq \emptyset$. Since A is normal

$\mathrm{ap}(A) = \{\pi \in P_n; a_{i,\pi(i)} = 0\}$. If $\pi \in \mathrm{ap}(A)$ then all constituent cyclic permutations of π can be identified as cycles in the digraph Z_A. We say that a cycle in a digraph is *odd* (*even*) if its length is odd (even). If $\pi \in \mathrm{ap}^-(A)$ then at least one of its constituent cycles is of odd parity and therefore its corresponding cycle in Z_A is even (Lemma 1.6.41). Also conversely, if there is an even cycle, say $(i_1, i_2, \ldots, i_k, i_1)$ in Z_A then the corresponding cyclic permutation $\sigma : i_1 \longrightarrow i_2 \longrightarrow \cdots \longrightarrow i_k \longrightarrow i_1$ is of odd parity and when complemented by loops (i, i) for $i \in N - \{i_1, i_2, \ldots, i_k\}$, the obtained permutation is odd, since loops are even cyclic permutations. We can summarize:

Theorem 1.6.44 *The problem of deciding whether all optimal permutations for an assignment problem are of the same parity is polynomially equivalent to the problem of deciding whether a digraph contains an even cycle ("Even Cycle Problem"). Once an even cycle in Z_A is known, optimal permutations of both parities can readily be identified.*

Remark 1.6.45 The computational complexity of the Even Cycle Problem was unresolved for almost 30 years until 1999 when an $O(n^3)$ algorithm was published [124].

Note that the problem of finding $\mathrm{maper}^+(A)$ and $\mathrm{maper}^-(A)$ has still unresolved computational complexity [29].

We close this subsection by a max-algebraic analogue of the *van der Waerden Conjecture*. Recall that an $n \times n$ matrix $A = (a_{ij})$ is called *doubly stochastic*, if all $a_{ij} \geq 0$ and all row and column sums of A equal 1.

Theorem 1.6.46 [20] (Max-algebraic van der Waerden Conjecture) *Among all doubly stochastic $n \times n$ matrices the max-algebraic permanent obtains its minimum for the matrix $A = (a_{ij})$, where $a_{ij} = \frac{1}{n}$ for all $i, j \in N$.*

Proof We have $\mathrm{maper}(A) = \max_{\pi \in P_n} \sum_{1 \leq i \leq n} a_{i,\pi(i)} = 1$. Assume that there is a doubly stochastic matrix $X = (x_{ij})$ with $\max_{\pi \in P_n} \sum_{1 \leq i \leq n} x_{i,\pi(i)} < 1$. Then we get for all permutations π: $\sum_{1 \leq i \leq n} x_{i,\pi(i)} < 1$. This holds in particular for the permutations π_k which map i to $i + k$ modulo n for $i = 1, 2, \ldots, n$ and $k = 0, 1, \ldots, n - 1$. Thus we get

$$n = \sum_{i=1}^{n} \sum_{j=1}^{n} x_{ij} = \sum_{k=0}^{n-1} \sum_{i=1}^{n} x_{i,\pi_k(i)} < n,$$

a contradiction. Therefore the matrix A yields the least optimal value for the max-algebraic permanent. $\qquad\square$

1.7 Exercises

Exercise 1.7.1 Evaluate the following expressions:

(a) $14 \otimes 3^2 \oplus 3 \otimes 5^8$ (all operations are max-algebraic). [The result is 43]

(b) $\begin{pmatrix} 4 & -1 & 5 \\ 0 & 3 & -2 \end{pmatrix} \otimes \begin{pmatrix} 7 & 1 \\ -3 & 4 \\ 5 & 3 \end{pmatrix}$. $[\begin{pmatrix} 11 & 8 \\ 3 & 7 \end{pmatrix}]$

(c) $3 \otimes A^2 \oplus A^3$, where $A = \begin{pmatrix} 2 & 0 \\ -1 & 3 \end{pmatrix}$. $[\begin{pmatrix} 7 & 6 \\ 5 & 9 \end{pmatrix}]$

(d) $A \otimes A^*, A^* \otimes A$, where $A = \begin{pmatrix} 3 & 2 \\ 1 & 5 \\ -3 & 0 \end{pmatrix}$. $[A \otimes A^* = \begin{pmatrix} 0 & 2 & 6 \\ 3 & 0 & 5 \\ -2 & -4 & 0 \end{pmatrix}, A^* \otimes A = \begin{pmatrix} 0 & 4 \\ 1 & 0 \end{pmatrix}]$

Exercise 1.7.2 Prove that $(A \oplus B)^* = A^* \oplus' B^*$ and $(A \otimes B)^* = B^* \otimes' A^*$ hold for any matrices A and B of compatible sizes. Use this to find $A \otimes' A^*, A^* \otimes' A$ for the matrix A of Exercise 1.7.1(d).

Exercise 1.7.3 About each of the matrices below decide whether it is definite and whether it is increasing. If it is definite then find also its weak transitive closure.

(a) $\begin{pmatrix} -2 & -1 \\ 3 & 0 \end{pmatrix}$. [Not increasing; not definite, positive cycle $(1, 2, 1)$]

(b) $\begin{pmatrix} -1 & 2 \\ -3 & -4 \end{pmatrix}$. [Not increasing; not definite, there is no zero cycle]

(c) $\begin{pmatrix} 0 & 2 \\ -3 & -4 \end{pmatrix}$. [Definite but not increasing, $\begin{pmatrix} 0 & 2 \\ -3 & -1 \end{pmatrix}$]

(d) $\begin{pmatrix} 3 & 2 \\ -5 & 0 \end{pmatrix}$. [Increasing; not definite, positive cycle $(1, 1)$]

(e) $\begin{pmatrix} 0 & 1 \\ -2 & 0 \end{pmatrix}$. [Definite and increasing (hence strongly definite), $\begin{pmatrix} 0 & 1 \\ -2 & 0 \end{pmatrix}$]

(f) $\begin{pmatrix} 0 & 2 & -4 & 1 \\ -3 & 0 & -2 & 0 \\ -5 & 1 & 0 & 1 \\ -4 & -2 & -3 & 0 \end{pmatrix}$. [Definite and increasing (hence strongly definite),

$\begin{pmatrix} 0 & 2 & 0 & 2 \\ -3 & 0 & -2 & 0 \\ -2 & 1 & 0 & 1 \\ -4 & -2 & -3 & 0 \end{pmatrix}$]

(g) $\begin{pmatrix} 0 & 2 & -4 & 1 \\ -3 & 0 & -2 & 0 \\ -5 & 2 & 0 & 1 \\ -4 & -2 & -1 & 0 \end{pmatrix}$. [Increasing; not definite, positive cycle $(2, 4, 3)$]

Exercise 1.7.4 (Symmetric matrices) Let $A \in \mathbb{R}^{n \times n}$ be symmetric. Prove then that:

(a) $\lambda(A) = \max_{i,j} a_{ij}$.
(b) There is a symmetric matrix B in normal form such that $ap(A) = ap(B)$. [See [19]]

(c) If A is also diagonally dominant then $\lambda(A) = \max_i a_{ii}$ and a best nondiagonal permutation has the form $(k, l) \circ \mathrm{id}$. Deduce then that both maper$^+(A)$ and maper$^-(A)$ can be found in $O(n^2)$ time. [See [29]]

Exercise 1.7.5 (Monge matrices) A matrix $A \in \mathbb{R}^{n \times n}$ is called Monge if $a_{ij} + a_{kl} \geq a_{il} + a_{kj}$ for all i, j, k, l such that $1 \leq i \leq k \leq n$ and $1 \leq j \leq l \leq n$. Prove that

(a) Every Monge matrix is diagonally dominant.
(b) If A is Monge and normal then a best nondiagonal permutation has the form $(k, k+1) \circ \mathrm{id}$.
     Deduce then that both maper$^+(A)$ and maper$^-(A)$ can be found in $O(n)$ time. [See [29]]

Exercise 1.7.6 (Matrix sums) For each of the following relations prove or disprove that it holds for all matrices $A, B \in \mathbb{R}^{n \times n}$:

(a) maper$(A \oplus B) \geq$ maper$(A) \oplus$ maper(B). [true; take $\pi \in ap(A)$ and show that $w(\pi, A) \leq$ maper$(A \oplus B)$]
(b) maper$(A \oplus B) \leq$ maper$(A) \oplus$ maper(B). [false]
(c) $\lambda(A \oplus B) \geq \lambda(A) \oplus \lambda(B)$. [true; take σ critical in A and show that $\mu(\sigma, A) \leq \lambda(A \oplus B)$]
(d) $\lambda(A \oplus B) \leq \lambda(A) \oplus \lambda(B)$. [false]

Exercise 1.7.7 (Matrix products) For each of the following relations prove or disprove that it holds for all matrices $A, B \in \mathbb{R}^{n \times n}$:

(a) maper$(A \otimes B) \geq$ maper$(A) \otimes$ maper(B). [true]
(b) maper$(A \otimes B) \leq$ maper$(A) \otimes$ maper(B). [false]
(c) $\lambda(A \otimes B) \geq \lambda(A) \otimes \lambda(B)$. [false]
(d) $\lambda(A \otimes B) \leq \lambda(A) \otimes \lambda(B)$. [false]
(e) $\lambda(A \otimes B) = \lambda(A \otimes B)$. [true]

Exercise 1.7.8 (AA^* products) Let $A \in \mathbb{R}^{n \times n}$ and P be a matrix product formed as follows: Write the letters A and A^* alternatingly starting by any of them, insert the product signs $\otimes$ and $\otimes'$ alternatingly between them and insert brackets so that a meaningful algebraic expression is obtained. Prove that if the total number of letters is odd then P is equal to the first symbol; if the total number is even then P is equal to the product of the first two letters. [See [60]]

Exercise 1.7.9 Two cross city line trains arrive at the central railway station C. One arrives at platform 1 from suburb A after a 40 minute journey, the other one at platform 7 from suburb B, journey time 30 minutes. Two trains connecting to both these trains leave from platforms 3 and 10 at 10.20 and 10.25, respectively. Find the latest times at which the cross city line trains should depart from A and B so that the passengers can board the connecting trains. Describe this problem as a problem of solving a max-algebraic system of simultaneous equations. Take into account times for changing the trains between platforms given in the following table:

Platform	3	10
1	6	15
7	8	4

$$\left[\begin{pmatrix} 46 & 38 \\ 55 & 34 \end{pmatrix} \otimes x = \begin{pmatrix} 80 \\ 85 \end{pmatrix}, \text{departures: } 9.30, 9.42\right]$$

Exercise 1.7.10 INDULGE produces milk chocolate bars in department D1 and drinking chocolate in department D2. Production runs in stages. D1 also simultaneously prepares milk (pasteurization etc.) for use by both departments in the next stage and similarly, D2 also prepares cocoa powder for both departments. At every stage each department prepares sufficient amount of milk and powder for both departments to run the next stage. The milk preparation takes 2 hours, cocoa powder 5, production of bars 3 and drinking chocolate 6 hours. Set up max-algebraic equations for starting times of the departments in stages $2, 3, \ldots$ depending on the starting times of the first stage. Then find the starting times of stages $2, 3, \ldots$ if (a) both departments start to work at the same time, (b) D1 starts 3 hours earlier than D2, (c) D1 starts 5 hours later than D2. You may assume that at the beginning of the first stage there are sufficient amounts of both cocoa powder and milk in stock to run the first stage.

$$\left[x(r+1) = \begin{pmatrix} 3 & 5 \\ 2 & 6 \end{pmatrix} \otimes x(r) \quad (r = 0, 1, \ldots);\right.$$

(a) $(0,0)^T$, $(5,6)^T$, $(11,12)^T$, $(17,18)^T, \ldots$; (b) $(0,3)^T$, $(8,9)^T$, $(14,15)^T$, $(20,21)^T, \ldots$; (c) $(5,0)^T$, $(8,7)^T$, $(12,13)^T$, $(18,19)^T, \ldots]$

Exercise 1.7.11 The matrix

$$A = \begin{pmatrix} 2 & 4 & 3 \\ 1 & 1 & 5 \\ 0 & 1 & 0 \end{pmatrix}$$

is the technological matrix of an MMIPP with starting vector $x = (0,0,0)^T$. Generate the starting time vectors of the first stages until periodicity is reached. Describe the periodic part by a formula. (This question is revisited in Exercise 9.4.2.)
$[(4,5,1)^T, (9,6,6)^T, (11,11,9)^T, (15,14,12)^T, (18,17,15)^T; \lambda(A) = 3;$

$$x(r+1) = 3 \otimes x(r) = (15 + 3(r-4), 14 + 3(r-4), 12 + 3(r-4))^T (r \geq 4)]$$

Exercise 1.7.12 The same task as in Exercise 1.7.11 but for the production matrix

$$A = \begin{pmatrix} 4 & 1 & 3 \\ 3 & 0 & 3 \\ 5 & 2 & 4 \end{pmatrix}.$$

$[(4,3,5)^T$, $(8,8,9)^T$, $(12,12,13)^T$; $\lambda(A) = 4$; $x(r+1) = 4 \otimes x(r) = (8 + 3(r-2), 8 + 3(r-2), 9 + 3(r-2))^T (r \geq 2)]$

Chapter 2
Max-algebra: Two Special Features

The aim of this chapter is to highlight two special features of max-algebra which make it unique as a modelling and solution tool: the ability to efficiently describe *all* solutions to some problems where it would otherwise be awkward or impossible to do so; and the potential to describe combinatorial problems algebraically.

First we show an example of a problem where max-algebra can help to efficiently find all solutions and, consequently, find a solution satisfying additional requirements (Sect. 2.1).

Then in Sect. 2.2 we show that using max-algebra a number of combinatorial and combinatorial optimization problems can be formulated in algebraic terms. Based on this max-algebra may, to some extent, be considered "an algebraic encoding" of combinatorics [27].

This chapter may be skipped without loss of continuity in reading this book.

2.1 Bounded Mixed-integer Solution to Dual Inequalities: A Mathematical Application

2.1.1 Problem Formulation

A special feature of max-algebra is the ability to efficiently describe the set of all solutions to some problems in contrast to standard approaches, using which we can usually find one solution. Finding all solutions may be helpful for identifying solutions that satisfy specific additional requirements. As an example consider the systems of the form

$$x_i - x_j \geq b_{ij} \quad (i, j = 1, \ldots, n) \tag{2.1}$$

where $B = (b_{ij}) \in \mathbb{R}^{n \times n}$. In [55] the matrix of the left-hand side coefficients of this system is called the *dual network matrix*. It is the transpose of the constraint matrix of a circulation problem in a network (such as the maximum flow or minimum-cost

P. Butkovič, *Max-linear Systems: Theory and Algorithms*,
Springer Monographs in Mathematics 151,
DOI 10.1007/978-1-84996-299-5_2, © Springer-Verlag London Limited 2010

flow problem) and inequalities of the form (2.1) therefore appear as dual inequalities for this type of problems. These facts motivate us to call (2.1) the *system of dual inequalities* (SDI). The aim of this section is to show that using standard max-algebraic techniques it is possible to generate the set of all solutions to (2.1) (which is of size $n^2 \times n$) using n generators. This description enables us then to find, or to prove that it does not exist, a *bounded mixed-integer solution* to the system of dual inequalities, that is, a vector $x = (x_1, \ldots, x_n)^T$ satisfying:

$$\left. \begin{array}{ll} x_i - x_j \geq b_{ij}, & (i, j \in N) \\ u_j \geq x_j \geq l_j, & (j \in N) \\ x_j \text{ integer}, & (j \in J) \end{array} \right\} \qquad (2.2)$$

where $u = (u_1, \ldots, u_n)^T, l = (l_1, \ldots, l_n)^T \in \mathbb{R}^n$ and $J \subseteq N = \{1, \ldots, n\}$ are given. We will refer to this problem as BMISDI. Note that without loss of generality u_j and l_j may be assumed to be integer for $j \in J$. This type of a system of inequalities has been studied for instance in [55] where it has been proved that a related mixed-integer feasibility question is *NP*-complete.

We will show that, in general, the application of max-algebra leads to a pseudopolynomial algorithm for solving BMISDI. However, an explicit solution is described in the case when B is integer (but still a mixed-integer solution is wanted). This implies that BMISDI can be solved using $O(n^3)$ operations when B is an integer matrix. Note that when $J = \emptyset$ then BMISDI is polynomially solvable since it is a set of constraints of a linear program. When $J = N$ and B is integer then BMISDI is also polynomially solvable since the matrix of the system is totally unimodular [120].

2.1.2 All Solutions to SDI and All Bounded Solutions

The system of inequalities

$$x_i - x_j \geq b_{ij} \quad (i, j \in N)$$

is equivalent to

$$\max_{j \in N} (b_{ij} + x_j) \leq x_i \quad (i \in N).$$

In max-algebraic notation this reads

$$\sum_{j \in N}^{\oplus} b_{ij} \otimes x_j \leq x_i \quad (i \in N)$$

or in the compact form

$$B \otimes x \leq x. \qquad (2.3)$$

Recall that using the notation introduced in Sect. 1.6.2 the set of finite solutions to (2.3) is $V_0^*(B)$.

The next theorem is straightforwardly deduced from Theorem 1.6.18.

Theorem 2.1.1 *If $B \in \mathbb{R}^{n \times n}$ then*

1. $V_0^*(B) \neq \emptyset$ *if and only if* $\lambda(B) \leq 0$.
2. *If* $V_0^*(B) \neq \emptyset$ *then*

$$V_0^*(B) = \left\{ \Delta(B) \otimes z; z \in \mathbb{R}^n \right\}.$$

We can now use Theorems 2.1.1 and 1.6.25 to describe all bounded solutions to SDI.

Corollary 2.1.2 *The set of all solutions x to SDI satisfying $x \leq u$ is*

$$\left\{ \Delta(B) \otimes z; z \leq (\Delta(B))^* \otimes' u \right\}$$

and if this set is nonempty then the vector $\Delta(B) \otimes ((\Delta(B))^* \otimes' u)$ is the greatest element of this set. Hence the inequality

$$l \leq \Delta(B) \otimes \left((\Delta(B))^* \otimes' u \right)$$

is necessary and sufficient for the existence of a solution to SDI satisfying $l \leq x \leq u$.

2.1.3 Solving BMISDI

We start with another corollary to Theorem 2.1.1.

Corollary 2.1.3 *A necessary condition for BMISDI to have a solution is that $\lambda(B) \leq 0$. If this condition is satisfied then BMISDI is equivalent to finding a vector $z \in \mathbb{R}^n$ such that*

$$l \leq \Delta(B) \otimes z \leq u$$

and

$$(\Delta(B) \otimes z)_j \in \mathbb{Z} \quad \text{for } j \in J.$$

In the rest of this subsection we will assume without loss of generality (Theorem 2.1.1) that $\lambda(B) \leq 0$.

Theorem 2.1.4 *Let $A \in \mathbb{R}^{n \times n}, b \in \mathbb{R}^n$ and $J \subseteq N$. Let $\tilde{b}$ be defined by*

$$\tilde{b}_j = \lfloor b_j \rfloor \quad \text{for } j \in J,$$
$$\tilde{b}_j = b_j \quad \text{for } j \notin J.$$

Then the following are equivalent:

1. *There exists a $z \in \mathbb{R}^n$ such that $l \leq A \otimes z \leq b$ and*

$$(A \otimes z)_j \in \mathbb{Z} \quad \text{for } j \in J.$$

2. *There exists a $z \in \mathbb{R}^n$ such that $l \leq A \otimes z \leq \tilde{b}$ and*

$$(A \otimes z)_j \in \mathbb{Z} \quad for \in J.$$

3. *There exists a $z \in \mathbb{R}^n$ such that $l \leq A \otimes z \leq A \otimes (A^* \otimes' \tilde{b})$ and*

$$(A \otimes z)_j \in \mathbb{Z} \quad for \ j \in J.$$

Proof 1. $\Longleftrightarrow$ 2. is trivial, 2. $\Longleftrightarrow$ 3. follows from Theorem 1.6.25, Corollary 1.6.26 and Lemma 1.1.1. $\qquad\square$

Theorem 2.1.4 enables us to compile the following algorithm.

Algorithm 2.1.5 BMISDI
   Input: $B \in \mathbb{R}^{n \times n}$, $u, l \in \mathbb{R}^n$ and $J \subseteq N$.
   Output: x satisfying (2.2) or an indication that no such vector exists.

1. $A := \Delta(B), x := u$
2. $x_j := \lfloor x_j \rfloor$ for $j \in J$
3. $z := A^* \otimes' x, x := A \otimes z$
4. If $l \not\leq x$ then stop (no solution)
5. If $l \leq x$ and $x_j \in \mathbb{Z}$ for $j \in J$ then stop else go to 2.

Theorem 2.1.6 [30] *The algorithm BMISDI is correct and requires $O(n^3 + n^2 L)$ operations of addition, maximum, minimum, comparison and integer part, where*

$$L = \sum\nolimits_{j \in J} \left(u_j - l_j \right).$$

Proof If the algorithm terminates at step 4 then there is no solution by the repeated use of Theorem 2.1.4.

The sequence of vectors x constructed by this algorithm is nonincreasing by Corollary 1.6.26 and hence $x = A \otimes z \leq u$ if it terminates at step 5. The remaining requirements of (2.2) are satisfied explicitly due to the conditions in step 5.

Computational complexity: The calculation of $\Delta(B)$ is $O(n^3)$ by Theorem 1.6.22. Each run of the loop between steps 2 and 5 is $O(n^2)$. In every iteration at least one component of $x_j, j \in J$ decreases by one and the statement now follows from the fact that all x_j range between l_j and u_j. $\qquad\square$

Example 2.1.7 Let

$$B = \begin{pmatrix} -2 & 2.7 & -2.1 \\ -3.8 & -1 & -5.2 \\ 1.6 & 3.5 & -3 \end{pmatrix},$$

$u = (5.2, 0.8, 7.4)^T$ and $J = \{1, 3\}$ (l is not specified). The algorithm BMISDI will find:

$$A = \Delta(B) = \begin{pmatrix} 0 & 2.7 & -2.1 \\ -3.6 & 0 & -5.2 \\ 1.6 & 4.3 & 0 \end{pmatrix},$$

$$x = (5, 0.8, 7)^T,$$

$$z = A^* \otimes' x = \begin{pmatrix} 0 & 3.6 & -1.6 \\ -2.7 & 0 & -4.3 \\ 2.1 & 5.2 & 0 \end{pmatrix} \otimes' x = \begin{pmatrix} 4.4 \\ 0.8 \\ 6 \end{pmatrix}$$

and

$$x = A \otimes z = (4.4, 0.8, 6)^T.$$

Now $x_1 \notin \mathbb{Z}$ so the algorithm continues by another iteration: $x = (4, 0.8, 6)^T$,

$$z = A^* \otimes' x = (4, 0.8, 6)^T$$

and

$$x = A \otimes z = (4, 0.8, 6)^T,$$

which is a solution (provided that $l \leq x$ since otherwise there is no solution) to the BMISDI since $x_1, x_3 \in \mathbb{Z}$.

2.1.4 Solving BMISDI for Integer Matrices

In this subsection we prove that a solution to the BMISDI can be found explicitly if B is integer. The following will be useful (the proof below is a simplification of the original proof due to [132]):

Theorem 2.1.8 [30] *Let $A \in \mathbb{Z}^{n \times n}$, $b \in \mathbb{R}^n$ and $A \otimes x = b$ for some $x \in \mathbb{R}^n$. Let $J \subseteq N$ and $\tilde{b}$ be defined by*

$$\tilde{b}_k = \lfloor b_k \rfloor \quad for \in J,$$

$$\tilde{b}_k = b_k \quad for\ k \notin J.$$

Then there exists an $\tilde{x} \in \mathbb{R}^n$ such that

$$A \otimes \tilde{x} \leq \tilde{b}$$

and

$$(A \otimes \tilde{x})_k = \tilde{b}_k \quad for\ k \in J.$$

Proof Without loss of generality assume that $b_k \notin \mathbb{Z}$ for some $k \in J$, then the set

$$S = \{s \in N; a_{ks} + x_s > \lfloor b_k \rfloor \text{ for some } k \in J\}$$

is nonempty and $x_s \notin \mathbb{Z}$ for every $s \in S$ since A is integer. Let $\tilde{x} \in \mathbb{R}^n$ be defined by $\tilde{x}_j = \lfloor x_j \rfloor$ for $j \in S$ and $\tilde{x}_j = x_j$ otherwise. Clearly $\tilde{x} \leq x$ and so $A \otimes \tilde{x} \leq A \otimes x$ by Lemma 1.1.1. Hence $\max_{j \in N}(a_{kj} + \tilde{x}_j) \leq b_k = \tilde{b}_k$ for all $k \notin J$. At the same time $\max_{j \in N}(a_{kj} + \tilde{x}_j) = \lfloor b_k \rfloor = \tilde{b}_k$ for all $k \in J$. □

For the main application, Theorem 2.1.10 below, it will be convenient to deduce from the statement of Theorem 2.1.8 a property of the greatest solution $\overline{x}$ to $A \otimes x \leq \tilde{b}$ (Corollary 1.6.26):

Corollary 2.1.9 *Under the assumptions of Theorem 2.1.8 and using the same notation, if $\overline{x} = A^* \otimes' \tilde{b}$ then*

$$A \otimes \overline{x} \leq \tilde{b}$$

and

$$(A \otimes \overline{x})_k = \tilde{b}_k \quad for \, k \in J.$$

Proof The inequality follows from Corollary 1.6.26. Let $\tilde{x}$ be the vector described in Theorem 2.1.8. By Theorem 1.6.25 we have $\tilde{x} \leq \overline{x}$ implying that

$$\tilde{b}_k = (A \otimes \tilde{x})_k \leq (A \otimes \overline{x})_k \leq \tilde{b}_k \quad for \, k \in J$$

which concludes the proof. □

Finally, we are prepared to use max-algebra and explicitly describe a solution to BMISDI in the case when B is an integer matrix:

Theorem 2.1.10 *Let $B \in \mathbb{Z}^{n \times n}$, $\lambda(B) \leq 0$, $A = \Delta(B)$, $b = A \otimes (A^* \otimes' u)$ and $\tilde{b}$ be defined by*

$$\tilde{b}_k = \lfloor b_k \rfloor \quad for \, k \in J$$

and

$$\tilde{b}_k = b_k \quad for \, k \notin J.$$

Then the BMISDI has a solution if and only if

$$l \leq A \otimes \left(A^* \otimes' \tilde{b} \right),$$

and $\hat{x} = A \otimes (A^* \otimes' \tilde{b})$ is then the greatest solution (that is, $y \leq \hat{x}$ for any solution y).

Proof Note first that A is an integer matrix and we therefore may apply Corollary 2.1.9 to A.

"If": By Corollary 1.6.26 $\hat{x} \leq \tilde{b} \leq b \leq u$. Let us take in Corollary 2.1.9 (and Theorem 2.1.8) $x = A^* \otimes' u$. Then $\hat{x} = A \otimes \bar{x}$ and so $\hat{x}_k \in \mathbb{Z}$ for $k \in J$.

"Only if": Let y be a solution. Then $y = A \otimes w \leq u$ for some $w \in \mathbb{R}^n$, thus by Theorem 1.6.25

$$w \leq A^* \otimes' u$$

and so

$$y = A \otimes w \leq A \otimes \left(A^* \otimes' u \right) = b.$$

Since $y_k \in \mathbb{Z}$ for $k \in J$ we also have

$$A \otimes w = y \leq \tilde{b}.$$

Hence by Theorem 1.6.25

$$w \leq A^* \otimes' \tilde{b}$$

and by Lemma 1.1.1 then

$$l \leq y = A \otimes w \leq A \otimes \left(A^* \otimes' \tilde{b} \right) = \hat{x}.$$

We also have $\hat{x} \leq \tilde{b} \leq b \leq u$ by Corollary 1.6.26 and $\hat{x}_k \in \mathbb{Z}$ for $k \in J$ by Corollary 2.1.9 as above, hence $\hat{x}$ is the greatest solution.   □

Example 2.1.11 Let

$$B = \begin{pmatrix} -2 & 2 & -2 \\ -3 & -1 & -4 \\ 1 & 3 & -3 \end{pmatrix},$$

$u = (3.5, 0.8, 5.7)^T$ and $J = \{1, 3\}$ (l is not specified). Then we have:

$$A = \Delta(B) = \begin{pmatrix} 0 & 2 & -2 \\ -3 & 0 & -4 \\ 1 & 3 & 0 \end{pmatrix},$$

$$A^* \otimes' u = \begin{pmatrix} 0 & 3 & -1 \\ -2 & 0 & -3 \\ 2 & 4 & 0 \end{pmatrix} \otimes' u = \begin{pmatrix} 3.5 \\ 0.8 \\ 4.8 \end{pmatrix},$$

$$b = A \otimes \left(A^* \otimes' u \right) = \begin{pmatrix} 3.5 \\ 0.8 \\ 4.8 \end{pmatrix},$$

$$\tilde{b} = \begin{pmatrix} 3 \\ 0.8 \\ 4 \end{pmatrix}$$

and

$$\hat{x} = A \otimes \left(A^* \otimes' \tilde{b} \right) = (3, 0.8, 4)^T.$$

By Theorem 2.1.10 $\hat{x}$ is the greatest solution to the BMISDI provided that $l \leq \hat{x}$ (otherwise there is no solution).

2.2 Max-algebra and Combinatorial Optimization

There is a number of combinatorial and combinatorial optimization problems closely related to max-algebra. In some cases max-algebra provides an efficient and elegant algebraic encoding of these problems. Although computational advantages do not necessarily follow from the max-algebraic formulation, for some problems this connection may help to deduce useful information [27].

2.2.1 Shortest/Longest Distances: Two Connections

Perhaps the most striking example is the *shortest-distances problem* which is one of the best known combinatorial optimization problems:

Given an $n \times n$ matrix A of direct distances between n places, find the matrix $\tilde{A}$ of shortest distances (that is, the matrix of the lengths of shortest paths between any pair of places).

It is known that the shortest-distances matrix exists if and only if there are no negative cycles in D_A. For the shortest-distances problem we may assume without loss of generality that all diagonal elements of A are 0.

We could continue from this and show a link to min-algebra; however, to be consistent with the rest of the book we shall formulate these results in max-algebraic terms, similarly as in Example 1.2.3. Hence the considered combinatorial optimization problem is:

Given an $n \times n$ matrix A of direct distances between n places, find the matrix $\tilde{A}$ of longest distances (that is, the matrix of the lengths of longest paths between any pair of places).

We may assume that all diagonal elements of A are 0 and that there are no positive cycles in D_A, thus A is strongly definite. We have seen in Sect. 1.6.2 that $\Gamma(A)$ is exactly $\tilde{A}$. By Proposition 1.6.12 then $\tilde{A} = A^{n-1}$. We have:

Theorem 2.2.1 *If $A \in \mathbb{R}^{n \times n}$ is a strongly definite direct-distances matrix then all matrices A^j ($j \geq n - 1$) are equal to the longest-distances matrix for D_A. Hence, the kth column ($k = 1, \ldots, n$) of A^j ($j \geq n - 1$) is the vector of longest distances to node k in D_A.*

One benefit of this result is that the longest- (and similarly shortest-) distances matrix for a strongly definite direct-distances matrix A can be found simply by repeated max-algebraic squaring of A, that is,

$$A^2, A^4, A^8, A^{16}, \ldots$$

until a power A^j ($j \geq n - 1$) is reached (see Sect. 1.6.2).

However, there exists another max-algebraic interpretation of the longest-distances problem. We have seen in Proposition 1.6.17 that for a strongly definite matrix A every column v of A^j ($j \geq n - 1$) is an eigenvector of A, that is,

$$A \otimes v = v.$$

Corollary 2.2.2 *If $A \in \mathbb{R}^{n \times n}$ is a strongly definite direct-distances matrix then every vector of longest-distances to a node in D_A is a max-algebraic eigenvector of A corresponding to the eigenvalue 0.*

2.2.2 Maximum Cycle Mean

The maximum cycle mean of a matrix (denoted $\lambda(A)$ for a matrix A), has been defined in Sect. 1.6.1. As already mentioned, the problem of calculating $\lambda(A)$ was studied independently in combinatorial optimization [106, 109]. At the same time the maximum cycle mean is very important in max-algebra. It is

- the eigenvalue of every matrix,
- the greatest eigenvalue of every matrix,
- the only eigenvalue whose corresponding eigenvectors may be finite.

Moreover, every eigenvalue of a matrix is the maximum cycle mean of some principal submatrix of that matrix.

All these and other aspects of the maximum cycle mean are proved in Chap. 4. Let us mention here a dual feature of the maximum cycle mean (see Corollary 4.5.6 and Theorem 1.6.29):

Theorem 2.2.3 *If $A \in \overline{\mathbb{R}}^{n \times n}$ then*

(a) $\lambda(A)$ *is the greatest eigenvalue of A, that is*

$$\lambda(A) = \max \left\{ \lambda \in \overline{\mathbb{R}}; \ A \otimes x = \lambda \otimes x, x \in \overline{\mathbb{R}}^n, x \neq \varepsilon \right\}$$

and, dually

(b)

$$\lambda(A) = \inf \left\{ \lambda \in \overline{\mathbb{R}}; \ A \otimes x \leq \lambda \otimes x, x \in \mathbb{R}^n \right\}.$$

2.2.3 The Job Rotation Problem

Characteristic maxpolynomials of matrices in max-algebra (Sect. 5.3) are related to the following *job rotation problem*. Suppose that a company with n employees requires these workers to swap their jobs (possibly on a regular basis) in order to avoid exposure to monotonous tasks (for instance manual workers at an assembly line, guards in a gallery or ride operators in a theme park). It may also be required that to maintain stability of service only a certain number of employees, say k $(k < n)$, actually swap their jobs. With each pair old job−new job a quantity may be associated expressing the cost (for instance for additional training) or the preference of the worker for this particular change. So the aim may be to select k employees and to suggest a schedule of the job swaps between them so that the sum of the parameters corresponding to these changes is either minimum or maximum. This task leads to finding a $k \times k$ principal submatrix of A for which the optimal assignment problem value is minimal or maximal (some entries can be set to $+\infty$ or $-\infty$ to avoid an assignment to the same or infeasible job). More formally, we deal with the *best principal submatrix problem* (BPSM):

Given a real $n \times n$ matrix A, for every $k \leq n$ find a $k \times k$ principal submatrix of A whose optimal assignment problem value is maximal.

Note that solving the assignment problem for all $\binom{n}{k}$ principal submatrices for each k would be computationally difficult since $\sum_{k=1}^{n} \binom{n}{k} = 2^n - 1$. No polynomial method for solving BPSM seems to be known, although its modification obtained after removing the word principal is known [73] and is polynomially solvable. This can also be seen from the following simple observation: Let $\widetilde{A}$ be the $(2n - k) \times (2n - k)$ matrix obtained from an $n \times n$ matrix $A \in \mathbb{R}^{n \times n}$ by adding $n - k$ rows and $n - k$ columns $(k < n)$ so that the entries in the intersection of these columns are $-\infty$ and the remaining new entries are zero, see Fig. 2.1. If the assignment problem is solved for $\widetilde{A}$ then every permutation selects $2n - k$ entries from $\widetilde{A}$. If A is finite then any optimal (maximizing) permutation avoids selecting entries from the intersection of the new columns and rows. But as it selects $n - k$ elements from the new rows and $n - k$ different elements from the new columns, it will select exactly $2n - k - 2(n - k) = k$ elements from A. No two of these k elements are from the same row or from the same column and so they represent a selection of k independent entries from a $k \times k$ submatrix of A. Their sum is maximum as the only elements taken from outside A are zero. So the best $k \times k$ submatrix problem can readily be solved as the classical assignment problem for a special matrix of order $2n - k$.

Unfortunately no similar trick seems to exist, that would enable us to find a best *principal* submatrix.

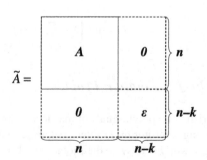

Fig. 2.1 Solving the best submatrix problem

Let us denote by δ_k the optimal value in the assignment problem for a best principal submatrix of order k ($k = 1, \ldots, n$). It will be proved in Sect. 5.3 that $\delta_1, \ldots, \delta_n$ are coefficients of the max-algebraic characteristic polynomial of A. It is not known whether the problem of finding all these quantities is an *NP*-complete or polynomially solvable problem (see Chap. 11). However, in Sect. 5.3.3 we will present a polynomial algorithm, based on the max-algebraic interpretation, for finding some and in some cases all these coefficients. Note that there is an indication that the problem of finding all coefficients is likely to be polynomially solvable as the following result suggests:

Theorem 2.2.4 [20] *If the entries of $A \in \mathbb{R}^{n \times n}$ are polynomially bounded, then the best principal submatrix problem for A and all k, $k \leq n$, can be solved by a randomized polynomial algorithm.*

2.2.4 Other Problems

In the table below (where SD stands for "strongly definite") is an overview of combinatorial or combinatorial optimization problems that can be formulated as max-algebraic problems [27]. The details of most of these links will be presented in the subsequent chapters.

Max-algebra	Combinatorics (0-1 entries)	Combinatorial Optimization
maper(A)	Term rank	Optimal value to the assignment problem
$A \otimes x = b$		
$\exists x$	Set covering	
$\exists! x$	Minimal set covering	
$\Gamma(A)$ if A SD	Transitive closure	Longest distances matrix
$A \otimes x = \lambda \otimes x$		
λ		Maximum cycle mean
x		Balancing coefficients
x if A SD	Connectivity to a node	Longest distances
x if A SD		Scaling to normal form
GM regularity	$\nexists$ even directed cycle 0-1 sign-nonsingularity	All optimal permutations of the same parity
Strong regularity	Digraph acyclic	Unique optimal permutation
Characteristic polynomial	$\exists$ exact cycle cover $\exists$ principal submatrix with > 0 permanent	Best principal submatrix (JRP)

2.3 Exercises

Exercise 2.3.1 The assignment problem for $A = (a_{ij}) \in \mathbb{R}^{n \times n}$ can be described as a (conventional) linear program

$$f(x) = \sum_{i,j \in N} a_{ij} x_{ij} \longrightarrow \max$$

s.t.

$$\sum_{j \in N} a_{ij} x_{ij} = 1, \quad i \in N,$$

$$\sum_{i \in N} a_{ij} x_{ij} = 1, \quad j \in N,$$

$$x_{ij} \geq 0.$$

Its dual is

$$g(u, v) = \sum_{i \in N} u_i + \sum_{j \in N} v_j \longrightarrow \min$$

s.t.

$$u_i + v_j \geq a_{ij}, \quad i, j \in N.$$

Show using max-algebra that $f^{\max} = g^{\min} = \mathrm{maper}(A)$.

(Hint: First show that $f \leq g$ and then prove the rest by using the results on the eigenproblem for strongly definite matrices.)

Exercise 2.3.2 A matrix $A = (a_{ij}) \in \mathbb{R}^{n \times n}$ is called pyramidal if $a_{ij} \geq a_{rs}$ whenever $\max(i, j) < \max(r, s)$. Prove that $\delta_k = \mathrm{maper}(A_k)$, where A_k is the principal submatrix of A determined by the first k row and column indices. [See [37].]

Chapter 3
One-sided Max-linear Systems and Max-algebraic Subspaces

Recall that one-sided max-linear systems are systems of equations of the form

$$A \otimes x = b \tag{3.1}$$

where $A \in \overline{\mathbb{R}}^{m \times n}$ and $b \in \overline{\mathbb{R}}^m$. They are closely related to systems of inequalities

$$A \otimes x \leq b. \tag{3.2}$$

Both were studied already in the first papers on max-algebra [57, 144] and the theory has further evolved in the 1960's and 1970's [149, 150], and later [24, 27].

It should be noted that one-sided max-linear systems can be solved more easily than their linear-algebraic counterparts. Also, unlike in conventional linear algebra, systems of inequalities (3.2) always have a solution $x \in \overline{\mathbb{R}}^n$ and the task of finding a solution to (3.1) is strongly related to the same task for the systems of inequalities. Note that, in contrast, the two-sided systems studied in Chap. 7 are much more difficult to solve.

In this chapter we will pay attention to two approaches for solving the one-sided systems, combinatorial and algebraic. Since the solvability question is essentially deciding whether a vector (b) is in a subspace (generated by the columns of A), later in this chapter we present a general theory of max-algebraic subspaces including the concepts of generators, independence and bases. We also briefly discuss unsolvable systems.

3.1 The Combinatorial Method

Let $A = (a_{ij}) \in \overline{\mathbb{R}}^{m \times n}$ and $b = (b_1, \ldots, b_m)^T \in \overline{\mathbb{R}}^m$. The set of solutions to (3.1) will be denoted by $S(A, b)$ or just S if no confusion can arise, that is,

$$S(A, b) = \left\{ x \in \overline{\mathbb{R}}^n ; A \otimes x = b \right\},$$

P. Butkovič, *Max-linear Systems: Theory and Algorithms*,
Springer Monographs in Mathematics 151,
DOI 10.1007/978-1-84996-299-5_3, © Springer-Verlag London Limited 2010

and $A_1, \ldots, A_n$ will stand for the columns of A.

We start with trivial cases. If $b = \varepsilon$ then

$$S(A, b) = \left\{ x = (x_1, \ldots, x_n)^T \in \overline{\mathbb{R}}^n; x_j = \varepsilon \text{ if } A_j \neq \varepsilon, j \in N \right\},$$

in particular $S(A, b) = \overline{\mathbb{R}}^n$ if $A = \varepsilon$. If $A = \varepsilon$ and $b \neq \varepsilon$ then $S(A, b) = \emptyset$. Hence we assume in what follows that $A \neq \varepsilon$ and $b \neq \varepsilon$.

If $b_k = \varepsilon$ for some $k \in M$ then for any $x \in S(A, b)$ we have $x_j = \varepsilon$ if $a_{kj} \neq \varepsilon, j \in N$; consequently the kth equation may be removed from the system together with every column A_j where $a_{kj} \neq \varepsilon$ (if any) and setting the corresponding $x_j = \varepsilon$. Hence there is no loss of generality to assume that $b \in \mathbb{R}^m$ (however, we will not always make this assumption).

If $b \in \mathbb{R}^m$ and A has an ε row then $S(A, b) = \emptyset$. If $A_j = \varepsilon, j \in N$ then x_j may take on any value in a solution x. Hence we may also suppose without loss of generality that A is doubly $\mathbb{R}$-astic.

Let A be column $\mathbb{R}$-astic and $b \in \mathbb{R}^m$. A key role is played by the vector $\overline{x} = (\overline{x}_1, \ldots, \overline{x}_n)^T$ where

$$\overline{x}_j = \left(\max_{i \in M} a_{ij} \otimes b_i^{-1} \right)^{-1}$$

for $j \in N$. Obviously, $\overline{x} \in \mathbb{R}^n$ and

$$\overline{x}_j = \min \left\{ b_i \otimes a_{ij}^{-1}; i \in M, a_{ij} \in \mathbb{R} \right\}$$

for $j \in N$. Where appropriate we will denote $\overline{x} = \overline{x}(A, b)$. We will also denote

$$M_j(A, b) = \left\{ i \in M; \overline{x}_j = b_i \otimes a_{ij}^{-1} \right\}$$

for $j \in N$. We will abbreviate $M_j(A, b)$ by M_j if no confusion can arise.

The combinatorial method follows from the next theorem.

Theorem 3.1.1 [57, 149] *Let $A \in \overline{\mathbb{R}}^{m \times n}$ be doubly $\mathbb{R}$-astic and $b \in \mathbb{R}^m$. Then*

(a) $A \otimes \overline{x}(A, b) \leq b$,
(b) $x \leq \overline{x}(A, b)$ *for every $x \in S(A, b)$,*
(c) $x \in S(A, b)$ *if and only if $x \leq \overline{x}(A, b)$ and*

$$\bigcup_{j:x_j = \overline{x}_j} M_j = M, \tag{3.3}$$

(d) $(A \otimes \overline{x})_i = b_i$ *for at least one $i \in M$.*

Proof (a) Let $k \in M, j \in N$ and suppose that $a_{kj} \in \mathbb{R}$. Then

$$a_{kj} \otimes \overline{x}_j \leq a_{kj} \otimes b_k \otimes a_{kj}^{-1} = b_k.$$

This inequality follows immediately if $a_{kj} = \varepsilon$. Hence

$$\sum_{j \in N}^{\oplus} \left(a_{kj} \otimes \overline{x}_j \right) \leq b_k \quad \text{for all } k \in M$$

and the statement follows.

(b) Let $x \in S(A, b), i \in M, j \in N$. Then $a_{ij} \otimes x_j \leq b_i$ thus $x_j^{-1} \geq a_{ij} \otimes b_i^{-1}$ and so $x_j^{-1} \geq \max_{i \in M} a_{ij} \otimes b_i^{-1}$. Therefore

$$x_j \leq \left(\max_{i \in M} a_{ij} \otimes b_i^{-1} \right)^{-1} = \overline{x}_j.$$

(c) Suppose first $x \in S(A, b)$. We only need to prove $M \subseteq \bigcup_{j:x_j=\overline{x}_j} M_j$. Let $k \in M$. Since $b_k = a_{kj} \otimes x_j > \varepsilon$ for some $j \in N$ and $x_j^{-1} \geq \overline{x}_j^{-1} \geq a_{ij} \otimes b_i^{-1}$ for every $i \in M$, we have $x_j^{-1} = a_{kj} \otimes b_k^{-1} = \max_{i \in M} a_{ij} \otimes b_i^{-1}$. Hence $k \in M_j$ and $x_j = \overline{x}_j$.

Suppose now $x \leq \overline{x}(A, b)$ and that (3.3) holds. Let $k \in M, j \in N$. Then $a_{kj} \otimes x_j \leq b_k$ if $a_{kj} = \varepsilon$. If $a_{kj} \neq \varepsilon$ then

$$a_{kj} \otimes x_j \leq a_{kj} \otimes \overline{x}_j \leq a_{kj} \otimes b_k \otimes a_{kj}^{-1} = b_k. \tag{3.4}$$

Therefore $A \otimes x \leq b$. At the same time $k \in M_j$ for some $j \in N$ satisfying $x_j = \overline{x}_j$. For this j both inequalities in (3.4) are equalities and thus $A \otimes x = b$.

(d) If $(A \otimes \overline{x})_i < b_i$ for all $i \in M$ then $A \otimes (\alpha \otimes \overline{x}) \leq b$ for some $\alpha > 0$ and so (due to the finiteness of $\overline{x}$) $\alpha \otimes \overline{x}$ would be a greater solution to $A \otimes x \leq b$ than $\overline{x}$, a contradiction with (b). $\qquad\square$

It follows that $\overline{x} = \overline{x}(A, b)$ is always a solution to $A \otimes x \leq b$, and $A \otimes x = b$ has a solution if and only if $\overline{x}(A, b)$ is a solution. Because of the special role of $\overline{x}$, this vector is called the *principal solution* to $A \otimes x = b$ and $A \otimes x \leq b$ [60]. Note that the principal solution may not be a solution to $A \otimes x = b$. More precisely, we have:

Corollary 3.1.2 *Let $A \in \overline{\mathbb{R}}^{m \times n}$ be doubly $\mathbb{R}$-astic and $b \in \mathbb{R}^m$. Then the following three statements are equivalent:*

(a) $S(A, b) \neq \emptyset$,
(b) $\overline{x} \in S(A, b)$,
(c) $\bigcup_{j \in N} M_j = M$.

The combinatorial aspect of systems $A \otimes x = b$ will become even more apparent when we deduce a criterion for unique solvability:

Corollary 3.1.3 *Let $A \in \overline{\mathbb{R}}^{m \times n}$ be doubly $\mathbb{R}$-astic and $b \in \mathbb{R}^m$. Then $S(A, b) = \{\overline{x}\}$ if and only if*

(a) $\bigcup_{j \in N} M_j = M$ and

(b) $\bigcup_{j \in N'} M_j \neq M$ for any $N' \subseteq N, N' \neq N$.

Example 3.1.4 Consider the system

$$
\begin{pmatrix}
-2 & 2 & 2 \\
-5 & -3 & -2 \\
\varepsilon & \varepsilon & 3 \\
-3 & -3 & 2 \\
1 & 4 & \varepsilon
\end{pmatrix}
\otimes
\begin{pmatrix}
x_1 \\
x_2 \\
x_3
\end{pmatrix}
=
\begin{pmatrix}
3 \\
-2 \\
1 \\
0 \\
5
\end{pmatrix}.
$$

The matrix $(a_{ij} \otimes b_i^{-1})$ is

$$
\begin{pmatrix}
-5 & -1 & -1 \\
-3 & -1 & 0 \\
\varepsilon & \varepsilon & 2 \\
-3 & -3 & 2 \\
-4 & -1 & \varepsilon
\end{pmatrix}.
$$

Hence $\bar{x} = (3, 1, -2)^T$, $M_1 = \{2, 4\}$, $M_2 = \{1, 2, 5\}$, $M_3 = \{3, 4\}$. The vector $\bar{x}$ is a solution since

$$
\bigcup_{j=1,2,3} M_j = M. \tag{3.5}
$$

However, $M_2 \cup M_3 = M$ as well and no other union of the sets M_1, M_2, M_3 is equal to M. Therefore we may describe the whole solution set:

$$
S(A, b) = \left\{ (x_1, x_2, x_3)^T \in \bar{\mathbb{R}}^3; x_1 \le 3, x_2 = 1, x_3 = -2 \right\}.
$$

Note that if $a_{22} = -3$ is reduced, say to -4, then (3.5) still holds but none of the sets M_1, M_2, M_3 may be omitted without violating this equality. Therefore $\bar{x}$ is a unique solution to this (new) system. If we further reduce $a_{12} = 2$, say to 1 then (3.5) is not satisfied any more and the system has no solution.

It is easily seen that the principal solution to $A \otimes x = b$ can be found in $O(mn)$ time and the same effort is sufficient for checking that it actually is a solution to this system.

The previous statements already indicate that the task of solving one-sided max-linear systems is essentially a combinatorial problem. To make it even more visible, let us consider the following problems:

(UNIQUE) SOLVABILITY: Given $A \in \bar{\mathbb{R}}^{m \times n}$ and $b \in \bar{\mathbb{R}}^m$ does the system $A \otimes x = b$ have a (unique) solution?

(MINIMAL) SET COVERING [126]: Given a finite set M and subsets $M_1, \ldots, M_n$ of M, is

$$
\bigcup_{j \in N} M_j = M
$$

(is

$$\bigcup_{j \in N} M_j = M \quad \text{but} \quad \bigcup_{\substack{j \in N \\ j \neq k}} M_j \neq M$$

for any $k \in N$)?

Corollaries 3.1.2 and 3.1.3 show that for every linear system it is possible to straightforwardly find a finite set and a collection of its subsets so that SOLVABIL-ITY is equivalent to SET COVERING and UNIQUE SOLVABILITY is equivalent to MINIMAL SET COVERING.

This correspondence is two-way, as the statements below suggest. Let us assume without loss of generality that M and its subsets $M_1, \ldots, M_n$ are given. Define $A = (a_{ij}) \in \mathbb{R}^{m \times n}$ as follows:

$$a_{ij} = \begin{cases} 1 & \text{if } i \in M_j \\ 0 & \text{else} \end{cases} \quad \text{for all } i \in M, \ j \in N,$$

$$b = 0.$$

The following are corollaries of Theorem 3.1.1.

Theorem 3.1.5 $\bigcup_{j \in N} M_j = M$ *if and only if* $A \otimes x = b$ *has a solution.*

Theorem 3.1.6 $\bigcup_{j \in N} M_j = M$ *and* $\bigcup_{j \in N'} M_j \neq M$ *for any* $N' \subseteq N, N' \neq N$ *if and only if* $A \otimes x = b$ *has a unique solution.*

We have demonstrated that every max-linear system is an algebraic representation of a set covering problem, and conversely. This has various consequences. For instance the task of finding a solution to $A \otimes x = b$ with the minimum number of components equal to $\bar{x}$ is polynomially equivalent to the minimum cardinality set cover problem and is therefore *NP*-complete [83]. Standard textbooks on combinatorial optimization such as [120] are recommended for more explanation on the set covering problem or for an explanation of *NP*-completeness.

Note that an interesting generalization of the combinatorial method to the infinite-dimensional case can be found in [5].

3.2 The Algebraic Method

In some theoretical and practical applications it may be helpful to express the principal solution algebraically rather than combinatorially. We start with inequalities. As already seen in Theorem 3.1.1, the systems of one-sided inequalities always have a solution and can be solved as easily as equations (unlike their linear-algebraic counterparts). The algebraic method slightly extends this result to any $A \in \overline{\mathbb{R}}^{m \times n}$

and $b \in \overline{\overline{\mathbb{R}}}^m$. Key statements are the following lemma and theorem; the reader is referred to p. 1 and Sect. 1.6.3 for the necessary definitions and conventions on $\pm\infty$. For consistency we will denote in this section a^{-1} (that is $-a$) for $a \in \overline{\overline{\mathbb{R}}}$ by a^*.

Lemma 3.2.1 *If $a, b \in \overline{\overline{\mathbb{R}}}$ then $x \in \overline{\overline{\mathbb{R}}}$ satisfies the inequality*

$$a \otimes x \leq b \tag{3.6}$$

if and only if

$$x \leq a^* \otimes' b. \tag{3.7}$$

Proof The statement holds when $a, b \in \mathbb{R}$ since $a^* \otimes' b = -a + b$. If $a = +\infty$ and $b = -\infty$ then $x = -\infty$ is the unique solution to (3.6) and (3.7) reads $x \leq -\infty$. In all other cases when $a, b \in \{-\infty, +\infty\}$ the solution set to (3.6) is $\overline{\overline{\mathbb{R}}}$ and (3.7) reads $x \leq +\infty$. $\square$

Theorem 3.2.2 [59] *If $A \in \overline{\overline{\mathbb{R}}}^{m \times n}, b \in \overline{\overline{\mathbb{R}}}^m$ and $x \in \overline{\overline{\mathbb{R}}}^n$ then*

$$A \otimes x \leq b \quad \text{if and only if} \quad x \leq A^* \otimes' b.$$

Proof The following are equivalent (Lemma 3.2.1 is used in the third equivalence):

$$A \otimes x \leq b,$$

$$\sum_{j \in N}^{\oplus} (a_{ij} \otimes x_j) \leq b_i \quad \text{for all } i \in M,$$

$$a_{ij} \otimes x_j \leq b_i \quad \text{for all } i \in M, j \in N,$$

$$x_j \leq (a_{ij})^* \otimes' b_i \quad \text{for all } i \in M, j \in N,$$

$$x_j \leq a_{ji}^* \otimes' b_i \quad \text{for all } i \in M, j \in N,$$

$$x_j \leq \sum_{i \in M}^{\oplus'} (a_{ji}^* \otimes' b_i) \quad \text{for all } j \in N,$$

$$x \leq A^* \otimes' b. \qquad \square$$

It follows from the definition of the principal solution $\bar{x}$ (p. 54) that $\bar{x} = A^* \otimes' b$ if A is doubly $\mathbb{R}$-astic and $b \in \mathbb{R}^m$. We will therefore extend this definition and call $A^* \otimes' b$ the principal solution for any $A \in \overline{\overline{\mathbb{R}}}^{m \times n}$ and $b \in \overline{\overline{\mathbb{R}}}^m$.

Corollary 3.2.3 *If $A \in \overline{\overline{\mathbb{R}}}^{m \times n}, b \in \overline{\mathbb{R}}^m$ and $c \in \overline{\mathbb{R}}^n$ then*

(a) *$\bar{x}$ is the greatest solution to $A \otimes x \leq b$, that is*

$$A \otimes (A^* \otimes' b) \leq b,$$

(b) $A \otimes x = b$ *has a solution if and only if $\overline{x}$ is a solution and*
(c)

$$A \otimes \left(A^* \otimes' (A \otimes c)\right) = A \otimes c.$$

Proof (a) $\overline{x}$ is a solution since it satisfies the condition of Theorem 3.2.2 and that theorem is also saying that $x \leq \overline{x}$ if $A \otimes x \leq b$, hence $\overline{x}$ is greatest.

(b) Suppose $A \otimes x = b$ for some $x \in \overline{\mathbb{R}}^n$. By Theorem 3.2.2 $x \leq \overline{x}$ and by Corollary 1.1.2 we then have

$$b = A \otimes x \leq A \otimes \overline{x} \leq b.$$

This implies $A \otimes \overline{x} = b$.

(c) The equation $A \otimes x = A \otimes c$ has a solution, thus by (b) $A^* \otimes' (A \otimes c)$ is a solution and the statement follows.   □

It will be useful to have an immediate generalization of these results to matrix inequalities:

Corollary 3.2.4 *If $A \in \overline{\mathbb{R}}^{m \times n}, B \in \overline{\mathbb{R}}^{m \times k}, C \in \overline{\mathbb{R}}^{n \times l}$ and $\overline{X} = A^* \otimes' B$ then*

(a) $\overline{X}$ *is the greatest solution to $A \otimes X \leq B$, that is*

$$A \otimes \left(A^* \otimes' B\right) \leq B,$$

(b) $A \otimes X = B$ *has a solution if and only if $\overline{X}$ is a solution and*
(c)

$$A \otimes \left(A^* \otimes' (A \otimes C)\right) = A \otimes C.$$

Proof This corollary follows immediately since $A \otimes X \leq B$ is equivalent to the system of one-sided max-linear systems:

$$A \otimes X_r \leq B_r \quad (r = 1, \dots, k)$$

where $X_1, \dots, X_k$ and $B_1, \dots, B_k$ are the columns of X and B, respectively.   □

3.3 Subspaces, Generators, Extremals and Bases

Being motivated by the results of the previous sections of this chapter we now present the theory of max-linear subspaces, independence and bases. The main benefit for the aims of this book is the result that every finitely generated subspace has an essentially unique basis. We will also show how to find a basis of a finitely generated subspace which will be of fundamental importance in Chap. 4 where we use this result for finding the bases of eigenspaces. Our presentation follows the lines of [43] and confirms the results of [69] developed for subspaces of $\mathbb{R}^n \cup \{\varepsilon\}$. Some of the results of this section have been proved in [60, 103, 105, 147].

Let $S \subseteq \overline{\mathbb{R}}^n$. The set S is called a *max-algebraic subspace* if

$$\alpha \otimes u \oplus \beta \otimes v \in S$$

for every $u, v \in S$ and $\alpha, \beta \in \overline{\mathbb{R}}$. The adjective "max-algebraic" will usually be omitted.

A vector $v = (v_1, \ldots, v_n)^T \in \overline{\mathbb{R}}^n$ is called a *max-combination* of S if

$$v = \sum_{x \in S}^{\oplus} \alpha_x \otimes x, \quad \alpha_x \in \overline{\mathbb{R}} \tag{3.8}$$

where only a finite number of α_x are finite. The set of all max-combinations of S is denoted by span(S). We set span($\emptyset$) = $\{\varepsilon\}$. It is easily seen that span(S) is a subspace. If span(S) = T then S is called a *set of generators* for T.

A vector $v \in S$ is called an *extremal in* S if $v = u \oplus w$ for $u, v \in S$ implies $v = u$ or $v = w$. Clearly, if $v \in S$ is an extremal in S and $\alpha \in \mathbb{R}$ then $\alpha \otimes v$ is also an extremal in S.

Note that terminology varies in the max-algebraic literature and, for instance, extremals are called vertices in [76, 105] and irreducible elements in [146].

Let $v = (v_1, \ldots, v_n)^T \in \overline{\mathbb{R}}^n, v \neq \varepsilon$. The *max-norm* or just *norm* of v is $\|v\| = \max(v_1, \ldots, v_n)$; v is called *scaled* if $\|v\| = 0$. The set S is called *scaled* if all its elements are scaled.

The set S is called *dependent* if v is a max-combination of $S - \{v\}$ for some $v \in S$. Otherwise S is *independent*. The set S is called *totally dependent* if every $v \in S$ is a max-combination of $S - \{v\}$. Note that $\emptyset$ is both independent and totally dependent and $\{\varepsilon\}$ is totally dependent.

Let $S, T \subseteq \overline{\mathbb{R}}^n$. The set S is called a *basis* of T if it is an independent set of generators for T. The set $\{e^i \in \overline{\mathbb{R}}^n; i = 1, \ldots, n\}$ defined by

$$e^i_j = \begin{cases} 0 & \text{if } j = i \\ \varepsilon & \text{if } j \neq i \end{cases}$$

is a basis of $\overline{\mathbb{R}}^n$; it will be called *standard*.

We start with two simple lemmas.

Lemma 3.3.1 *Let S be a set of generators of a subspace $T \subseteq \overline{\mathbb{R}}^n$ and let v be a scaled extremal in T. Then $v \in S$.*

Proof Let v be a max-combination (3.8). Since the number of finite α_x is finite and v is an extremal we deduce by induction that $v = \alpha_x \otimes x$ for some $\alpha_x \in \mathbb{R}$. But both v and x are scaled and therefore $v = x$ yielding $v \in S$. $\qquad\square$

Lemma 3.3.2 *The set of scaled extremals of a subspace is independent.*

Proof Let $E \neq \emptyset$ be the set of extremals of a subspace T and $v \in E$. By applying Lemma 3.3.1 to the subspace $T' = \text{span}(E - \{v\})$ we get $v \notin T'$ and the statement follows. $\qquad\square$

If $v = (v_1, \ldots, v_n)^T \in \overline{\mathbb{R}}^n$ then the *support* of v is defined by

$$\text{Supp}(v) = \{j \in N; v_j \in \mathbb{R}\}.$$

We will use the following notation. If $j \in \text{Supp}(v)$ then $v(j) = v_j^{-1} \otimes v$. For any $j \in N$ and $S \subseteq \overline{\mathbb{R}}^n$ we denote

$$S(j) = \{v(j); v \in S, j \in \text{Supp}(v)\}.$$

An element of $v \in S$ is called *minimal* in S if $u \leq v, u \in S$ imply $u = v$. If $S \subseteq \overline{\mathbb{R}}^n$ is a subspace, $v \in S$ and $j \in \text{Supp}(v)$ then we denote

$$D_j(v) = \{u \in S(j); u \leq v(j)\}.$$

The following will be important for the main results of this section.

Proposition 3.3.3 *Let* $S \subseteq \overline{\mathbb{R}}^n$. *Then the following are equivalent:*

(a) $v \in \text{span}(S)$.
(b) *For each* $j \in \text{Supp}(v)$ *there is an* $x^j \in S$ *such that* $j \in \text{Supp}(x^j)$ *and* $x^j(j) \in D_j(v)$.

Proof If (b) holds then $v = \sum_{j \in \text{Supp}(v)}^{\oplus} \alpha_j \otimes x^j$, where $\alpha_j = v_j \otimes \left(x_j^j\right)^{-1}$.

Let now $v \in \text{span}(S)$. Then for each $j \in \text{Supp}(v)$ there is an $x^j \in S$ with $\alpha_j \otimes x^j \leq v$ and $(\alpha_j \otimes x^j)_j \leq v_j$. Clearly, $\alpha_j = v_j \otimes \left(x_j^j\right)^{-1}$ and (b) follows. $\square$

The following immediate corollary is an analogue of Carathéodory's Theorem and was essentially proved in [76] and [103].

Corollary 3.3.4 *Let* $S \subseteq \overline{\mathbb{R}}^n$. *Then* $v \in \text{span}(S)$ *if and only if* $v \in \text{span}\{x^1, \ldots, x^k\}$ *for some* $x^1, \ldots, x^k \in S$ *where* $k \leq |\text{Supp}(v)|$.

We add another straightforward corollary that will be used later on.

Corollary 3.3.5 *Let* $T \subseteq \overline{\mathbb{R}}^n$ *be a subspace and* Q *be a set of generators for* T. *Let* $U \subseteq Q$ *and* $S = Q - U$. *Then* S *generates* T *if and only if each* $v \in Q$ *satisfies condition* (b) *of Proposition* 3.3.3.

The next statement provides two criteria for a vector to be an extremal.

Proposition 3.3.6 *Let* $T \subseteq \overline{\mathbb{R}}^n$ *be a subspace and* S *be a set of generators for* T. *Let* $v \in S, v \neq \varepsilon$. *Then the following are equivalent:*

(a) v *is an extremal in* T.
(b) $v(j)$ *is minimal in* $T(j)$ *for some* $j \in \text{Supp}(v)$.
(c) $v(j)$ *is minimal in* $S(j)$ *for some* $j \in \text{Supp}(v)$.

Proof (a) $\Longrightarrow$ (c): If $|\mathrm{Supp}(v)| = 1$ then $v(j)$ is minimal in $S(j)$. So suppose that $|\mathrm{Supp}(v)| > 1$ and $v(j)$ is not minimal in $S(j)$ for any $j \in \mathrm{Supp}(v)$. Then for each $j \in \mathrm{Supp}(v)$ there is an $x^j \in S(j)$ such that $x^j \leq v(j), x^j \neq v(j)$. Therefore $v = \sum_{j \in \mathrm{Supp}(v)}^{\oplus} v_j \otimes x^j$, and v is proportional with none of x^j. Hence v is not an extremal in T.

(c) $\Longrightarrow$ (b): Let $u \in T$ and assume that $j \in \mathrm{Supp}(v)$ and $u(j) \leq v(j)$. We need to show that $u(j) = v(j)$. By Proposition 3.3.3 the inequality $w(j) \leq u(j)$ holds for some $w \in S$. Thus $w(j) \leq u(j) \leq v(j)$ and by (c) it follows that $w(j) = u(j) = v(j)$.

(b) $\Longrightarrow$ (a): Let $v(j)$ be minimal in $T(j)$ for some $j \in \mathrm{Supp}(v)$ and suppose that $v = u \oplus w$ for some $u, w \in T$. Then both $u \leq v$ and $w \leq v$ and either $u_j = v_j$ or $w_j = v_j$, say (without loss of generality) $u_j = v_j$. Hence $u(j) \leq v(j)$ and it follows from (b) that $u(j) = v(j)$. Therefore also $u = v$ and (a) follows. $\square$

We can now easily deduce a corollary that shows the crucial role of extremals: they are generators.

Corollary 3.3.7 *Let $T \subseteq \overline{\mathbb{R}}^n$ be a subspace. If $D_j(v)$ has a minimal element for each $v \in T$ and each $j \in \mathrm{Supp}(v)$ then T is generated by its extremals.*

Proof Suppose that x^j is a minimal element of $D_j(v)$. Since, for $u \in T(j)$, the inequality $u \leq x^j$ implies $u \in D_j(v)$, x^j is also a minimal element of $T(j)$. The statement now follows by combining Propositions 3.3.3 and 3.3.6. $\square$

The following fundamental result was essentially proved in [147]. Here we slightly reformulate it: every set of generators S of a subspace T can be partitioned as $E \cup F$ where E is a set of extremals for T and the remainder F is redundant.

Theorem 3.3.8 *Let $T \subseteq \overline{\mathbb{R}}^n$ be a subspace and S be a set of scaled generators for T. Let E be a set of scaled extremals in T. Then*

(a) $E \subseteq S$.
(b) *Let $F = S - E$. Then for any $v \in F$ the set $S - \{v\}$ is (also) a set of generators for T.*

Proof Part (a) repeats Lemma 3.3.1.

To prove (b), let $v \in F$. Since v is not an extremal, by Proposition 3.3.6 for each $j \in \mathrm{Supp}(v)$ there is a $z^j \in T$ such that $z^j(j) < v(j)$. Since $T = \mathrm{span}(S)$, by Proposition 3.3.3 there is also an $y^j \in S$ satisfying $y^j(j) \leq z^j(j) < v(j)$. Obviously, $y^j \neq v$ and by applying Proposition 3.3.3 again we get that v is a max-combination of $\{y^j; j \in \mathrm{Supp}(v)\}$ where $y^j \in S$ are different from v. Thus in any max-combination involving v, this vector can be replaced by a max-combination of vectors in $S - \{v\}$ which completes the proof. $\square$

The following refinement of Theorem 3.3.8 will also be useful.

Theorem 3.3.9 *Let E be the set of scaled extremals in a subspace T. Let $S \subseteq T$ consist of scaled vectors. Then the following are equivalent:*

(a) *S is a minimal set of generators for T.*
(b) *$S = E$ and S generates T.*
(c) *S is a basis for T.*

Proof (a) $\Longrightarrow$ (b): By Theorem 3.3.8 we have $S = E \cup F$ where every element of F is redundant in S. But since S is a minimal set of generators, we have $F = \emptyset$. Hence $S = E$.

(b) $\Longrightarrow$ (c): E is independent and generating.

(c) $\Longrightarrow$ (a): By independence of S the span of a proper subset of S is strictly contained in span(S). $\square$

Theorem 3.3.9 shows that if a subspace has a (scaled) basis then it must be its set of (scaled) extremals, hence the basis is essentially unique. Note that a maximal independent set in a subspace T may not be a basis for T as is shown by the following example.

Example 3.3.10 Let $T \subseteq \overline{\mathbb{R}}^2$ consist of all $(x_1, x_2)^T$ with $x_1 \geq x_2 > \varepsilon$. If $0 > a > b > \varepsilon$ then $\{(0, a)^T, (0, b)^T\}$ is a maximal independent set in T but it does not generate T.

We now deduce a few corollaries of Theorem 3.3.9. The first one can be found in [76, 105] and [131].

Corollary 3.3.11 *If T is a finitely generated subspace then its set of scaled extremals is nonempty and it is the unique scaled basis for T.*

Proof Since T is finitely generated there exists a minimal set of generators S. By Theorem 3.3.9 $S = E$ and S is a basis. $\square$

The next corollaries are related to totally dependent sets.

Corollary 3.3.12 *If S is a nonempty scaled totally dependent set then S is infinite.*

Proof Suppose that S is finite and let $T = \text{span}(S)$. By Corollary 3.3.11 T contains scaled extremals, which by Theorem 3.3.8 are contained in S, given that $T = \text{span}(S)$. But then S is not totally dependent, a contradiction. $\square$

Corollary 3.3.13 *Let $T \subseteq \overline{\mathbb{R}}^n$ be a subspace. Then the following are equivalent:*

(a) *There is no extremal in T.*
(b) *There exists a totally dependent set of generators for T.*
(c) *Every set of generators for T is totally dependent.*

Proof Since there always is a set of generators for T (e.g. the set T itself), each of (b) and (c) is equivalent to (a) by Theorem 3.3.8. □

A subspace S in $\overline{\mathbb{R}}^n$ is called *open* if $S - \{\varepsilon\}$ is open in the Euclidean topology.

Corollary 3.3.14 *Let* $T \subseteq \mathbb{R}^n \cup \{\varepsilon\}, n > 1$, *be a subspace. If* $T - \{\varepsilon\}$ *is open then every generating set for* T *is totally dependent* (*and hence* T *has no basis*).

Proof It is sufficient to show that there is no scaled extremal in T since the result then follows from Theorem 3.3.8. Let $v \in T - \{\varepsilon\}$. Since T is open there exist vectors $w^p \in T$ $(p = k, l)$, where $w_p^p < v_p$ and $w_i^p = v_i$ for $i \neq p$. Hence $v = w^k \oplus w^l$ and $v \neq w^k, v \neq w^l$. Therefore there are no scaled extremals in T. □

An example of an open subspace is $T = \mathbb{R}^n \cup \{\varepsilon\}$. For this particular case Corollary 3.3.14 was proved in [69]. Another example consists of all vectors $(a, b)^T$ with $a, b \in \mathbb{R}, a > b$.

More geometric and topological properties of max-algebraic subspaces can be found in [43, 52–54, 87, 89] and [103].

3.4 Column Spaces

We have seen a number of corollaries of the key result, Theorem 3.3.9. We shall now link the first of these corollaries, Corollary 3.3.11, to the results of the previous sections of this chapter. As usual the *column space* of a matrix $A \in \overline{\mathbb{R}}^{m \times n}$ with columns $A_1, \ldots, A_n$ is the set

$$\text{Col}(A) = \left\{ \sum_{j \in N}^{\oplus} x_j \otimes A_j; x_j \in \overline{\mathbb{R}} \right\} = \left\{ A \otimes x; x \in \overline{\mathbb{R}}^n \right\}.$$

Since $\alpha \otimes A \otimes x \oplus \beta \otimes A \otimes y = A \otimes (\alpha \otimes x \oplus \beta \otimes y)$, we readily see that any column space is a subspace. Observe that by finding a solution to a system $A \otimes x = b$ we prove that $b \in \text{Col}(A)$. A natural task then is to find a basis of this subspace. Corollary 3.3.11 guarantees that such a basis exists and is unique up to scalar multiples of its elements. Note that for a formal proof we would have to first remove repeated columns as they would be indistinguishable in a set of columns, but they may be re-instated after deducing the uniqueness of the basis since the expression "multiples of a vector v" also covers vectors identical with v. We summarize:

Theorem 3.4.1 *For every* $A \in \overline{\mathbb{R}}^{m \times n}$ *there is a matrix* $B \in \overline{\mathbb{R}}^{m \times k}, k \leq n$, *consisting of some columns of* A *such that no two columns of* B *are equal and the set of column vectors of* B *is a basis of* $\text{Col}(A)$. *This matrix* B *is unique up to the order and scalar multiples of its columns.*

It remains to show how to find a basis of the column space of a matrix, say A. If a column, say A_k is a max-combination of the remaining columns and A' arises from A by removing A_k then $\text{Col}(A) = \text{Col}(A')$ since in every max-combination of the columns of A, the vector A_k may be replaced by a max-combination of the other columns, that is, columns of A'. By repeating this process until no column is a max-combination of the remaining columns, we arrive at a set that satisfies both requirements in the definition of a basis. Every check of linear independence is equivalent to solving an $m \times (n-1)$ one-sided system and can therefore be performed using $O(mn)$ operations, thus the whole process is $O(mn^2)$. Although asymptotically equally efficient, a method called the $\mathcal{A}$-test, essentially described in the following theorem, is more compact:

Theorem 3.4.2 [60] *Let $A \in \overline{\mathbb{R}}^{m \times n}$ be a matrix with columns $A_1, \ldots, A_n$ and $\mathcal{A}$ be the matrix arising from $A^* \otimes' A$ after replacing the diagonal entries by ε. Then for all $j \in N$ the vector A_j is equal to the jth column of $A \otimes \mathcal{A}$ if and only if A_j is a max-combination of the other columns of A. The elements of the jth column of $\mathcal{A}$ then provide the coefficients to express the max-combination.*

Proof See [60], Theorem 16-2.   □

Example 3.4.3 Let

$$A = \begin{pmatrix} 1 & 1 & 2 & \varepsilon & 5 \\ 1 & 0 & 4 & 1 & 5 \\ 1 & \varepsilon & -1 & 1 & 0 \end{pmatrix}.$$

Then

$$A^* \otimes' A = \begin{pmatrix} -1 & -1 & -1 \\ -1 & 0 & -\varepsilon \\ -2 & -4 & 1 \\ -\varepsilon & -1 & -1 \\ -5 & -5 & 0 \end{pmatrix} \otimes' \begin{pmatrix} 1 & 1 & 2 & \varepsilon & 5 \\ 1 & 0 & 4 & 1 & 5 \\ 1 & \varepsilon & -1 & 1 & 0 \end{pmatrix}$$

$$= \begin{pmatrix} 0 & -1 & -2 & 0 & -1 \\ 0 & \varepsilon & 1 & \varepsilon & 4 \\ -3 & \varepsilon & 0 & \varepsilon & 1 \\ 0 & \varepsilon & -2 & \varepsilon & -1 \\ -4 & \varepsilon & -3 & \varepsilon & 0 \end{pmatrix}.$$

Hence

$$A \otimes \mathcal{A} = \begin{pmatrix} 1 & 0 & 2 & 1 & 5 \\ 1 & \cdots & 2 & \cdots & 5 \\ 1 & \cdots & \cdots & \cdots & 0 \end{pmatrix}.$$

We deduce

$$A_1 = 0 \otimes A_2 \oplus -3 \otimes A_3 \oplus 0 \otimes A_4 \oplus -4 \otimes A_5$$

$$A_5 = -1 \otimes A_1 \oplus 4 \otimes A_2 \oplus 1 \otimes A_3 \oplus -1 \otimes A_4$$

and the basis of $\text{Col}(A)$ is $\{A_2, A_3, A_4\}$.

The number of vectors in any basis of a finitely generated subspace T is called the *dimension* of T, notation $\dim(T)$. Unlike in linear algebra, the dimensions of max-algebraic subspaces are unrelated to the numbers of components of the vectors in these subspaces. This has been observed in the early years of max-algebra and the following two statements describe the anomaly.

Theorem 3.4.4 [60] *Let $m \geq 3$ and $k \geq 2$. There exist k vectors in $\overline{\mathbb{R}}^m$, none of which is a max-combination of the others.*

Proof It is sufficient to find k such vectors for $m = 3$. Consider

$$A = \begin{pmatrix} 0 & 0 & \cdots & 0 \\ 1 & 2 & \cdots & k \\ -1 & -2 & \cdots & -k \end{pmatrix}$$

and apply the $\mathcal{A}$-test to A

$$A^* \otimes' A = \begin{pmatrix} 0 & -1 & 1 \\ 0 & -2 & 2 \\ \cdots & \cdots & \cdots \\ 0 & -k & k \end{pmatrix} \otimes' \begin{pmatrix} 0 & 0 & \cdots & 0 \\ 1 & 2 & \cdots & k \\ -1 & -2 & \cdots & -k \end{pmatrix}$$

$$= \begin{pmatrix} 0 & -1 & \cdots & -k+1 \\ -1 & 0 & \cdots & -k+2 \\ \cdots & \cdots & \cdots & \cdots \\ -k+1 & -k+2 & \cdots & 0 \end{pmatrix}.$$

Hence all entries in the first row of the matrix

$$A \otimes A = \begin{pmatrix} 0 & 0 & \cdots & 0 \\ 1 & 2 & \cdots & k \\ -1 & -2 & \cdots & -k \end{pmatrix} \otimes \begin{pmatrix} \varepsilon & -1 & \cdots & -k+1 \\ -1 & \varepsilon & \cdots & -k+2 \\ \cdots & \cdots & \cdots & \cdots \\ -k+1 & -k+2 & \cdots & \varepsilon \end{pmatrix}$$

are -1 yielding that no column of $A \otimes A$ is equal to the corresponding column in A. Using the $\mathcal{A}$-test we deduce that none of the columns of A is a max-combination of the others.   □

Theorem 3.4.5 [60] *Every real $2 \times n$ matrix, $n \geq 2$, has two columns such that all other columns are a max-combination of these two columns.*

Proof Let $A = (a_{ij}) \in \mathbb{R}^{2 \times n}$. We may assume without loss of generality that the order of the columns is such that

$$a_{11} \otimes a_{21}^{-1} \leq a_{12} \otimes a_{22}^{-1} \leq \cdots \leq a_{1n} \otimes a_{2n}^{-1}. \tag{3.9}$$

It is sufficient to prove that the system

$$
\begin{pmatrix} a_{11} & a_{1n} \\ a_{21} & a_{2n} \end{pmatrix} \otimes x = \begin{pmatrix} a_{1k} \\ a_{2k} \end{pmatrix}
$$

has a solution for every $k = 1, \ldots, n$. From (3.9) we deduce for every k:

$$
a_{11} \otimes a_{1k}^{-1} \leq a_{21} \otimes a_{2k}^{-1},
$$

$$
a_{1n} \otimes a_{1k}^{-1} \geq a_{2n} \otimes a_{2k}^{-1},
$$

which imply $2 \in M_1$ and $1 \in M_2$ and the statement now follows by Corollary 3.1.2. $\square$

These results indicate that the question of a dimension in max-algebra is more complicated than that in conventional linear algebra. We will return to this in Chap. 6.

3.5 Unsolvable Systems

If a system $A \otimes x = b$ has no solution then the question of a best approximation of b by the mapping $x \longmapsto A \otimes x$ arises. For this we need to introduce the concept of a distance between two vectors. We shall consider the distance based on the Chebyshev norm for which a quick answer follows from our previous results. If $x = (x_1, \ldots, x_n)^T, y = (y_1, \ldots, y_n)^T \in \mathbb{R}^n$ then the *Chebyshev distance* of x and y is $\xi(x, y) = \max_{j \in N} |x_j - y_j|$. Max-algebraically,

$$
\xi(x, y) = \sum_{j \in N}^{\oplus} \left(x_j \otimes y_j^{-1} \oplus x_j^{-1} \otimes y_j \right).
$$

It is easily verified that

$$
\xi(\alpha \otimes x, y) \leq |\alpha| \otimes \xi(x, y) \tag{3.10}
$$

for any $\alpha \in \mathbb{R}$.

For the approximation of b by $A \otimes x$ we distinguish two important cases:

Case 1 When x has to satisfy the condition $A \otimes x \leq b$ (recall that this system always has a solution). In MMIPP (see p. 9) b corresponds to required completion times and $A \otimes x$ is the actual completion times vector. Thus the approximation using a Chebyshev distance of $A \otimes x$ and b subject to $A \otimes x \leq b$ can be described as "minimal earliness subject to zero tardiness" [60].

Case 2 When x is unrestricted, $x \in \mathbb{R}^n$.

The following two theorems show that the principal solution plays a key role in the answers to both questions. Recall that $\overline{x}(A, b)$ is finite if A is doubly $\mathbb{R}$-astic and b finite.

Theorem 3.5.1 [60] *Let $A \in \overline{\mathbb{R}}^{m \times n}$ be doubly $\mathbb{R}$-astic, $b \in \mathbb{R}^m$, $\overline{x} = \overline{x}(A, b)$ and*

$$Q = \left\{ x \in \overline{\mathbb{R}}^n; A \otimes x \leq b \right\}.$$

Then

$$\xi (A \otimes \overline{x}, b) = \min_{x \in Q} \xi (A \otimes x, b).$$

Proof It follows from Theorem 3.1.1 that $x \in Q$ if and only if $x \leq \overline{x}$. By Corollary 1.1.2 then

$$A \otimes x \leq A \otimes \overline{x} \leq b$$

for every $x \in Q$.   □

Theorem 3.5.2 [60] *Let $A \in \overline{\mathbb{R}}^{m \times n}$ be doubly $\mathbb{R}$-astic, $b \in \mathbb{R}^m$, $\overline{x} = \overline{x}(A, b)$, $\mu^2 = \xi(A \otimes \overline{x}, b)$ and $y = \mu \otimes \overline{x}$. Then*

$$\xi (A \otimes y, b) = \min_{x \in \mathbb{R}^n} \xi (A \otimes x, b).$$

Proof Since $A \otimes \overline{x} \leq b$ and $(A \otimes \overline{x})_i = b_i$ for some $i \in M$ (Theorem 3.1.1) we have $\xi(A \otimes y, b) = \mu$.

Suppose $\xi(A \otimes z, b) < \xi(A \otimes y, b)$ for some $z \in \mathbb{R}^n$ and let $\rho = \xi(A \otimes z, b)$. Then $\rho < \mu$ and

$$A \otimes z \leq \rho \otimes b.$$

Hence

$$A \otimes \left(\rho^{-1} \otimes z \right) \leq b$$

and so by Theorem 3.5.1 and (3.10)

$$\mu^2 = \xi (A \otimes \overline{x}, b) \leq \xi \left(A \otimes \left(\rho^{-1} \otimes z \right), b \right) \leq \left| \rho^{-1} \right| \otimes \xi (A \otimes z, b) = \rho^2.$$

It follows that $\mu \leq \rho$, a contradiction, hence the statement.   □

There are other ways of approximating b using $A \otimes x$, for instance by permuting the components of $A \otimes x$ [42]. For more types of approximation see e.g. [47].

3.6 Exercises

Exercise 3.6.1 Describe the solution set to the system $A \otimes x = b$, where

$$
A = \begin{pmatrix} 3 & 2 & 4 \\ 6 & 7 & 6 \\ 2 & 4 & 8 \\ 0 & 2 & 3 \\ 3 & 1 & 8 \end{pmatrix}, \qquad b = \begin{pmatrix} -p \\ 1 \\ 1 \\ -4 \\ 1 \end{pmatrix}
$$

in terms of the real parameter p. [No solution for $p < 2$ or $p > 3$; $(-5, \leq -6, -7)^T$ for $p = 2$; unique solution $(-3 - p, -6, -7)^T$ if $2 < p < 3$; $(\leq -6, -6, -7)^T$ for $p = 3$]

Exercise 3.6.2 As in the previous question but for $A \otimes x \leq b$.

$$
\left[x \leq \begin{pmatrix} -3 & -6 & -2 & 0 & -3 \\ -2 & -7 & -4 & -2 & -1 \\ -4 & -6 & -8 & -3 & -8 \end{pmatrix} \otimes' \begin{pmatrix} -p \\ 1 \\ 1 \\ -4 \\ 1 \end{pmatrix} = \begin{pmatrix} \max(-p - 3, -1) \\ \max(-p - 2, \ 0) \\ \max(-p - 4, -5) \end{pmatrix} \right]
$$

Exercise 3.6.3 Find the scaled basis of the column space of the matrix

$$
A = \begin{pmatrix} 3 & -2 & 0 & 3 & 2 \\ 1 & 1 & -2 & 6 & 3 \\ 4 & 3 & 1 & 8 & 0 \end{pmatrix}.
$$

$[\{(-1, -3, 0)^T, (-5, -2, 0)^T, (-1, 0, -3)^T\}.]$

Exercise 3.6.4 For A and b with $p = 0$ of Exercise 3.6.1 find the Chebyshev best approximation of b by $A \otimes x$ over the set $\{x \in \overline{\mathbb{R}}^n; A \otimes x \leq b\}$ and then over $\mathbb{R}^n$.

$$
\left[\begin{pmatrix} -2 \\ 1 \\ 1 \\ -4 \\ 1 \end{pmatrix} \text{ for } x = \begin{pmatrix} -5 \\ -6 \\ -7 \end{pmatrix} ; \begin{pmatrix} -1 \\ 2 \\ 2 \\ -3 \\ 2 \end{pmatrix} \text{ for } x = \begin{pmatrix} -4 \\ -5 \\ -6 \end{pmatrix} \right]
$$

Exercise 3.6.5 Find the Chebyshev best approximation of b by $A \otimes x$ over the set $\{x \in \overline{\mathbb{R}}^n; A \otimes x \leq b\}$ and then over $\mathbb{R}^n$ for $A = \begin{pmatrix} 3 & 1 \\ 2 & 5 \end{pmatrix}$ and $b = \begin{pmatrix} 2 \\ 0 \end{pmatrix}$.

$$
\left[\begin{pmatrix} 1 \\ 0 \end{pmatrix} \text{ for } x = \begin{pmatrix} -2 \\ -5 \end{pmatrix} ; \begin{pmatrix} 3/2 \\ 1/2 \end{pmatrix} \text{ for } = \begin{pmatrix} -3/2 \\ -9/2 \end{pmatrix} \right]
$$

Exercise 3.6.6 Let $A \in \mathbb{R}^{m \times 2}$. Prove that there exist positions $(k, 1)$ and $(l, 2)$ in A such that for any b, for which $A \otimes x = b$ has a solution, $(k, 1)$ is a column maximum in column 1 of $(diag(b))^{-1} \otimes A$ and $(l, 2)$ is a column maximum in column 2 of this matrix, respectively. [See [42]]

Exercise 3.6.7 Prove that the following problem is *NP*-complete. Given $A \in \overline{\mathbb{R}}^{m \times n}$ and $b \in \overline{\mathbb{R}}^{m}$, decide whether it is possible to permute the components of b so that for the obtained vector b' the system $A \otimes x = b'$ has a solution. [See [31]]

Chapter 4
Eigenvalues and Eigenvectors

This chapter provides an account of the max-algebraic eigenvalue-eigenvector theory for square matrices over $\overline{\mathbb{R}}$. The algorithms presented and proved here enable us to find all eigenvalues and bases of all eigenspaces of an $n \times n$ matrix in $O(n^3)$ time. These results are of fundamental importance for solving the reachability problems in Chap. 8 and elsewhere.

We start with definitions and basic properties of the eigenproblem, then continue by proving one of the most important results in max-algebra, namely that for every matrix the maximum cycle mean is the greatest eigenvalue, which motivates us to call it the principal eigenvalue. We then show how to describe the corresponding (principal) eigenspace. Next we present the Spectral Theorem, that enables us to find all eigenvalues of a matrix. It also makes it possible to characterize matrices with finite eigenvectors. Finally, we discuss how to efficiently describe all eigenvectors of a matrix.

4.1 The Eigenproblem: Basic Properties

Given $A \in \overline{\mathbb{R}}^{n \times n}$, the task of finding the vectors $x \in \overline{\mathbb{R}}^n, x \neq \varepsilon$ (*eigenvectors*) and scalars $\lambda \in \overline{\mathbb{R}}$ (*eigenvalues*) satisfying

$$A \otimes x = \lambda \otimes x \tag{4.1}$$

is called the (max-algebraic) *eigenproblem*. For some applications it may be sufficient to find one eigenvalue-eigenvector pair; however, in this chapter we show that all eigenvalues can be found and all eigenvectors can efficiently be described for any matrix.

The eigenproblem is of key importance in max-algebra. It has been studied since the 1960's [58] in connection with the analysis of the steady-state behavior of production systems (see Sect. 1.3.3). Full solution of the eigenproblem in the case of irreducible matrices has been presented in [60] and [98], see also [11, 61] and [144]. A general spectral theorem for reducible matrices has appeared in [84] and [12], and

P. Butkovič, *Max-linear Systems: Theory and Algorithms*,
Springer Monographs in Mathematics 151,
DOI 10.1007/978-1-84996-299-5_4, © Springer-Verlag London Limited 2010

partly in [48]. An application of the max-algebraic eigenproblem to the conventional eigenproblem and in music theory can be found in [79].

For $A \in \overline{\mathbb{R}}^{n \times n}$ and $\lambda \in \overline{\mathbb{R}}$ we denote by $V(A, \lambda)$ the set consisting of ε and all eigenvectors of A corresponding to λ, and by $\Lambda(A)$ the set of all eigenvalues of A, that is

$$V(A, \lambda) = \left\{ x \in \overline{\mathbb{R}}^n; A \otimes x = \lambda \otimes x \right\}$$

and

$$\Lambda(A) = \left\{ \lambda \in \overline{\mathbb{R}}; V(A, \lambda) \neq \{\varepsilon\} \right\}.$$

We also denote by $V(A)$ the set consisting of ε and all eigenvectors of A, that is

$$V(A) = \bigcup_{\lambda \in \Lambda(A)} V(A, \lambda).$$

Finite eigenvectors are of special significance for both theory and applications and we denote:

$$V^+(A, \lambda) = V(A, \lambda) \cap \mathbb{R}^n$$

and

$$V^+(A) = V(A) \cap \mathbb{R}^n.$$

We start by presenting basic properties of eigenvalues and eigenvectors. The set $\{\alpha \otimes x; x \in S\}$ for $\alpha \in \overline{\mathbb{R}}$ and $S \subseteq \overline{\mathbb{R}}^n$ will be denoted $\alpha \otimes S$.

Proposition 4.1.1 *Let $A, B \in \overline{\mathbb{R}}^{n \times n}, \alpha \in \mathbb{R}, \lambda, \mu \in \overline{\mathbb{R}}$ and $x, y \in \overline{\mathbb{R}}^n$. Then*

(a) $V(\alpha \otimes A) = V(A)$,
(b) $\Lambda(\alpha \otimes A) = \alpha \otimes \Lambda(A)$,
(c) $V(A, \lambda) \cap V(B, \mu) \subseteq V(A \oplus B, \lambda \oplus \mu)$,
(d) $V(A, \lambda) \cap V(B, \mu) \subseteq V(A \otimes B, \lambda \otimes \mu)$,
(e) $V(A, \lambda) \subseteq V(A^k, \lambda^k)$ *for all integers $k \geq 0$,*
(f) $x \in V(A, \lambda) \Longrightarrow \alpha \otimes x \in V(A, \lambda)$,
(g) $x, y \in V(A, \lambda) \Longrightarrow x \oplus y \in V(A, \lambda)$.

Proof If $A \otimes x = \lambda \otimes x$ then $(\alpha \otimes A) \otimes x = (\alpha \otimes \lambda) \otimes x$ which proves (a) and (b). If $A \otimes x = \lambda \otimes x$ and $B \otimes x = \mu \otimes x$ then

$$(A \oplus B) \otimes x = A \otimes x \oplus B \otimes x$$

$$= \lambda \otimes x \oplus \mu \otimes x$$

$$= (\lambda \oplus \mu) \otimes x$$

and

$$(A \otimes B) \otimes x = A \otimes (B \otimes x)$$

$$= A \otimes \mu \otimes x$$
$$= \mu \otimes A \otimes x$$
$$= \mu \otimes \lambda \otimes x$$

which prove (c) and (d). Statement (e) follows by a repeated use of (d) and setting $A = B$.

If $A \otimes x = \lambda \otimes x$ then $A \otimes (\alpha \otimes x) = \lambda \otimes (\alpha \otimes x)$ which proves (f).

Finally, if $A \otimes x = \lambda \otimes x$ and $A \otimes y = \lambda \otimes y$ then

$$A \otimes (x \oplus y) = A \otimes x \oplus A \otimes y$$
$$= \lambda \otimes (x \oplus y)$$

and (g) follows. $\qquad \square$

It follows from Proposition 4.1.1 that $V(A, \lambda)$ is a subspace for every $\lambda \in \Lambda(A)$; it will be called an *eigenspace* (corresponding to the eigenvalue λ).

Remark 4.1.2 By (c) and (e) of Proposition 4.1.1 we have: If $A \in \overline{\mathbb{R}}^{n \times n}$ and $\varepsilon < \lambda(A) \leq 0$ then $V(A) \subseteq V(\Gamma(A))$. In particular,

$$V(A_\lambda, 0) \subseteq V(\Gamma(A_\lambda), 0).$$

The next statement summarizes spectral properties that are unaffected by a simultaneous permutation of the rows and columns.

Proposition 4.1.3 *Let* $A, B \in \overline{\mathbb{R}}^{n \times n}$ *and* $B = P^{-1} \otimes A \otimes P$, *where* P *is a permutation matrix. Then*

(a) *A is irreducible if and only if B is irreducible.*
(b) *The sets of cycle lengths in D_A and D_B are equal.*
(c) *A and B have the same eigenvalues.*
(d) *There is a bijection between $V(A)$ and $V(B)$ described by:*

$$V(B) = \left\{ P^{-1} \otimes x; x \in V(A) \right\}.$$

Proof To prove (a) and (b) note that B is obtained from A by simultaneous permutations of the rows and columns. Hence D_B differs from D_A by the numbering of the nodes only and the statements follow. For (c) and (d) we observe that $B \otimes z = \lambda \otimes z$ if and only if $A \otimes P \otimes z = \lambda \otimes P \otimes z$, that is, $z \in V(B)$ if and only if $z = P^{-1} \otimes x$ for some $x \in V(A)$. $\qquad \square$

Remark 4.1.4 The eigenvectors as defined by (4.1) are also called right eigenvectors in contrast to left eigenvectors that are defined by the equation

$$y^T \otimes A = y^T \otimes \lambda.$$

By the rules for transposition we have that y is a left eigenvector of A if and only if y is a right eigenvector of A^T (corresponding to the same eigenvalue), and hence the task of finding left eigenvectors for A is converted to the task of finding right eigenvectors for A^T.

4.2 Maximum Cycle Mean is the Principal Eigenvalue

When solving the eigenproblem a crucial role is played by the concepts of the maximum cycle mean and that of a definite matrix. The aim of this section is to prove that the maximum cycle mean is an eigenvalue of every square matrix over $\overline{\mathbb{R}}$. We will first solve the extreme case when $\lambda(A) = \varepsilon$ and then we prove that the columns of $\Gamma(A_\lambda)$ with zero diagonal entries are eigenvectors corresponding to $\lambda(A)$ if $\lambda(A) > \varepsilon$.

Recall that the maximum cycle mean of $A = (a_{ij}) \in \overline{\mathbb{R}}^{n \times n}$ is

$$\lambda(A) = \max \frac{a_{i_1 i_2} + a_{i_2 i_3} + \cdots + a_{i_{k-1} i_k} + a_{i_k i_1}}{k}$$

where the maximization is taken over all (elementary) cycles $(i_1, \ldots, i_k, i_1)$ in D_A ($k = 1, \ldots, n$), see Lemma 1.6.2. Due to the convention $\max \emptyset = \varepsilon$, it follows from this definition that $\lambda(A) = \varepsilon$ if and only if D_A is acyclic.

Lemma 4.2.1 Let $A = (a_{ij}) \in \overline{\mathbb{R}}^{n \times n}$ have columns $A_1, A_2, \ldots, A_n$. If $\lambda(A) = \varepsilon$ then $\Lambda(A) = \{\varepsilon\}$, at least one column of A is ε and the eigenvectors of A are exactly the vectors $(x_1, \ldots, x_n)^T \in \overline{\mathbb{R}}^n, x \neq \varepsilon$ such that $x_j = \varepsilon$ whenever $A_j \neq \varepsilon$ ($j \in N$). Hence $V(A, \varepsilon) = \{G \otimes z; z \in \overline{\mathbb{R}}^n\}$, where $G \in \overline{\mathbb{R}}^{n \times n}$ has columns $g_1, g_2, \ldots$ and for all $j \in N$:

$$g_j = \begin{cases} e^j, & \text{if } A_j = \varepsilon, \\ \varepsilon, & \text{if } A_j \neq \varepsilon. \end{cases}$$

Proof Suppose $\lambda(A) = \varepsilon$ and $A \otimes x = \lambda \otimes x$ for some $\lambda \in \mathbb{R}, x \neq \varepsilon$. Hence

$$\max_{j=1,\ldots,n} (a_{ij} + x_j) = \lambda + x_i \quad (i = 1, \ldots, n).$$

For every $i \in N$ there is a $j \in N$ such that

$$a_{ij} + x_j = \lambda + x_i.$$

Thus if, say $x_{i_1} > \varepsilon$, and $i = i_1$ then there are $i_2, i_3, \ldots$ such that

$$a_{i_1 i_2} + x_{i_2} = \lambda + x_{i_1}$$
$$a_{i_2 i_3} + x_{i_3} = \lambda + x_{i_2}$$

$$\ldots.$$

where $x_{i_1}, x_{i_2}, x_{i_3}, \ldots > \varepsilon$. This process will eventually cycle. Let us assume without loss of generality that the cycle is $(i_1, \ldots, i_k, i_{k+1} = i_1)$. Hence the last equation in the above system is

$$a_{i_k i_1} + x_{i_1} = \lambda + x_{i_k}.$$

In all these equations both sides are finite. If we add them up and simplify, we get

$$a_{i_1 i_2} + a_{i_2 i_3} + \cdots + a_{i_{k-1} i_k} + a_{i_k i_1} = k\lambda$$

showing that a cycle in D_A exists, a contradiction to $\lambda(A) = \varepsilon$. Therefore $\Lambda(A) \cap \mathbb{R} = \emptyset$. At the same time A has an ε column by Lemma 1.5.3. If the jth column is ε then $A \otimes x = \lambda(A) \otimes x$ for any vector x whose components are all ε, except for the jth which may be of any finite value. Hence $\Lambda(A) = \{\varepsilon\}$ and the rest of the lemma follows. $\qquad \square$

Since Lemma 4.2.1 completely solves the case $\lambda(A) = \varepsilon$, we may now assume that we deal with matrices whose maximum cycle mean is finite. Recall that the matrix $A_\lambda = (\lambda(A))^{-1} \otimes A$ is definite for any $A \in \overline{\mathbb{R}}^{n \times n}$ whenever $\lambda(A) > \varepsilon$ (Theorem 1.6.5).

Proposition 4.2.2 *Let $A \in \overline{\mathbb{R}}^{n \times n}$ and $\lambda(A) > \varepsilon$. Then*

$$V(A) = V(\lambda(A)^{-1} \otimes A).$$

Proof The statement follows from part (a) of Proposition 4.1.1. $\qquad \square$

Thus by Lemma 4.2.1, Proposition 4.1.1 (parts (a) and (b)) and Proposition 4.2.2 the task of finding all eigenvalues and eigenvectors of a matrix has been reduced to the same task for definite matrices.

Recall that $\Gamma(A)$ was defined in Sect. 1.6.2 as the series $A \oplus A^2 \oplus A^3 \oplus \cdots$ and that

$$\Gamma(A) = A \oplus A^2 \oplus \cdots \oplus A^n$$

if and only if $\lambda(A) \leq 0$ (Proposition 1.6.10).

Let us denote the columns of $\Gamma(A) = (\gamma_{ij})$ by $g_1, \ldots, g_n$. Recall that if A is definite then the values γ_{ij} $(i, j \in N)$ represent the weights of heaviest $i - j$ paths in D_A (Sect. 1.6.2). The significance of $\Gamma(A)$ for matrices with $\lambda(A) \leq 0$ is indicated by the fact that for such matrices

$$A \otimes \Gamma(A) = A^2 \oplus \cdots \oplus A^{n+1} \leq \Gamma(A)$$

due to (1.20), thus yielding

$$A \otimes g_j \leq g_j \quad \text{for every } j \in N. \tag{4.2}$$

An important point of the max-algebraic eigenproblem theory is that in (4.2) actually equality holds whenever A is definite and $j \in N_c(A)$:

Lemma 4.2.3 *Let* $A = (a_{ij}) \in \overline{\mathbb{R}}^{n \times n}$. *If* A *is definite,* $g_1, \ldots, g_n$ *are the columns of* $\Gamma(A)$ *and* $j \in N_c(A)$ *then* $A \otimes g_j = g_j$.

Proof Let $j \in N_c(A)$ and $i \in N$. Then by (4.2)

$$\max_{r=1,\ldots,n} (a_{ir} + \gamma_{rj}) \leq \gamma_{ij}$$

and we need to prove that actually equality holds. We may assume without loss of generality $\gamma_{ij} > \varepsilon$ (otherwise the wanted equality follows). Let $(i, k, \ldots, j)$ be a heaviest $i - j$ path. If $k = j$ then $\gamma_{ij} = a_{ij} = a_{ij} + \gamma_{jj}$. If $k \neq j$ then $\gamma_{ij} = a_{ik} + \gamma_{kj}$. In each case there is an r such that $a_{ir} + \gamma_{rj} = \gamma_{ij}$. $\square$

Before we summarize our results in the main statement of this section, we give a practical description of the set of critical nodes $N_c(A)$. Since there are no cycles of weight more than 0 in D_A for definite matrices A but at least one has weight 0, we have then that for a definite matrix A at least one diagonal entry in $\Gamma(A)$ is 0 and all diagonal entries are 0 or less since the kth diagonal entry is the greatest weight of a cycle in D_A containing node k.

It also follows for any definite matrix A that zero diagonal entries in $\Gamma(A)$ exactly correspond to critical nodes, that is, we have

$$N_c(A) = \{ j \in N; \gamma_{jj} = 0 \}. \tag{4.3}$$

By Lemma 4.2.3 zero is an eigenvalue of every definite matrix. Hence Proposition 4.1.1 (part 2), Lemmas 4.2.1, 4.2.2, 1.6.6 and 4.2.3 and (4.3) imply:

Theorem 4.2.4 $\lambda(A)$ *is an eigenvalue for any matrix* $A \in \overline{\mathbb{R}}^{n \times n}$. *If* $\lambda(A) > \varepsilon$ *then up to* n *eigenvectors of* A *corresponding to* $\lambda(A)$ *can be found among the columns of* $\Gamma(A_\lambda)$. *More precisely, every column of* $\Gamma(A_\lambda)$ *with zero diagonal entry is an eigenvector of* A *with corresponding eigenvalue* $\lambda(A)$.

In view of Theorem 4.2.4 we will call $\lambda(A)$ the *principal eigenvalue* of A.

Note that when the result of Theorem 4.2.4 is generalized to matrices over linearly ordered commutative groups then the concept of radicability of the underlying group (see Sect. 1.4) is crucial, since otherwise it is not possible to guarantee the existence of the maximum cycle mean. Therefore in groups that are not radicable, such as the additive group of integers, an eigenvalue of a matrix may not exist.

4.3 Principal Eigenspace

The results of the previous section enable us to present a complete description of all eigenvectors corresponding to the principal eigenvalue. Such eigenvectors will be called *principal* and $V(A, \lambda(A))$ will be called the *principal eigenspace* of A. Our aim in this section is to describe bases of $V(A, \lambda(A))$.

The columns of $\Gamma(A_\lambda)$ with zero diagonal entry are principal eigenvectors by Theorem 4.2.4. We will call them the *fundamental eigenvectors* [60] of A (FEV). Clearly, every max-combination of fundamental eigenvectors is also a principal eigenvector.

We will use Theorem 4.2.4 and

- prove that there are no principal eigenvectors other than max-combinations of fundamental eigenvectors,
- identify fundamental eigenvectors that are multiples of the others, and
- prove that by removing fundamental eigenvectors that are multiples of the others we produce a basis of the principal eigenspace, that is, none of the remaining columns is a max-combination of the others.

We start with a technical lemma.

Lemma 4.3.1 [65] *Let* $A \in \overline{\mathbb{R}}^{n \times n}$, $\lambda(A) > \varepsilon$ *and* $g_1, \ldots, g_n$ *be the columns of* $\Gamma(A_\lambda) = (\gamma_{ij})$. *If* $x = (x_1, \ldots, x_n)^T \in V(A, \lambda(A))$ *and* $x_i > \varepsilon$ ($i \in N$) *then there is an* $s \in N_c(A)$ *such that*

$$x_i = x_s + \gamma_{is}.$$

Proof Let $A_\lambda = (d_{ij})$ and $i \in N$, $x_i > \varepsilon$. Then $A_\lambda \otimes x = x$ by Proposition 4.1.1 (parts (a) and (b)) and $N_c(A) = N_c(A_\lambda)$ by Lemma 1.6.6. This implies that there is a sequence of indices $i_1 = i, i_2, \ldots$ such that

$$x_{i_1} = d_{i_1 i_2} + x_{i_2}$$
$$x_{i_2} = d_{i_2 i_3} + x_{i_3} \tag{4.4}$$

$$\ldots$$

This sequence will eventually cycle. Let us assume that the cycle is

$$(i_r, \ldots, i_k, i_{k+1} = i_r).$$

For this subsequence we have

$$x_{i_r} = d_{i_r i_{r+1}} + x_{i_{r+1}}$$

$$\ldots$$

$$x_{i_k} = d_{i_k i_r} + x_{i_r}.$$

In all these equations both sides are finite. If we add them up and simplify, we get

$$d_{i_r i_{r+1}} + \cdots + d_{i_k i_r} = 0$$

and hence $i_k \in N_c(A_\lambda) = N_c(A)$.

If we add up the first $k - 1$ equations in (4.4) and simplify, we get

$$x_{i_1} = d_{i_1 i_2} + \cdots + d_{i_{k-1} i_k} + x_{i_k}.$$

Since $d_{i_1 i_2} + \cdots + d_{i_{k-1} i_k}$ is the weight of an $i_1 - i_k$ path in D_{A_λ} and $\gamma_{i_1 i_k}$ is the weight of a heaviest $i_1 - i_k$ path, we have

$$x_{i_1} \leq \gamma_{i_1 i_k} + x_{i_k}.$$

At the same time $x \in V(\Gamma(A_\lambda))$ (see Remark 4.1.2) and so

$$x_{i_1} = \sum_{j \in N}^{\oplus} \gamma_{i_1 j} \otimes x_j \geq \gamma_{i_1 i_k} + x_{i_k}.$$

Hence i_k is the sought s.   □

We are ready to prove that there are no principal eigenvectors other than max-combinations of fundamental eigenvectors:

Lemma 4.3.2 *Suppose that $A = (a_{ij}) \in \overline{\mathbb{R}}^{n \times n}$, $\lambda(A) > \varepsilon$ and $g_1, \ldots, g_n$ are the columns of $\Gamma(A_\lambda) = (\gamma_{ij})$. If $x = (x_1, \ldots, x_n)^T \in V(A, \lambda(A))$ then*

$$x = \sum_{j \in N_c(A)}^{\oplus} x_j \otimes g_j.$$

Proof Let $x = (x_1, \ldots, x_n)^T \in V(A, \lambda(A))$. We have

$$A_\lambda \otimes x = x \tag{4.5}$$

by Proposition 4.1.1 (parts (a) and (b)) and $N_c(A) = N_c(A_\lambda)$ by Lemma 1.6.6. This implies (see Remark 4.1.2) that $x \in V(\Gamma(A_\lambda), 0)$, yielding

$$x = \sum_{j \in N}^{\oplus} x_j \otimes g_j \geq \sum_{j \in N_c(A)}^{\oplus} x_j \otimes g_j.$$

We need to prove that the converse inequality holds too, that is, for every $i \in N$ there is an $s \in N_c(A)$ such that

$$x_i \leq x_s + \gamma_{is}.$$

If $x_i = \varepsilon$ then this is trivially true. If $x_i > \varepsilon$ then it follows from Lemma 4.3.1.   □

Clearly, when considering all possible max-combinations of a set of fundamental eigenvectors (or, indeed, of any vectors), we may remove from this set fundamental eigenvectors that are multiples of some other. To be more precise, we say that two fundamental eigenvectors g_i and g_j are *equivalent* if $g_i = \alpha \otimes g_j$ for some $\alpha \in \mathbb{R}$ and *nonequivalent* otherwise. We characterize equivalent fundamental eigenvectors using the equivalence of eigennodes in the next statement (note that the relation $i \sim j$ has been defined in Sect. 1.6.1):

Theorem 4.3.3 [60] *Suppose that $A = (a_{ij}) \in \overline{\mathbb{R}}^{n \times n}$, $\lambda(A) > \varepsilon$ and $g_1, \ldots, g_n$ are the columns of $\Gamma(A_\lambda) = (\gamma_{ij})$. If $i, j \in N_c(A)$ then $g_i = \alpha \otimes g_j$ for some $\alpha \in \mathbb{R}$ if and only if $i \sim j$.*

Proof Recall that $N_c(A) = N_c(A_\lambda)$ by Lemma 1.6.6.

Let $i, j \in N_c(A_\lambda)$. If $g_i = \alpha \otimes g_j$, $\alpha \in \mathbb{R}$ then $\gamma_{ji} = \alpha \otimes \gamma_{jj} = \alpha$ and $\gamma_{ij} = \alpha^{-1} \otimes \gamma_{ii} = \alpha^{-1}$. Hence the heaviest $i - j$ path extended by the heaviest $j - i$ path is a cycle of weight $\alpha^{-1} \otimes \alpha = 0$, thus $i \sim j$. Conversely, let $i \sim j$ and α be the weight of the $j - i$ subpath of the critical cycle containing both i and j. Then for any $k \in N$ we have $\gamma_{ki} = \alpha \otimes \gamma_{kj}$ since $\geq$ follows from the definition of γ_{ki} and $>$ would imply $\alpha^{-1} \otimes \gamma_{ki} > \gamma_{kj}$. But α^{-1} is the weight of the $i - j$ subpath of the critical cycle containing both i and j and thus $\alpha^{-1} \otimes \gamma_{ki}$ is the weight of a $k - j$ path which is a contradiction with the maximality of γ_{kj}. Hence $g_i = \alpha \otimes g_j$. □

Note that if $i \sim j$ then we also write $g_i \sim g_j$.

From the last two theorems we can readily deduce:

Corollary 4.3.4 [60] *Suppose that $A = (a_{ij}) \in \overline{\mathbb{R}}^{n \times n}$, $\lambda(A) > \varepsilon$ and $g_1, \ldots, g_n$ are the columns of $\Gamma(A_\lambda)$. Then*

$$V(A, \lambda(A)) = \left\{ \sum_{j \in N_c^*(A)}^{\oplus} \alpha_j \otimes g_j ; \alpha_j \in \overline{\mathbb{R}}, j \in N_c^*(A) \right\}$$

where $N_c^*(A)$ is any maximal set of nonequivalent eigennodes of A.

Clearly, any set $N_c^*(A)$ in Corollary 4.3.4 can be obtained by taking exactly one g_k for each equivalence class in $(N_c(A), \sim)$. The results on bases in Chap. 3 enable us now to easily describe bases of principal eigenspaces and, consequently, to define the principal dimension.

Theorem 4.3.5 [6] *Suppose that $A = (a_{ij}) \in \overline{\mathbb{R}}^{n \times n}$, $\lambda(A) > \varepsilon$ and $g_1, \ldots, g_n$ are the columns of $\Gamma(A_\lambda)$. Then $V(A, \lambda(A))$ is a nontrivial subspace and we obtain a basis of $V(A, \lambda(A))$ by taking exactly one g_k for each equivalence class in $(N_c(A), \sim)$.*

Proof $V(A, \lambda(A))$ is a subspace by Proposition 4.1.1 (parts (f) and (g)). It is nontrivial due to (4.3) and Lemma 4.2.3. By Corollary 3.3.11 it remains to prove that every $g_k, k \in N_c(A)$, is an extremal.

Let $k \in N_c(A)$ be fixed and suppose that $g_k = u \oplus v$ where $u, v \in V(A, \lambda(A))$. Then by Lemma 4.3.2 we have:

$$u = \sum_{j \in N_c^*(A)}^{\oplus} \alpha_j \otimes g_j$$

and

$$v = \sum_{j \in N_c^*(A)}^{\oplus} \beta_j \otimes g_j$$

where $N_c^*(A)$ is a fixed maximal set of nonequivalent eigennodes of A and $\alpha_j, \beta_j \in \mathbb{R}$. We may assume without loss of generality that $g_k \in N_c^*(A)$ and thus $g_k \nsim g_h$ for any $h \in N_c^*(A), h \neq k$. Hence

$$g_k = \sum_{j \in N_c^*(A)}^{\oplus} \delta_j \otimes g_j$$

where $\delta_j = \alpha_j \oplus \beta_j$. Clearly $\delta_k \leq 0$. Suppose $\delta_k < 0$ then

$$g_k = \sum_{\substack{j \in N_c^*(A) \\ j \neq k}}^{\oplus} \delta_j \otimes g_j.$$

It follows that

$$0 = \gamma_{kk} = \sum_{\substack{j \in N_c^*(A) \\ j \neq k}}^{\oplus} \delta_j \otimes \gamma_{kj} = \delta_h \otimes \gamma_{kh}$$

for some $h \in N_c^*(A), h \neq k$. At the same time

$$\gamma_{hk} = \sum_{\substack{j \in N_c^*(A) \\ j \neq k}}^{\oplus} \delta_j \otimes \gamma_{hj} \geq \delta_h \otimes \gamma_{hh} = \delta_h.$$

Therefore

$$\gamma_{kh} \otimes \gamma_{hk} \geq \delta_h^{-1} \otimes \delta_h = 0.$$

The last inequality is in fact equality since there are no positive cycles in $D_{\Gamma(A_\lambda)}$, implying that $k \sim h$, a contradiction. Hence $\delta_k = 0$. Then (without loss of generality) $\alpha_k = 0$ implying $u \geq g_k = u \oplus v$ and thus $u = g_k$. $\qquad\square$

The dimension of the principal eigenspace of A will be called the *principal dimension* of A and will be denoted pd(A). It follows from Theorems 4.3.3 and 4.3.5 that pd(A) is equal to the number of critical components of $C(A)$ or, equivalently, to the size of any basis of the column space of the matrix consisting of fundamental eigenvectors of A. Since this basis can be found in $O(n^3)$ time (Sect. 3.4), pd(A) can be found with the same computational effort.

Remark 4.3.6 It is easily seen that $\lambda(A^T) = \lambda(A)$, $\Gamma(A^T) = (\Gamma(A))^T$ and $N_c(A^T) = N_c(A)$. Hence an analogue of Theorem 4.3.5 in terms of rows of $\Gamma(A_\lambda)$ for left principal eigenvectors immediately follows. See also Remark 4.1.4.

Example 4.3.7 Consider the matrix

$$
A = \begin{pmatrix}
7 & 9 & 5 & 5 & 3 & 7 \\
7 & 5 & 2 & 7 & 0 & 4 \\
8 & 0 & 3 & 3 & 8 & 0 \\
7 & 2 & 5 & 7 & 9 & 5 \\
4 & 2 & 6 & 6 & 8 & 8 \\
3 & 0 & 5 & 7 & 1 & 2
\end{pmatrix}.
$$

The maximum cycle mean is 8, attained by three critical cycles: $(1, 2, 1)$, $(5, 5)$ and $(4, 5, 6, 4)$. Thus $\lambda(A) = 8$, $\mathrm{pd}(A) = 2$ and

$$
\Gamma(A_\lambda) = \begin{pmatrix}
0 & 1 & -1 & 0 & 1 & 1 \\
-1 & 0 & 2 & -1 & 0 & 0 \\
0 & 1 & -1 & 0 & 1 & 1 \\
-1 & 0 & -1 & 0 & 1 & 1 \\
-2 & -1 & -2 & -1 & 0 & 0 \\
-2 & -1 & -2 & -1 & 0 & 0
\end{pmatrix}.
$$

Critical components have node sets $\{1, 2\}$ and $\{4, 5, 6\}$. Hence the first and second columns of $\Gamma(A_\lambda)$ are multiples of each other and similarly the fourth, fifth and sixth columns. For the basis of $V(A, \lambda(A))$ we may take for instance the first and fourth columns.

Example 4.3.8 Consider the matrix

$$
A = \begin{pmatrix}
0 & 3 & & \\
1 & -1 & & \\
& & 2 & \\
& & & 1
\end{pmatrix},
$$

where the missing entries are ε. Then $\lambda(A) = 2$, $N_c(A) = \{1, 2, 3\}$, critical components have node sets $\{1, 2\}$ and $\{3\}$, $\mathrm{pd}(A) = 2$. We can compute

$$
\Gamma(A_\lambda) = \begin{pmatrix}
0 & 1 & & \\
-1 & 0 & & \\
& & 0 & \\
& & & -1
\end{pmatrix},
$$

hence a basis of the principal eigenspace is

$$
\{g_2, g_3\} = \left\{ (1, 0, \varepsilon, \varepsilon)^T , (\varepsilon, \varepsilon, 0, \varepsilon)^T \right\}.
$$

4.4 Finite Eigenvectors

The aim in this chapter is to show how to find all eigenvalues and describe all eigenvectors of a matrix. To achieve this goal, in this section we will study the set of finite eigenvectors. We will show how to efficiently describe all finite eigenvectors.

We will continue to use the notation $\Gamma(A_\lambda) = (\gamma_{ij})$ if $\lambda(A) > \varepsilon$. Recall that $N_c(A) = N_c(A_\lambda)$ by Lemma 1.6.6.

We will present the main results of this section in the following order:

- A proof that the maximum cycle mean is the only possible eigenvalue corresponding to finite eigenvectors.
- Criteria for the existence of finite eigenvectors.
- Description of all finite eigenvectors.
- A proof that irreducible matrices have only finite eigenvectors.

The first result shows that $\lambda(A)$ is the only possible eigenvalue corresponding to finite eigenvectors. Note that if $A = \varepsilon$ then every finite vector of a suitable dimension is an eigenvector of A and all correspond to the unique eigenvalue $\lambda(A) = \varepsilon$.

Theorem 4.4.1 [60] *Let* $A = (a_{ij}) \in \overline{\mathbb{R}}^{n \times n}$. *If* $A \neq \varepsilon$ *and* $V^+(A) \neq \emptyset$ *then* $\lambda(A) > \varepsilon$ *and* $A \otimes x = \lambda(A) \otimes x$ *for every* $x \in V^+(A)$.

Proof Let $x = (x_1, \ldots, x_n)^T \in V^+(A)$. We have

$$\max_{j=1,\ldots,n} \left(a_{ij} + x_j \right) = \lambda + x_i \quad (i = 1, \ldots, n)$$

for some $\lambda \in \overline{\mathbb{R}}$. Since $A \neq \varepsilon$ the LHS is finite for at least one i and thus $\lambda > \varepsilon$.

For every $i \in N$ there is a $j \in N$ such that

$$a_{ij} + x_j = \lambda + x_i.$$

Hence, if $i = i_1$ is any fixed index then there are indices $i_2, i_3, \ldots$ such that

$$a_{i_1 i_2} + x_{i_2} = \lambda + x_{i_1},$$
$$a_{i_2 i_3} + x_{i_3} = \lambda + x_{i_2},$$

$$\ldots$$

This process will eventually cycle. Let us assume without loss of generality that the cycle is $(i_1, \ldots, i_k, i_{k+1} = i_1)$, otherwise we remove the necessary first elements of this sequence. Hence the last equation in the above system is

$$a_{i_k i_1} + x_{i_1} = \lambda + x_{i_k}.$$

In all these equations both sides are finite. If we add them up and simplify, we get

$$\lambda = \frac{a_{i_1 i_2} + a_{i_2 i_3} + \cdots + a_{i_{k-1} i_k} + a_{i_k i_1}}{k}.$$

At the same time, if $\sigma = (i_1, \ldots, i_k, i_{k+1} = i_1)$ is an arbitrary cycle in D_A then it satisfies the system of inequalities obtained from the above system of equations after replacing $=$ by $\leq$. Hence

$$\lambda \geq \frac{a_{i_1 i_2} + a_{i_2 i_3} + \cdots + a_{i_{k-1} i_k} + a_{i_k i_1}}{k} = \mu(\sigma, A).$$

It follows that $\lambda = \max_\sigma \mu(\sigma, A) = \lambda(A)$. $\square$

Theorem 4.4.1 opens the possibility of answering questions such as the existence and description of finite eigenvectors.

Lemma 4.4.2 Let $A \in \overline{\mathbb{R}}^{n \times n}$. If $A \neq \varepsilon$ and $x = (x_1, \ldots, x_n)^T \in V^+(A)$ then for every $i \in N$ there is an $s \in N_c(A)$ such that

$$x_i = x_s + \gamma_{is},$$

where $\Gamma(A_\lambda) = (\gamma_{ij})$.

Proof Since $\lambda(A) > \varepsilon$ and $x \in V(A, \lambda(A))$ by Theorem 4.4.1, the statement follows immediately from Lemma 4.3.1. $\square$

We are ready to formulate the first criterion for the existence of finite eigenvectors.

Theorem 4.4.3 *Suppose that* $A \in \overline{\mathbb{R}}^{n \times n}$, $\lambda(A) > \varepsilon$ *and* $g_1, \ldots, g_n$ *are the columns of* $\Gamma(A_\lambda) = (\gamma_{ij})$. *Then*

$$V^+(A) \neq \emptyset \iff \sum_{j \in N_c(A)}^{\oplus} g_j \in \mathbb{R}^n.$$

Proof Suppose $\sum_{j \in N_c(A)}^{\oplus} g_j \in \mathbb{R}^n$. Every g_j ($j \in N_c(A)$) is in $V(A, \lambda(A))$ by Lemma 4.2.3 and $\sum_{j \in N_c(A)}^{\oplus} g_j \in V(A)$ by Proposition 4.1.1. Hence $\sum_{j \in N_c(A)}^{\oplus} g_j \in V^+(A)$.

On the other hand, by Lemma 4.4.2, if $x = (x_1, \ldots, x_n)^T \in V^+(A)$ then for every $i \in N$ there is an $s \in N_c(A)$ such that $\gamma_{is} \in \mathbb{R}$ and so $\sum_{j \in N_c(A)}^{\oplus} g_j \in \mathbb{R}^n$. $\square$

We can now easily deduce a classical result:

Corollary 4.4.4 [60] *Suppose* $A \in \overline{\mathbb{R}}^{n \times n}$, $A \neq \varepsilon$. *Then* $V^+(A) \neq \emptyset$ *if and only if the following are satisfied:*

(a) $\lambda(A) > \varepsilon$.
(b) *In* D_A *there is*

$$(\forall i \in N)(\exists j \in N_c(A)) i \to j.$$

Proof By Theorem 4.4.1, $A \neq \varepsilon$ and $V^+(A) \neq \emptyset$ implies $\lambda(A) > \varepsilon$. Observe that

$$\sum_{j \in N_c(A)}^{\oplus} g_j \in \mathbb{R}^n \Longleftrightarrow \sum_{j \in N_c(A)}^{\oplus} \gamma_{ij} \in \mathbb{R} \quad \text{for all } i \in N.$$

Hence by Theorem 4.4.3 $V^+(A) \neq \emptyset$ if and only if

$$(\forall i \in N)(\exists j \in N_c(A))\gamma_{ij} \in \mathbb{R}.$$

However, γ_{ij} is the greatest weight of an $i - j$ path in D_{A_λ} or ε, if there is no such path, and the statement follows. $\square$

The description of all finite eigenvectors can now easily be deduced:

Theorem 4.4.5 *Let* $A \in \overline{\mathbb{R}}^{n \times n}$. *If* $\lambda(A) > \varepsilon$, $g_1, \ldots, g_n$ *are the columns of* $\Gamma(A_\lambda)$ *and* $V^+(A) \neq \emptyset$ *then*

$$V^+(A) = \left\{ \sum_{j \in N_c^*(A)}^{\oplus} \alpha_j \otimes g_j; \alpha_j \in \mathbb{R} \right\}, \tag{4.6}$$

where $N_c^*(A)$ *is any maximal set of nonequivalent eigennodes of* A.

Proof $\supseteq$ follows from Lemma 4.2.3, Proposition 4.1.1 and Theorem 4.4.3 immediately. $\subseteq$ follows from Lemma 4.3.2. $\square$

Remark 4.4.6 Note that (4.6) requires $\alpha_j \in \mathbb{R}$ and, in general, g_j may or may not be in $V^+(A)$. Therefore the subspace $V^+(A) \cup \{\varepsilon\}$ may or may not be finitely generated and hence, in general, there is no guarantee that it has a basis.

Example 4.4.7 Consider the matrix

$$A = \begin{pmatrix} 0 & 3 & & \\ 1 & -1 & & \\ & & 2 & \\ & & 0 & 1 \end{pmatrix},$$

where the missing entries are ε. Then $\lambda(A) = 2$, $N_c(A) = \{1, 2, 3\}$, critical components have node sets $\{1, 2\}$ and $\{3\}$, $\text{pd}(A) = 2$. A finite eigenvector exists since an eigennode is accessible from every node (unlike in the slightly different Example 4.3.8). We can compute

$$\Gamma(A_\lambda) = \begin{pmatrix} 0 & 1 & & \\ -1 & 0 & & \\ & & 0 & \\ & & -2 & -1 \end{pmatrix},$$

hence a basis of the principal eigenspace is $\{(1, 0, \varepsilon, \varepsilon)^T, (\varepsilon, \varepsilon, 0, -2)^T\}$. All finite eigenvectors are max-combinations of the vectors in the basis provided that both coefficients are finite. However, $V^+(A) \cup \{\varepsilon\}$ has no basis.

The following classical complete solution of the eigenproblem for irreducible matrices is now easy to prove:

Theorem 4.4.8 (Cuninghame-Green [60]) *Every irreducible matrix $A \in \overline{\mathbb{R}}^{n \times n}$ ($n > 1$) has a unique eigenvalue equal to $\lambda(A)$ and*

$$V(A) - \{\varepsilon\} = V^+(A) = \left\{ \sum_{j \in N_c^*(A)}^{\oplus} \alpha_j \otimes g_j; \alpha_j \in \mathbb{R} \right\},$$

where $g_1, \ldots, g_n$ are the columns of $\Gamma(A_\lambda)$ and $N_c^*(A)$ is any maximal set of nonequivalent eigennodes of A.

Proof Let A be irreducible, thus $\lambda(A) > \varepsilon$. Also, $\Gamma(A_\lambda)$ is finite by Proposition 1.6.10. Every eigenvector of A is also an eigenvector of $\Gamma(A_\lambda)$ with eigenvalue 0 (Remark 4.1.2) but the product of a finite matrix and a vector $x \neq \varepsilon$ is finite. Hence an irreducible matrix can only have finite eigenvectors and thus its only eigenvalue is $\lambda(A)$ by Theorem 4.4.1.

On the other hand, due to the finiteness of all columns of $\Gamma(A_\lambda)$, by Theorem 4.4.3, $V^+(A) \neq \emptyset$ and the rest follows from Theorem 4.4.5.   □

Remark 4.4.9 Note that every 1×1 matrix A over $\overline{\mathbb{R}}$ is irreducible and $V(A) - \{\varepsilon\} = V^+(A) = \mathbb{R}$.

The fact that $\lambda(A)$ is the unique eigenvalue of an irreducible matrix A was already proved in [58] and then independently in [144] for finite matrices. Since then it has been rediscovered in many papers worldwide. The description of $V^+(A)$ for irreducible matrices as given in Corollary 4.4.4 was also proved in [98].

Note that for an irreducible matrix A we have:

$$V(A) = V^+(A) \cup \{\varepsilon\} = \{\Gamma(A_\lambda) \otimes z; z \in \overline{\mathbb{R}}^n, z_j = \varepsilon \quad \text{for all } j \notin N_c(A)\}.$$

Remark 4.4.10 Since $\Gamma(A_\lambda)$ is finite for an irreducible matrix A, the generators of $V^+(A)$ are all finite if A is irreducible. Hence $V^+(A) \cup \{\varepsilon\} = V(A)$ has a basis in this case, which coincides with the basis of $V(A)$.

Example 4.4.11 Consider the irreducible matrix

$$A = \begin{pmatrix} 0 & 3 & & 0 \\ 1 & -1 & 0 & \\ & 0 & 2 & \\ & & 0 & 1 \end{pmatrix},$$

where the missing entries are ε. Then $\lambda(A) = 2$, $N_c(A) = \{1, 2, 3\}$, critical components have node sets $\{1, 2\}$ and $\{3\}$, $\mathrm{pd}(A) = 2$. We can compute

$$\Gamma(A_\lambda) = \begin{pmatrix} 0 & 1 & -4 & -2 \\ -1 & 0 & -5 & -3 \\ -3 & -2 & 0 & -5 \\ -5 & -4 & -2 & -1 \end{pmatrix},$$

hence a basis of the principal eigenspace is

$$\left\{ (1, 0, -2, -4)^T, \ (-4, -5, 0, -2)^T \right\}.$$

4.5 Finding All Eigenvalues

Our next step is to describe all eigenvalues of square matrices over $\overline{\mathbb{R}}$. The information about principal eigenvectors obtained in the previous sections will be substantially used.

We have already seen in Sect. 1.5 that if $A, B \in \overline{\mathbb{R}}^{n \times n}$ are equivalent ($A \equiv B$),
then D_A can be obtained from D_B by a renumbering of the nodes and that
$B = P^{-1} \otimes A \otimes P$ for some permutation matrix P. Hence if $A \equiv B$ then A is irreducible if and only if B is irreducible. We also know by Proposition 4.1.3 that $V(A)$
and $V(B)$ are essentially the same (the eigenvectors of A and B only differ by the
order of their components).

It follows from Theorem 4.4.8 that a matrix with a nonfinite eigenvector cannot
be irreducible. The following lemma provides an alternative and somewhat more
detailed explanation of this simple but remarkable property. It may also be useful
for a good understanding of the structure of the set $V(A)$ for a general matrix A.

Lemma 4.5.1 *Let* $A = (a_{ij}) \in \overline{\mathbb{R}}^{n \times n}$ *and* $\lambda \in \Lambda(A)$. *If* $x \in V(A, \lambda) - V^+(A, \lambda)$,
$x \neq \varepsilon$, *then* $n > 1$,

$$A \equiv \begin{pmatrix} A^{(11)} & \varepsilon \\ A^{(21)} & A^{(22)} \end{pmatrix},$$

$\lambda = \lambda(A^{(22)})$, *and hence* A *is reducible.*

Proof Permute the rows and columns of A simultaneously so that the vector aris-
ing from x by the same permutation of its components is $x' = \begin{pmatrix} x^{(1)} \\ x^{(2)} \end{pmatrix}$, where
$x^{(1)} = \varepsilon \in \overline{\mathbb{R}}^p$ and $x^{(2)} \in \mathbb{R}^{n-p}$ for some p ($1 \leq p < n$). Denote the obtained matrix
by A' (thus $A \equiv A'$) and let us write blockwise

$$A' = \begin{pmatrix} A^{(11)} & A^{(12)} \\ A^{(21)} & A^{(22)} \end{pmatrix},$$

where $A^{(11)}$ is $p \times p$. The equality $A' \otimes x' = \lambda \otimes x'$ now yields blockwise:

$$A^{(12)} \otimes x^{(2)} = \varepsilon,$$

$$A^{(22)} \otimes x^{(2)} = \lambda \otimes x^{(2)}.$$

Since $x^{(2)}$ is finite, it follows from Theorem 4.4.4 that $\lambda = \lambda(A^{(22)})$; also clearly $A^{(12)} = \varepsilon$. $\qquad\square$

We already know (Theorem 4.4.8) that all eigenvectors of an irreducible matrix are finite. We now can prove that only irreducible matrices have this property.

Theorem 4.5.2 Let $A = (a_{ij}) \in \overline{\mathbb{R}}^{n \times n}$. Then $V(A) - \{\varepsilon\} = V^+(A)$ if and only if A is irreducible.

Proof It remains to prove the "only if" part since the "if" part follows from Theorem 4.4.8. If A is reducible then $n > 1$ and $A \equiv \left(\begin{smallmatrix} A^{(11)} & \varepsilon \\ A^{(21)} & A^{(22)} \end{smallmatrix} \right)$, where $A^{(22)}$ is irreducible. By setting $\lambda = \lambda(A^{(22)}), x^{(2)} \in V^+(A_{22}), x = \left(\begin{smallmatrix} \varepsilon \\ x^{(2)} \end{smallmatrix} \right) \in \overline{\mathbb{R}}^n$ we see that $x \in V(A) - V^+(A), x \neq \varepsilon$. $\qquad\square$

Theorem 4.5.2 does not exclude the possibility that a reducible matrix has finite eigenvectors. The following spectral theory will, as a by-product, enable us to characterize all situations when this occurs.

Every matrix $A = (a_{ij}) \in \overline{\mathbb{R}}^{n \times n}$ can be transformed in linear time by simultaneous permutations of the rows and columns to a *Frobenius normal form* (FNF) [11, 18, 126]

$$\begin{pmatrix} A_{11} & \varepsilon & \cdots & \varepsilon \\ A_{21} & A_{22} & \cdots & \varepsilon \\ \cdots & \cdots & \cdots & \cdots \\ A_{r1} & A_{r2} & \cdots & A_{rr} \end{pmatrix} \qquad (4.7)$$

where $A_{11}, \ldots, A_{rr}$ are irreducible square submatrices of A. The diagonal blocks are determined uniquely up to a simultaneous permutation of their rows and columns: however, their order is not determined uniquely. Since any such form is essentially determined by strongly connected components of D_A, an FNF can be found in $O(|V| + |E|)$ time [18, 142]. It will turn out later in this section that the FNF is a particularly convenient form for studying spectral properties of matrices. Since these are essentially preserved by simultaneous permutations of the rows and columns (Proposition 4.1.3) we will often assume, without loss of generality, that the matrix under consideration already is in an FNF.

If A is in an FNF then the corresponding partition of the node set N of D_A will be denoted as $N_1, \ldots, N_r$ and these sets will be called *classes* (of A). It follows that each of the induced subgraphs $D_A[N_i]$ $(i = 1, \ldots, r)$ is strongly connected and an arc from N_i to N_j in D_A exists only if $i \geq j$. Clearly, every A_{jj} has a unique eigenvalue $\lambda(A_{jj})$. As a slight abuse of language we will, for simplicity, also say that $\lambda(A_{jj})$ is the eigenvalue of N_j.

Fig. 4.1 Condensation
digraph (6 classes)

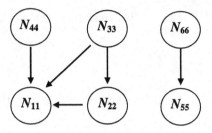

If A is in an FNF, say (4.7), then the *condensation digraph*, notation C_A, is the
digraph

$$(\{N_1, \ldots, N_r\}, \{(N_i, N_j); (\exists k \in N_i)(\exists l \in N_j)a_{kl} > \varepsilon\}).$$

Observe that C_A is acyclic.

Recall that the symbol $N_i \to N_j$ means that there is a directed path from a node
in N_i to a node in N_j in C_A (and therefore from each node in N_i to each node in N_j
in D_A).

If there are neither outgoing nor incoming arcs from or to an induced subgraph
$C_A[\{N_{i_1}, \ldots, N_{i_s}\}]$ $(1 \le i_1 < \cdots < i_s \le r)$ and no proper subdigraph has this property then the submatrix

$$\begin{pmatrix} A_{i_1 i_1} & \varepsilon & \cdots & \varepsilon \\ A_{i_2 i_1} & A_{i_2 i_2} & \cdots & \varepsilon \\ \cdots & \cdots & \cdots & \cdots \\ A_{i_s i_1} & A_{i_s i_2} & \cdots & A_{i_s i_s} \end{pmatrix}$$

is called an *isolated superblock* (or just *superblock*). The nodes of C_A (that is,
classes of A) with no incoming arcs are called the *initial classes*, those with no
outgoing arcs are called the *final classes*. Note that an isolated superblock may have
several initial and final classes.

For instance the condensation digraph for the matrix

$$\begin{pmatrix} A_{11} & \varepsilon & \varepsilon & \varepsilon & \varepsilon & \varepsilon \\ * & A_{22} & \varepsilon & \varepsilon & \varepsilon & \varepsilon \\ * & * & A_{33} & \varepsilon & \varepsilon & \varepsilon \\ * & \varepsilon & \varepsilon & A_{44} & \varepsilon & \varepsilon \\ \varepsilon & \varepsilon & \varepsilon & \varepsilon & A_{55} & \varepsilon \\ \varepsilon & \varepsilon & \varepsilon & \varepsilon & * & A_{66} \end{pmatrix} \tag{4.8}$$

can be seen in Fig. 4.1 (note that in (4.8) and elsewhere $*$ indicates a submatrix
different from ε). It consists of two superblocks and six classes including three
initial and two final ones.

Lemma 4.5.3 *If $x \in V(A)$, $N_i \to N_j$ and $x[N_j] \ne \varepsilon$ then $x[N_i]$ is finite. In particular, $x[N_j]$ is finite.*

Proof Suppose that $x \in V(A, \lambda)$ for some $\lambda \in \overline{\mathbb{R}}$. Fix $s \in N_j$ such that $x_s > \varepsilon$. Since $N_i \to N_j$ we have that for every $r \in N_i$ there is a positive integer q such that $b_{rs} > \varepsilon$ where $B = A^q = (b_{ij})$. Since $x \in V(B, \lambda^q)$ by Proposition 4.1.1 we also have $\lambda^q \otimes x_r \geq b_{rs} \otimes x_s > \varepsilon$. Hence $x_r > \varepsilon$. $\qquad \square$

We are now able to describe all eigenvalues of any square matrix over $\overline{\mathbb{R}}$.

Theorem 4.5.4 (Spectral Theorem) *Let* (4.7) *be an FNF of a matrix* $A = (a_{ij}) \in \overline{\mathbb{R}}^{n \times n}$. *Then*

$$\Lambda(A) = \left\{ \lambda(A_{jj}); \lambda(A_{jj}) = \max_{N_i \to N_j} \lambda(A_{ii}) \right\}.$$

Proof Note that

$$\lambda(A) = \max_{i=1,\dots,r} \lambda(A_{ii}) \tag{4.9}$$

for a matrix A in FNF (4.7).

First we prove the inclusion $\supseteq$. Suppose

$$\lambda(A_{jj}) = \max\{\lambda(A_{ii}); N_i \to N_j\}$$

for some $j \in R = \{1, \dots, r\}$. Denote

$$S_2 = \{i \in R; N_i \to N_j\},$$

$$S_1 = R - S_2$$

and

$$M_p = \bigcup_{i \in S_p} N_i \quad (p = 1, 2).$$

Then $\lambda(A_{jj}) = \lambda(A[M_2])$ and

$$A \equiv \begin{pmatrix} A[M_1] & \varepsilon \\ * & A[M_2] \end{pmatrix}.$$

If $\lambda(A_{jj}) = \varepsilon$ then at least one column, say the lth in A is ε. We set x_l to any real number and $x_j = \varepsilon$ for $j \neq l$. Then $x \in V(A, \lambda(A_{jj}))$.

If $\lambda(A_{jj}) > \varepsilon$ then $A[M_2]$ has a finite eigenvector by Theorem 4.4.4, say $\tilde{x}$. Set $x[M_2] = \tilde{x}$ and $x[M_1] = \varepsilon$. Then $x = (x[M_1], x[M_2]) \in V(A, \lambda(A_{jj}))$.

Now we prove $\subseteq$. Suppose that $x \in V(A, \lambda), x \neq \varepsilon$, for some $\lambda \in \overline{\mathbb{R}}$.

If $\lambda = \varepsilon$ then A has an ε column, say the kth, thus $a_{kk} = \varepsilon$. Hence the 1×1 submatrix (a_{kk}) is a diagonal block in an FNF of A. In the corresponding decomposition of N one of the sets, say N_j, is $\{k\}$. The set $\{i; N_i \to N_j\} = \{j\}$ and the theorem statement follows.

If $\lambda > \varepsilon$ and $x \in V^+(A)$ then $\lambda = \lambda(A)$ (cf. Theorem 4.4.1) and the statement now follows from (4.9).

If $\lambda > \varepsilon$ and $x \notin V^+(A)$ then similarly as in the proof of Lemma 4.5.1 permute the rows and columns of A simultaneously so that

$$x = \begin{pmatrix} x^{(1)} \\ x^{(2)} \end{pmatrix},$$

where $x^{(1)} = \varepsilon \in \overline{\mathbb{R}}^p$, $x^{(2)} \in \mathbb{R}^{n-p}$ for some p $(1 \leq p < n)$. Hence

$$A \equiv \begin{pmatrix} A^{(11)} & \varepsilon \\ A^{(21)} & A^{(22)} \end{pmatrix}$$

and we can assume without loss of generality that both $A^{(11)}$ and $A^{(22)}$ are in an FNF and therefore also

$$\begin{pmatrix} A^{(11)} & \varepsilon \\ A^{(21)} & A^{(22)} \end{pmatrix}$$

is in an FNF. Let

$$A^{(11)} = \begin{pmatrix} A_{i_1 i_1} & \varepsilon & \cdots & \varepsilon \\ A_{i_2 i_1} & A_{i_2 i_2} & \cdots & \varepsilon \\ \cdots & \cdots & \cdots & \cdots \\ A_{i_s i_1} & A_{i_s i_2} & \cdots & A_{i_s i_s} \end{pmatrix}$$

and

$$A^{(22)} = \begin{pmatrix} A_{i_{s+1} i_{s+1}} & \varepsilon & \cdots & \varepsilon \\ A_{i_{s+2} i_{s+1}} & A_{i_{s+2} i_{s+2}} & \cdots & \varepsilon \\ \cdots & \cdots & \cdots & \cdots \\ A_{i_q i_{s+1}} & A_{i_q i_{s+2}} & \cdots & A_{i_q i_q} \end{pmatrix}.$$

We have

$$\lambda = \lambda(A^{(22)}) = \lambda(A_{jj}) = \max_{i=s+1,\ldots,q} \lambda(A_{ii}),$$

where $j \in \{s+1, \ldots, q\}$. It remains to say that if $N_i \to N_j$ then $i \in \{s+1, \ldots, q\}$. $\square$

The Spectral Theorem has been proved in [84] and, independently, also in [12]. Spectral properties of reducible matrices have also been studied in [10] and [145]. Significant correlation exists between the max-algebraic spectral theory and that for nonnegative matrices in linear algebra [13, 128], see also [126]. For instance the Frobenius normal form and accessibility between classes play a key role in both theories. The maximum cycle mean corresponds to the Perron root for irreducible (nonnegative) matrices and finite eigenvectors in max-algebra correspond to positive eigenvectors in the spectral theory of nonnegative matrices. However there are also differences, see Remark 4.6.8.

Let A be in the FNF (4.7). If

$$\lambda(A_{jj}) = \max_{N_i \to N_j} \lambda(A_{ii})$$

then A_{jj} (and also N_j or just j) will be called *spectral*. Thus $\lambda(A_{jj}) \in \Lambda(A)$ if j is spectral but not necessarily the other way round.

Corollary 4.5.5 *All initial classes of C_A are spectral.*

Proof Initial classes have no predecessors and so the condition of the theorem is satisfied. $\qquad\square$

Recall that $\lambda(A) = \min\{\lambda; (\exists x \in \mathbb{R}^n) A \otimes x \leq \lambda \otimes x\}$ if $\lambda(A) > \varepsilon$ (Theorem 1.6.29). In contrast we have:

Corollary 4.5.6

$$\lambda(A) = \max \Lambda(A)$$
$$= \max \left\{\lambda; \left(\exists x \in \overline{\mathbb{R}}^n, x \neq \varepsilon\right) A \otimes x = \lambda \otimes x\right\}$$

for every matrix $A \in \overline{\mathbb{R}}^{n \times n}$.

Proof If A is in an FNF, say (4.7), then $\lambda(A) = \max_{i=1,\dots,r} \lambda(A_{ii}) \geq \lambda(A_{jj})$ for all j. $\qquad\square$

We easily deduce two more useful statements:

Corollary 4.5.7 $1 \leq |\Lambda(A)| \leq n$ *for every $A \in \overline{\mathbb{R}}^{n \times n}$.*

Proof Follows from the previous corollary and from the fact that the number of classes of A is at most n. $\qquad\square$

Corollary 4.5.8 $V(A) = V(A, \lambda(A))$ *if and only if all initial classes have the same eigenvalue $\lambda(A)$.*

Proof The eigenvalues of all initial classes are in $\Lambda(A)$ since all initial classes are spectral, hence all must be equal to $\lambda(A)$ if $\Lambda(A) = \{\lambda(A)\}$. On the other hand, if all initial classes have the same eigenvalue $\lambda(A)$, and λ is the eigenvalue of any spectral class then

$$\lambda \geq \lambda(A) = \max_i \lambda(A_{ii})$$

since there is a path from some initial class to this class and thus $\lambda = \lambda(A)$. $\qquad\square$

Figure 4.2 shows a condensation digraph with 14 classes including two initial classes and four final ones. The integers indicate the eigenvalues of the corresponding classes. The six bold classes are spectral, the others are not.

Note that the unique eigenvalues of all classes (that is, of diagonal blocks of an FNF) can be found in $O(n^3)$ time by applying Karp's algorithm (see Sect. 1.6) to

Fig. 4.2 Condensation
digraph

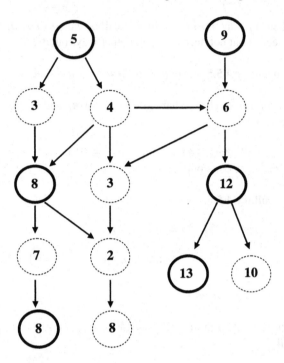

each block. The condition for identifying all spectral submatrices in an FNF provided in Theorem 4.5.4 enables us to find them in $O(r^2) \leq O(n^2)$ time by applying
standard reachability algorithms to C_A.

Example 4.5.9 Consider the matrix

$$
A = \begin{pmatrix}
0 & 3 & & & & \\
1 & 1 & & & & \\
& & 4 & & & \\
& & 0 & 3 & 1 & \\
& & & -1 & 2 & \\
& & & 1 & & 5
\end{pmatrix},
$$

where the missing entries are ε. Then $\lambda(A_{11}) = 2$, $\lambda(A_{22}) = 4$, $\lambda(A_{33}) = 3$,
$\lambda(A_{44}) = 5$, $r = 4$; $\Lambda(A) = \{2, 5\}$, $\lambda(A) = 5$, initial classes are N_1 and N_4 and
there are no other spectral classes. Final classes are N_1 and N_2.

We will now use the Spectral Theorem to prove two results, Theorems 4.5.10
and 4.5.14, whose proofs are easier when the Spectral Theorem is available. The
first of them has been known for certain types of matrices for some time [65, 102]:
however, using Theorem 4.5.4 we are able to prove it conveniently for any matrix:

Theorem 4.5.10 *Let $A \in \overline{\mathbb{R}}^{n \times n}$. Then*

$$\lambda(A^k) = (\lambda(A))^k$$

holds for all integers $k \geq 0$.

Proof The proof is trivial if $n = 1$ or $k = 0$, so assume $n \geq 2, k \geq 1$.

Suppose first that A is irreducible. Let $x \in V^+(A) = V(A, \lambda(A)) - \{\varepsilon\}$. By Proposition 4.1.1 we have $x \in V(A^k, \lambda(A^k))$ and thus by Theorem 4.4.1 $(\lambda(A))^k = \lambda(A^k)$. It also follows that $(\lambda(A))^k$ is the greatest principal eigenvalue of a diagonal block in any FNF of (possibly reducible) A^k.

Now suppose that A is reducible and without loss of generality let A be in the FNF (4.7). Then $\lambda(A) = \lambda(A_{ii})$ for some i, $1 \leq i \leq r$. The matrix A^k is again lower blockdiagonal and has diagonal blocks $A_{11}^k, \ldots, A_{ii}^k, \ldots, A_{rr}^k$. These blocks may or may not be irreducible. However $(\lambda(A))^k = (\lambda(A_{ii}))^k$ is the greatest principal eigenvalue of a diagonal block in any FNF of A_{ii}^k (by the first part of this proof since A_{ii} is irreducible) and therefore also in any FNF of A^k. This completes the proof. $\square$

For the second result we need two lemmas.

Lemma 4.5.11 *Let $A \in \overline{\mathbb{R}}^{n \times n}$. Then $\varepsilon \in \Lambda(A)$ if and only if A has an ε column.*

Proof If $A \otimes x = \varepsilon$ and $x_k \neq \varepsilon$ then the kth column of A is ε. A similar argument is used for the converse. $\square$

Lemma 4.5.12 *Let $A \in \overline{\mathbb{R}}^{n \times n}$ be irreducible. If $A \otimes x \leq \lambda \otimes x, x \neq \varepsilon, \lambda \in \overline{\mathbb{R}}$ then $x \in \mathbb{R}^n$.*

Proof The statement is trivial for $n = 1$. Let $n > 1$, then $\lambda(A) > \varepsilon$. Without loss of generality we assume that A is definite. Then we have

$$\Gamma(A) \otimes x = A \otimes x \oplus A^2 \otimes x \oplus \cdots \oplus A^n \otimes x$$
$$\leq \lambda \otimes x \oplus \lambda^2 \otimes x \oplus \cdots \oplus \lambda^n \otimes x$$
$$= (\lambda \oplus \cdots \oplus \lambda^n) \otimes x.$$

The LHS is finite since $\Gamma(A)$ is finite (Proposition 1.6.10) and $x \neq \varepsilon$, hence both λ and x are finite. $\square$

Corollary 4.5.13 *Let $A \in \overline{\mathbb{R}}^{n \times n}$ be irreducible. Then*

$$\lambda(A) = \min\{\lambda; (\exists x \in \mathbb{R}^n) A \otimes x \leq \lambda \otimes x\}$$
$$= \min\left\{\lambda; (\exists x \in \overline{\mathbb{R}}^n, x \neq \varepsilon) A \otimes x \leq \lambda \otimes x\right\}.$$

Proof The statement is trivial for $n = 1$. If $n > 1$ then $\lambda(A) > \varepsilon$ and the first equality follows from Theorem 1.6.29. The second follows from Lemma 4.5.12. $\square$

We now make another use of Theorem 4.5.4 and prove a more general version of Theorem 1.6.29:

Theorem 4.5.14 *If $A \in \overline{\mathbb{R}}^{n \times n}$ then*

$$\min \left\{ \lambda; \left(\exists x \in \mathbb{R}^n, x \neq \varepsilon \right) A \otimes x \leq \lambda \otimes x \right\} = \min \Lambda(A).$$

Proof Without loss of generality let A be in the FNF (4.7) and as before $R = \{1, \dots, r\}$. Let

$$L = \inf \left\{ \lambda; \left(\exists x \in \mathbb{R}^n, x \neq \varepsilon \right) A \otimes x \leq \lambda \otimes x \right\}.$$

Clearly $L \leq \min \Lambda(A)$ since for x we may take any eigenvector of A. If $\varepsilon \in \Lambda(A)$ then using $x \in V(A, \varepsilon) - \{\varepsilon\}$ we deduce that $L = \varepsilon$. We will therefore assume in the rest of the proof that $\varepsilon \notin \Lambda(A)$.

Let $x \in \mathbb{R}^n, x \neq \varepsilon, \lambda \in \mathbb{R}$ and $A \otimes x \leq \lambda \otimes x$. We need to show that $\lambda \geq \min \Lambda(A)$. Observe that $\lambda > \varepsilon$ since otherwise $x \in V(A, \varepsilon) - \{\varepsilon\}$, a contradiction with $\varepsilon \notin \Lambda(A)$. Let us denote

$$K = \{k \in R; x[N_k] \neq \varepsilon\}.$$

Take any $k \in K$. We have

$$A[N_k] \otimes x[N_k] \leq (A \otimes x)[N_k] \leq \lambda \otimes x[N_k].$$

Then $x[N_k]$ is finite by Lemma 4.5.12 and so $\lambda \geq \lambda(A[N_k])$ by Theorem 1.6.18.

If $a_{st} = \varepsilon$ for all $s \in N_i, i \in R$ and $t \in N_k$, then N_k is spectral and the statement follows.

If $a_{st} > \varepsilon$ for some $s \in N_i, i \in R$ and $t \in N_k$, then $x_s \geq \lambda^{-1} \otimes a_{st} \otimes x_t > \varepsilon$. Therefore $i \in K$ and again, as above, by Lemma 4.5.12 $x[N_i]$ is finite. C_A is acyclic and finite, hence after a finite number of repetitions we will reach an $i \in R$ such that N_i is initial, and hence also spectral, yielding $\lambda(A[N_i]) > \varepsilon$ (since $\varepsilon \notin \Lambda(A)$) and $\lambda(A[N_i]) \geq \min \Lambda(A)$.

At the same time

$$A[N_i] \otimes x[N_i] \leq (A \otimes x)[N_i] \leq \lambda \otimes x[N_i].$$

Therefore $x[N_i]$ is finite by Lemma 4.5.12 and by Theorem 1.6.18 we have:

$$\lambda \geq \lambda(A[N_i]),$$

from which the statement follows. $\square$

4.6 Finding All Eigenvectors

Our final effort in this chapter is to show how to efficiently describe all eigenvectors of a matrix.

Let $A \in \overline{\mathbb{R}}^{n \times n}$ be in the FNF (4.7), $N_1, \ldots, N_r$ be the classes of A and $R = \{1, \ldots, r\}$. For the following discussion suppose that $\lambda \in \Lambda(A)$ is a fixed eigenvalue, $\lambda > \varepsilon$, and denote $I(\lambda) = \{i \in R; \lambda(N_i) = \lambda, N_i \text{ spectral}\}$.

We denote by $g_1, \ldots, g_n$ the columns of $\Gamma(\lambda^{-1} \otimes A) = (\gamma_{ij})$. Note that $\lambda(\lambda^{-1} \otimes A) = \lambda^{-1} \otimes \lambda(A)$ may be positive since $\lambda \leq \lambda(A)$ and thus $\Gamma(\lambda^{-1} \otimes A)$ may include entries equal to $+\infty$ (Proposition 1.6.10). However, for $i \in I(\lambda)$ we have

$$\lambda \left(\lambda^{-1} \otimes A_{ii} \right) = \lambda^{-1} \otimes \lambda(A_{ii}) \leq 0$$

by Theorem 4.5.4 and hence $\Gamma(\lambda^{-1} \otimes A_{ii})$ is finite for $i \in I(\lambda)$.

Let us denote

$$N_c(\lambda) = \bigcup_{i \in I(\lambda)} N_c(A_{ii}) = \left\{ j \in N; \gamma_{jj} = 0, j \in \bigcup_{i \in I(\lambda)} N_i \right\}.$$

Two nodes i and j in $N_c(\lambda)$ are called λ-*equivalent* (notation $i \sim_\lambda j$) if i and j belong to the same cycle whose mean is λ. Note that if $\lambda = \lambda(A)$ then $\sim_\lambda$ coincides with $\sim$.

Theorem 4.6.1 [44] *Suppose $A \in \overline{\mathbb{R}}^{n \times n}$ and $\lambda \in \Lambda(A), \lambda > \varepsilon$. Then $g_j \in \overline{\mathbb{R}}^n$ (that is, g_j does not contain $+\infty$) for all $j \in N_c(\lambda)$ and a basis of $V(A, \lambda)$ can be obtained by taking one g_j for each $\sim_\lambda$ equivalence class.*

Proof Let us denote $M = \bigcup_{i \in I(\lambda)} N_i$. By Lemma 4.1.3 we may assume without loss of generality that A is of the form

$$\begin{pmatrix} \bullet & \varepsilon \\ \bullet & A[M] \end{pmatrix}.$$

Hence $\Gamma(\lambda^{-1} \otimes A)$ is

$$\begin{pmatrix} \bullet & \varepsilon \\ \bullet & C \end{pmatrix}$$

where $C = \Gamma((\lambda(A[M]))^{-1} \otimes A[M])$, and the statement now follows by Proposition 1.6.10 and Theorem 4.3.5 since $\lambda = \lambda(A[M])$ and thus $\sim_\lambda$ equivalence for A is identical with $\sim$ equivalence for $A[M]$. $\qquad \square$

Corollary 4.6.2 *A basis of $V(A, \lambda)$ for $\lambda \in \Lambda(A), \lambda > \varepsilon$, can be found using $O(k^3)$ operations, where $k = |I(\lambda)|$ and we have*

$$V(A, \lambda) = \{\Gamma(\lambda^{-1} \otimes A) \otimes z; z \in \overline{\mathbb{R}}^n, z_j = \varepsilon \text{ for all } j \notin N_c(\lambda)\}.$$

Consequently, the bases of all eigenspaces can be found in $O(n^3)$ operations.

Using Lemma 4.2.1 and Corollary 4.6.2 we get:

Corollary 4.6.3 *If $A \in \overline{\mathbb{R}}^{n \times n}$, $\lambda \in \Lambda(A)$ and the dimension of $V(A, \lambda)$ is r_λ then there is a column $\mathbb{R}$-astic matrix $G_\lambda \in \overline{\mathbb{R}}^{n \times r_\lambda}$ such that*

$$V(A, \lambda) = \left\{ G_\lambda \otimes z; z \in \overline{\mathbb{R}}^{r_\lambda} \right\}.$$

It follows from the proofs of Lemma 4.5.1 and Theorem 4.5.4 that $V(A, \lambda)$ can also be found as follows: If $I(\lambda) = \{j\}$ then define

$$M_2 = \bigcup_{N_i \to N_j} N_i, \qquad M_1 = N - M_2.$$

Hence

$$V(A, \lambda) = \{x; x[M_1] = \varepsilon, x[M_2] \in V^+(A[M_2])\}.$$

If the set $I(\lambda)$ consists of more than one index then the same process has to be repeated for each nonempty subset of $I(\lambda)$, that is, for each $J \subseteq I(\lambda)$, $J \neq \emptyset$, we set $S = \bigcup_{j \in J} N_j$ and

$$M_2 = \bigcup_{N_i \to S} N_i, \qquad M_1 = N - M_2.$$

Obviously, this is not a practical way of finding all eigenvectors as considering all subsets would be computationally infeasible, but it enables us to conveniently prove another criterion for the existence of finite eigenvectors:

Theorem 4.6.4 [10] *$V^+(A) \neq \emptyset$ if and only if $\lambda(A)$ is the eigenvalue of all final classes (in all superblocks).*

Proof The set M_1 in the above construction must be empty to obtain a finite eigenvector, hence a class in S must be reachable from every class of its superblock. This is only possible if S is the set of all final classes since no class is reachable from a final class (other than the final class itself). Conversely, if all final classes have the same eigenvalue $\lambda(A)$ then for $\lambda = \lambda(A)$ the set S contains all the final classes, they are reachable from all classes of their superblocks, and consequently $M_1 = \emptyset$, yielding a finite eigenvector.   □

Corollary 4.6.5 *$V^+(A) = \emptyset$ if and only if a final class has eigenvalue less than $\lambda(A)$.*

Example 4.6.6 For the matrix A of Example 4.5.9 each of the two eigenspaces has dimension 1. Since

$$\Gamma((A_{11})_\lambda) = \begin{pmatrix} 0 & 1 \\ -1 & 0 \end{pmatrix}$$

$V(A, 2)$ is the set of multiples of $(1, 0, \varepsilon, \varepsilon, \varepsilon, \varepsilon)^T$, similarly $V(A, 5)$ is the set of multiples of $(\varepsilon, \varepsilon, \varepsilon, \varepsilon, \varepsilon, 0)^T$. There are no finite eigenvectors since for the final class N_2 we have $\lambda(A_{22}) < 5$.

Remark 4.6.7 Note that a final class with eigenvalue less than $\lambda(A)$ may not be spectral and so $\Lambda(A) = \{\lambda(A)\}$ is possible even if $V^+(A) = \emptyset$. For instance in the case of

$$A = \begin{pmatrix} 1 & \varepsilon & \varepsilon \\ \varepsilon & 0 & \varepsilon \\ 0 & 0 & 1 \end{pmatrix}$$

we have $\lambda(A) = 1$, but $V^+(A) = \emptyset$.

Remark 4.6.8 Following the terminology of nonnegative matrices in linear algebra we say that a class is basic if its eigenvalue is $\lambda(A)$. It follows from Theorem 4.6.4 that $V^+(A) \neq \emptyset$ if basic classes and final classes coincide. Obviously this requirement is not necessary for $V^+(A) \neq \emptyset$, which is in contrast to the spectral theory of nonnegative matrices where for A to have a positive eigenvector it is necessary and sufficient that basic classes (that is, those whose eigenvalue is the Perron root) are exactly the final classes [126].

Remark 4.6.9 The principal eigenspace of any matrix may contain either finite eigenvectors only (for instance when the matrix is irreducible) or only nonfinite eigenvectors (see Remark 4.6.7), or both finite and non-finite eigenvectors, for instance when $A = I$.

4.7 Commuting Matrices Have a Common Eigenvector

The theory of commuting matrices in max-algebra seems to be rather modest at the time when this book goes to print: however, it is known that any two commuting matrices have a common eigenvector. This will be useful in the theory of two-sided max-linear systems (Chap. 7) and for solving some special cases of the generalized eigenproblem (Chap. 9).

Lemma 4.7.1 [70] *Let* $A, B \in \overline{\mathbb{R}}^{n \times n}$ *and* $A \otimes B = B \otimes A$. *If* $x \in V(A, \lambda)$, $\lambda \in \overline{\mathbb{R}}$, *then* $B \otimes x \in V(A, \lambda)$.

Proof We have $A \otimes x = \lambda \otimes x$ and thus

$$A \otimes (B \otimes x) = B \otimes (A \otimes x) = B \otimes \lambda \otimes x = \lambda \otimes (B \otimes x). \qquad \square$$

Theorem 4.7.2 (Schneider [107]) *If* $A, B \in \overline{\mathbb{R}}^{n \times n}$ *and* $A \otimes B = B \otimes A$ *then* $V(A) \cap V(B) \neq \{\varepsilon\}$, *more precisely, for every* $\lambda \in \Lambda(A)$ *there is a* $\mu \in \Lambda(B)$ *such that*

$$V(A, \lambda) \cap V(B, \mu) \neq \{\varepsilon\}.$$

Proof Let $\lambda \in \Lambda(A)$ and r_λ be the dimension of $V(A, \lambda)$. By Corollary 4.6.3 there is a matrix $G_\lambda \in \overline{\mathbb{R}}^{n \times r_\lambda}$ such that

$$V(A, \lambda) = \left\{ G_\lambda \otimes z; z \in \overline{\mathbb{R}}^{r_\lambda} \right\}.$$

Clearly, $A \otimes G_\lambda = \lambda \otimes G_\lambda$. It follows from Lemma 4.7.1 that all columns of $B \otimes G_\lambda$ are in $V(A, \lambda)$ and hence

$$B \otimes G_\lambda = G_\lambda \otimes C$$

for some $r_\lambda \times r_\lambda$ matrix C. Let $v \in V(C), v \neq \varepsilon$, thus $v \in V(C, \mu)$ for some $\mu \in \overline{\mathbb{R}}$, and set $u = G_\lambda \otimes v$. Then $u \neq \varepsilon$ since G_λ is column $\mathbb{R}$-astic and we have:

$$A \otimes u = A \otimes G_\lambda \otimes v = \lambda \otimes G_\lambda \otimes v = \lambda \otimes u$$

and

$$B \otimes u = B \otimes G_\lambda \otimes v = G_\lambda \otimes C \otimes v = \mu \otimes G_\lambda \otimes v = \mu \otimes u.$$

Hence $u \in V(A, \lambda) \cap V(B, \mu)$ and $u \neq \varepsilon$. $\square$

The proof of Theorem 4.7.2 is constructive and enables us to find a common eigenvector of commuting matrices: The system $B \otimes G_\lambda = G_\lambda \otimes C$ is a one-sided system for C and since a solution exists, the principal solution $\overline{C} = G_\lambda^* \otimes' (B \otimes G_\lambda)$ is a solution (Corollary 3.2.4).

Note that [107] contains more information on commuting matrices in max-algebra.

4.8 Exercises

Exercise 4.8.1 Find the eigenvalue, $\Gamma(A_\lambda)$ and the scaled basis of the unique eigenspace for each of the matrices below:

(a) $A = \begin{pmatrix} 3 & 6 \\ 2 & 1 \end{pmatrix}$. $[\lambda(A) = 4;$

$$\Gamma(A_\lambda) = \begin{pmatrix} 0 & 2 \\ -2 & 0 \end{pmatrix},$$

the scaled basis is $\{(0, -2)^T\}.]$

(b) $A = \begin{pmatrix} 0 & 0 \\ -1 & 0 \end{pmatrix}$. $[\lambda(A) = 0; \Gamma(A_\lambda) = A$, the scaled basis is $\{(0, -1)^T, (0, 0)^T\}.]$

(c) $A = \begin{pmatrix} 1 & 0 & 4 & 3 \\ 0 & 1 & -3 & 3 \\ 0 & 1 & 0 & 2 \\ -3 & -1 & 0 & 1 \end{pmatrix}$. $[\lambda(A) = 2;$

$$\Gamma(A_\lambda) = \begin{pmatrix} 0 & 1 & 2 & 2 \\ -2 & -1 & 0 & 1 \\ -2 & -1 & 0 & 0 \\ -4 & -3 & -2 & -1 \end{pmatrix},$$

the scaled basis is $\{(0, -2, -2, -4)^T\}$.]

(d) Find the eigenvalue, $\Gamma(A_\lambda)$ and the scaled basis of the unique eigenspace of the matrix

$$A = \begin{pmatrix} 4 & 4 & 3 & 8 & 1 \\ 3 & 3 & 4 & 5 & 4 \\ 5 & 3 & 4 & 7 & 3 \\ 2 & 1 & 2 & 3 & 0 \\ 6 & 6 & 4 & 8 & 1 \end{pmatrix}.$$

$[\lambda(A) = 5;$

$$\Gamma(A_\lambda) = \begin{pmatrix} 0 & -1 & 0 & 3 & -2 \\ 0 & 0 & 0 & 3 & -1 \\ 0 & -1 & 0 & 3 & -2 \\ -3 & -4 & -3 & 0 & -5 \\ 1 & 1 & 1 & 4 & 0 \end{pmatrix},$$

the scaled basis is $\{(-1, -1, -1, -4, 0)^T, (-2, -1, -2, -5, 0)^T\}$.]

Exercise 4.8.2 Find all eigenvalues and the scaled bases of all eigenspaces of the matrix

$$A = \begin{pmatrix} 3 & 2 & & & & & & & \\ 2 & 3 & & & & & & & \\ & & 4 & & & & & & \\ & & & 3 & 4 & & & & \\ & & & 6 & 1 & & & & \\ & & & 4 & 1 & 7 & 2 & & \\ & & & & & 3 & 0 & & \\ & & & & & & & 1 & 4 \\ 0 & & & & & & & 0 & 2 \end{pmatrix},$$

where the missing entries are ε. $[\Lambda(A) = \{3, 4, 7, 2\}$, the scaled basis of $V(A, 3)$ is

$$\left\{ (0, -1, \varepsilon, \varepsilon, \varepsilon, \varepsilon, \varepsilon, -2, -3)^T, (-1, 0, \varepsilon, \varepsilon, \varepsilon, \varepsilon, \varepsilon, -3, -4)^T \right\},$$

the scaled basis of $V(A, 4)$ is

$$\left\{(\varepsilon, \varepsilon, 0, \varepsilon, \varepsilon, \varepsilon, \varepsilon, \varepsilon, \varepsilon)^T\right\},$$

the scaled basis of $V(A, 7)$ is

$$\left\{(\varepsilon, \varepsilon, \varepsilon, \varepsilon, \varepsilon, 0, -4, \varepsilon, \varepsilon)^T\right\},$$

the scaled basis of $V(A, 2)$ is

$$\left\{(\varepsilon, \varepsilon, \varepsilon, \varepsilon, \varepsilon, \varepsilon, \varepsilon, 0, -2)^T\right\}.\right]$$

Exercise 4.8.3 In the matrix A below the sign $\times$ indicates a finite entry, all other off-diagonal entries are ε. Find all spectral indices and all eigenvalues of A, and decide whether this matrix has finite eigenvectors.

$$A = \begin{pmatrix} 4 & & & & & & & & \\ \times & 3 & & & & & & & \\ & \times & 5 & & & & & & \\ & & & 7 & & & & & \\ & & & & \times & 8 & & & \\ & & \times & \times & & & 2 & \\ & & & \times & \times & & & 4 \end{pmatrix}$$

[Spectral indices: 3, 5, 6, 7, $\Lambda(A) = \{5, 8, 2, 4\}$, no finite eigenvectors.]

Exercise 4.8.4 Prove that $\lambda(A) = \lambda(A^T)$, $\Gamma(A^T) = (\Gamma(A))^T$ and $N_c(A) = N_c(A^T)$ for every square matrix A. Then prove or disprove that $\Lambda(A) = \Lambda(A^T)$. [false]

Exercise 4.8.5 Prove or disprove each of the following statements:

(a) If $A \in \mathbb{Z}^{n \times n}$ then A has an integer eigenvector if and only if $\lambda(A) \in \mathbb{Z}$. [true]
(b) If $A \in \mathbb{R}^{n \times n}$ then A has an integer eigenvector if and only if $\lambda(A) \in \mathbb{Z}$. [false]
(c) If $A \in \mathbb{R}^{n \times n}$ then A has an integer eigenvalue and an integer eigenvector if and only if $A \in \mathbb{Z}^{n \times n}$. [false]

Exercise 4.8.6 We say that $T = (t_{ij}) \in \mathbb{R}^{n \times n}$ is triangular if it satisfies the condition $t_{ij} < \lambda(T)$ for all $i, j \in N$, $i \le j$. Prove the statement: If $A \in \mathbb{R}^{n \times n}$ then $\lambda(A) = \lambda(B)$ for every B equivalent to A if and only if A is not equivalent to a triangular matrix. [See [39]]

Exercise 4.8.7 Show that the maximum cycle mean and an eigenvector for $0 - 1$ matrices can be found using $O(n^2)$ operations. [See [33, 66]]

Exercise 4.8.8 Prove that the following problem is *NP*-complete: Given $A \in \overline{\mathbb{R}}^{n \times n}$ and $x \in \overline{\mathbb{R}}^n$, decide whether it is possible to permute the components of x so that the obtained vector is an eigenvector of A. [See [31]]

Exercise 4.8.9 Let A and B be square matrices of the same order. Prove then that the set of finite eigenvalues of $A \otimes B$ is the same as the set of finite eigenvalues of $B \otimes A$.

Chapter 5
Maxpolynomials. The Characteristic Maxpolynomial

The aim of this chapter is to study *max-algebraic polynomials*, that is, expressions of the form

$$p(z) = \sum_{r=0,\ldots,p}^{\oplus} c_r \otimes z^{j_r}, \tag{5.1}$$

where $c_r, j_r \in \mathbb{R}$. The number j_p is called the *degree* of $p(z)$ and $p+1$ is called its *length*.

We will consider (5.1) both as formal algebraic expressions with z as an indeterminate and as max-algebraic functions of z. We will abbreviate "max-algebraic polynomial" to "*maxpolynomial*". Note that j_r are not restricted to integers and so (5.1) covers expressions such as

$$8.3 \otimes z^{-7.2} \oplus (-2.6) \otimes z^{3.7} \oplus 6.5 \otimes z^{12.3}. \tag{5.2}$$

In conventional notation $p(z)$ has the form

$$\max_{r=0,\ldots,p} (c_r + j_r z)$$

and if considered as a function, it is piecewise linear and convex.

Each expression $c_r \otimes z^{j_r}$ will be called a *term* of the maxpolynomial $p(z)$. For a maxpolynomial of the form (5.1) we will always assume

$$j_0 < j_1 < \cdots < j_p,$$

where p is a nonnegative integer. If $c_p = 0 = j_0$ then $p(z)$ is called *standard*. Clearly, every maxpolynomial $p(z)$ can be written as

$$c \otimes z^j \otimes q(z), \tag{5.3}$$

where $q(z)$ is a standard maxpolynomial. For instance (5.2) is of degree 12.3 and length 3. It can be written as

$$6.5 \otimes z^{-7.2} \otimes q(z),$$

P. Butkovič, *Max-linear Systems: Theory and Algorithms*,
Springer Monographs in Mathematics 151,
DOI 10.1007/978-1-84996-299-5_5, © Springer-Verlag London Limited 2010

where $q(z)$ is the standard maxpolynomial

$$1.8 \oplus (-9.1) \otimes z^{10.9} \oplus z^{19.5}.$$

There are many similarities with conventional polynomial algebra, in particular (see Sect. 5.1) there is an analogue of the fundamental theorem of algebra, that is, every maxpolynomial factorizes to linear terms (although these terms do not correspond to "roots" in the conventional terminology). However, there are aspects that make this theory different. This is caused, similarly as in other parts of max-algebra, by idempotency of addition, which for instance yields the formula

$$(a \oplus b)^k = a^k \oplus b^k \tag{5.4}$$

for all $a, b, k \in \mathbb{R}$. This property has a significant impact on many results. Perhaps the most important feature that makes max-algebraic polynomial theory different is the fact that the functional equality $p(z) = q(z)$ does not imply equality between p and q as formal expressions. For instance $(1 \oplus z)^2$ is equal by (5.4) to $2 \oplus z^2$ but at the same time expands to $2 \oplus 1 \otimes z \oplus z^2$ by basic arithmetic laws. Hence the expressions $2 \oplus 1 \otimes z \oplus z^2$ and $2 \oplus z^2$ are identical as functions. This demonstrates the fact that some terms of maxpolynomials, do not actually contribute to the function value. In our example $1 \otimes z \leq 2 \oplus z^2$ for all $z \in \mathbb{R}$. This motivates the following definitions: A term $c_s \otimes z^{j_s}$ of a maxpolynomial $\sum_{r=0,\dots,p}^{\oplus} c_r \otimes z^{j_r}$ is called *inessential* if

$$c_s \otimes z^{j_s} \leq \sum_{r \neq s}^{\oplus} c_r \otimes z^{j_r}$$

holds for every $z \in \mathbb{R}$ and *essential* otherwise. Clearly, an inessential term can be removed from [reinstated in] a maxpolynomial *ad lib* when this maxpolynomial is considered as a function. Note that the terms $c_0 \otimes z^{j_0}$ and $c_p \otimes z^{j_p}$ are essential in any maxpolynomial $\sum_{r=0,\dots,p}^{\oplus} c_r \otimes z^{j_r}$.

Lemma 5.0.1 *If the term $c_s \otimes z^{j_s}$, $0 < s < p$, is essential in the maxpolynomial $\sum_{r=0,\dots,p}^{\oplus} c_r \otimes z^{j_r}$ then*

$$\frac{c_s - c_{s+1}}{j_{s+1} - j_s} > \frac{c_{s-1} - c_s}{j_s - j_{s-1}}.$$

Proof Since the term $c_s \otimes z^{j_s}$ is essential and the sequence $\{j_r\}_{r=0}^p$ is increasing there is an $\alpha \in \mathbb{R}$ such that

$$c_s + j_s \alpha > c_{s-1} + j_{s-1} \alpha$$

and

$$c_s + j_s \alpha > c_{s+1} + j_{s+1} \alpha.$$

Hence

$$\frac{c_s - c_{s+1}}{j_{s+1} - j_s} > \alpha > \frac{c_{s-1} - c_s}{j_s - j_{s-1}}. \qquad \square$$

We will first analyze general properties of maxpolynomials yielding an analogue of the fundamental theorem of algebra and we will also briefly study maxpolynomial equations. Then we discuss characteristic maxpolynomials of square matrices. Maxpolynomials, including characteristic maxpolynomials, were studied in [8, 20, 62, 65, 71]. The material presented in Sect. 5.1 follows the lines of [65] with kind permission of Academic Press.

5.1 Maxpolynomials and Their Factorization

One of the aims in this section is to seek factorization of maxpolynomials. We will see that unlike in conventional algebra it is always possible to factorize a maxpolynomial as a function (although not necessarily as a formal expression) into linear factors over $\mathbb{R}$ with a relatively small computational effort. We will therefore first study expressions of the form

$$\prod_{r=1,\ldots,p}^{\otimes} (\beta_r \oplus z)^{e_r} \tag{5.5}$$

where $\beta_r \in \overline{\mathbb{R}}$ and $e_r \in \mathbb{R}$ ($r = 1, \ldots, p$) and show how they can be multiplied out; this operation will be called *evolution*. We call expressions (5.5) a *product form* and will assume

$$\beta_1 < \cdots < \beta_p. \tag{5.6}$$

The constants β_r will be called *corners* of the product form (5.5). Note that (5.5) in conventional notation reads

$$\sum_{r=1,\ldots,p} e_r \max(\beta_r, z).$$

Hence, a factor $(\varepsilon \oplus z)^e$ is the same as the linear function ez of slope e. A factor $(\beta \oplus z)^e$, $\beta \in \mathbb{R}$, is constant $e\beta$ while $z \leq \beta$ and linear function ez if $z \geq \beta$. Therefore (5.5) is the function $b(z) + f(z)z$, where

$$b(z) = \sum_{z \leq \beta_s} e_s \beta_s, \ f(z) = \sum_{z > \beta_s} e_s.$$

Every product form is a piecewise linear function with constant slope between any two corners, and for $z < \beta_1$ and $z > \beta_p$. It follows that a product form is convex when all exponents e_r are positive. However, this function may, in general, be nonconvex and therefore we cannot expect each product form to correspond to a maxpolynomial as a function.

Let us first consider product forms

$$(z \oplus \beta_1) \otimes (z \oplus \beta_2) \otimes \cdots \otimes (z \oplus \beta_p), \tag{5.7}$$

that is, product forms where all exponents are 1 and all $\beta_r \in \mathbb{R}$ (and still $\beta_1 < \cdots < \beta_p$). Such product forms will be called *simple*.

We can multiply out any simple product form using basic arithmetic laws as in conventional algebra. This implies that the coefficient at z^k ($k = 0, \ldots, p$) of the obtained maxpolynomial is

$$\sum_{1 \leq i_1 < \cdots < i_r \leq p}^{\oplus} \beta_{i_1} \otimes \beta_{i_2} \otimes \cdots \otimes \beta_{i_r}, \tag{5.8}$$

where $r = p - k$. Note that (5.8) is 0 if $r = 0$. However, due to (5.6) this coefficient significantly simplifies, namely (5.8) is actually the same as

$$\beta_{k+1} \otimes \cdots \otimes \beta_p$$

when $k < p$ and 0 when $k = p$. Hence the maxpolynomial obtained by multiplying out a simple product form (5.7) is of length $p + 1$ and can be found as follows.

The constant term is $\beta_1 \otimes \cdots \otimes \beta_p$; the term involving z^k ($k \geq 1$) is obtained by replacing β_k in the term involving z^{k-1} by z.

We now generalize this procedure to an algorithm for any product form with positive exponents and finite corners. Product forms with these two properties are called *standard*.

Algorithm 5.1.1 EVOLUTION
Input: $\beta_1, \ldots, \beta_p, e_1, \ldots, e_p \in \mathbb{R}$ (parameters of a product form).
Output: Terms of the maxpolynomial obtained by multiplying out (5.5).
$t_0 := \beta_1^{e_1} \otimes \cdots \otimes \beta_p^{e_p}$
for $r = 1, \ldots, p$ do
$t_r := t_{r-1}$ after replacing $\beta_r^{e_r}$ by z^{e_r}

The general step of this algorithm can also be interpreted as follows:
$c_r := c_{r-1} \otimes (\beta_r^{e_r})^{-1}$ and $j_r := j_{r-1} + e_r$ with $c_0 := \beta_1^{e_1} \otimes \cdots \otimes \beta_p^{e_p}$ and $j_0 = 0$.
Alternatively, the sequence of pairs $\{(e_r, \beta_r)\}_{r=1}^{p}$ is transformed into the sequence

$$\{(c_r, j_r)\}_{r=0}^{p} = \left\{ \left(\sum_{s \geq r} e_s \beta_s, \sum_{s < r} e_s \right) \right\}_{r=1}^{p+1},$$

where the sum of an empty set is 0 by definition. Note that the algorithm EVOLUTION is formulated for general product forms but its correctness is guaranteed for standard product forms:

Theorem 5.1.2 *If the algorithm EVOLUTION is applied to standard product form (5.5) then the maxpolynomial $f(z) = \sum_{r=0,\ldots,p}^{\oplus} t_r$ is standard, has no inessential terms and is the same function as the product form.*

Proof Let $f(z) = \sum_{r=0,\ldots,p}^{\oplus} t_r$. Then $f(z)$ is standard since all terms involving z have positive exponents and one of the terms (t_0) is constant. The highest order term (t_p) has coefficient zero.

Let $r \in \{0, 1, \ldots, p\}$ and let z be any value satisfying $\beta_r < z < \beta_{r+1}$. Then

$$t_r = c_r \otimes z^{j_r} = z^{e_1} \otimes \cdots \otimes z^{e_r} \otimes \beta_{r+1}^{e_{r+1}} \otimes \cdots \otimes \beta_p^{e_p}$$

and $f(z) = c_r \otimes z^{j_r}$ because any other term has either some z's replaced by some β's ($\leq \beta_r < z$) or some β's ($\geq \beta_{r+1} > z$) replaced by z's and will therefore be strictly less than t_r. At the same time, if $\beta_r < z < \beta_{r+1}$, then the value of (5.5) is $c_r \otimes z^{j_r}$ for $r = 0, 1, \ldots, p$. We deduce that $f(z)$ and (5.5) are equal for all $z \in \mathbb{R}$ and hence $f(z)$ has no inessential terms. □

Example 5.1.3 Let us apply EVOLUTION to the product form $(1 \oplus z) \otimes (3 \oplus z)^2$. Here

$$\{(e_r, \beta_r)\}_{r=1}^{p} = \{(1, 1), (2, 3)\}.$$

We find

$$t_0 = 1^1 \otimes 3^2 = 7,$$

$$t_1 = z^1 \otimes 3^2 = 6 \otimes z,$$

$$t_2 = z^1 \otimes z^2 = z^{3\cdot}.$$

For the inverse operation (that will be called *resolution*) we first notice that if a standard maxpolynomial $p(z)$ was obtained by EVOLUTION then two consecutive terms of $p(z)$ are of the form

$$\cdots \oplus \beta_r^{e_r} \otimes \cdots \otimes \beta_p^{e_p} \otimes z^{e_1+\cdots+e_{r-1}} \oplus \beta_{r+1}^{e_{r+1}} \otimes \cdots \otimes \beta_p^{e_p} \otimes z^{e_1+\cdots+e_r} \oplus \cdots .$$

By cancelling the common factors we get $\beta_r^{e_r} \oplus z^{e_r}$ or, alternatively $(\beta_r \oplus z)^{e_r}$.

Example 5.1.4 Consider the maxpolynomial $7 \oplus 6 \otimes z \oplus z^3$. By cancelling the common factor for the first two terms we find $1 \oplus z$, for the next two terms we get $6 \oplus z^2 = (3 \oplus z)^2$. Hence the product form is $(1 \oplus z) \otimes (3 \oplus z)^2$.

This idea generalizes to nonstandard maxpolynomials as they can always be written in the form (5.3).

Example 5.1.5

$$10 \otimes z^{-1} \oplus 9 \oplus 3 \otimes z^2 = 3 \otimes z^{-1} \otimes \left(7 \oplus 6 \otimes z \oplus z^3\right)$$

$$= 3 \otimes z^{-1} \otimes (1 \oplus z) \otimes (3 \oplus z)^2.$$

In fact there is no need to transform a maxpolynomial to a standard one before we apply the idea of cancellation of common factors and we can straightforwardly formulate the algorithm:

Algorithm 5.1.6 RESOLUTION
Input: Maxpolynomial $\sum_{r=0,\ldots,p}^{\oplus} c_r \otimes z^{j_r}$.
Output: Product form $\prod_{r=1,\ldots,p}^{\otimes} (\beta_r \oplus z)^{e_r}$.
For each $r = 0, 1, \ldots, p-1$ cancel a common factor $c_{r+1} \otimes z^{j_r}$ of two consecutive terms $c_r \otimes z^{j_r}$ and $c_{r+1} \otimes z^{j_{r+1}}$ to obtain $c_r \otimes c_{r+1}^{-1} \oplus z^{j_{r+1}-j_r} = (\beta_{r+1} \oplus z)^{e_{r+1}}$.

Observe that $e_{r+1} = j_{r+1} - j_r$ and $\beta_{r+1} = \frac{c_r - c_{r+1}}{j_{r+1} - j_r}$ for $r = 0, 1, \ldots, p-1$. Again, this algorithm is formulated without specific requirements on the input and we need to identify the conditions under which it will work correctly.

It will be shown that the algorithm RESOLUTION works correctly if the sequence

$$\left\{ \frac{c_r - c_{r+1}}{j_{r+1} - j_r} \right\}_{r=0}^{p-1}$$

is increasing (in which case the sequence $\{\beta_r\}$ is increasing). A maxpolynomial satisfying this requirement is said to satisfy the *concavity condition*. Before we answer the question of the correctness of the algorithm RESOLUTION, we present an observation that will be useful:

Theorem 5.1.7 *The algorithms EVOLUTION and RESOLUTION are mutually inverse.*

Proof EVOLUTION maps

$$(e_r, \beta_r) \longrightarrow \left(\sum_{s \geq r} e_s \beta_s, \sum_{s < r} e_s \right),$$

while RESOLUTION maps

$$(c_r, j_r) \longrightarrow \left(j_{r+1} - j_r, \frac{c_r - c_{r+1}}{j_{r+1} - j_r} \right).$$

Hence EVOLUTION applied to the result of RESOLUTION produces

$$\left(\sum_{s \geq r} (j_{s+1} - j_s) \frac{c_s - c_{s+1}}{j_{s+1} - j_s}, \sum_{s < r} (j_{s+1} - j_s) \right)$$

$$= (c_r - c_p, j_r - j_0) = (c_r, j_r).$$

One can similarly deduce that RESOLUTION applied to the result of EVOLUTION produces (e_r, β_r).   □

This theorem finds an immediate use in the following key statement.

Theorem 5.1.8 *For a standard maxpolynomial $p(z)$ satisfying the concavity condition the algorithm RESOLUTION finds a standard product form $q(z)$ such that $p(z) = q(z)$ for all $z \in \mathbb{R}$.*

Proof Suppose that the maxpolynomial $p(z)$ satisfies the concavity condition. Then the sequence

$$\{\beta_r\}_{r=1}^{p} = \left\{ \frac{c_r - c_{r+1}}{j_{r+1} - j_r} \right\}_{r=0}^{p-1}$$

is increasing and finite and $e_r > 0$, since j_r are increasing. Hence the product form $q(z)$ produced by RESOLUTION is standard.

By an application of EVOLUTION to $q(z)$ we get a maxpolynomial $t(z)$ and $t(z) = q(z)$ for all $z \in \mathbb{R}$ by Theorem 5.1.2. At the same time $t(z) = p(z)$ for all $z \in \mathbb{R}$ by Theorem 5.1.7. Hence the statement. $\qquad \square$

Note that the computational complexity of RESOLUTION is $O(p)$.

Lemma 5.1.9 *Let $p(z)$ and $p'(z)$ be two maxpolynomials such that $p'(z) = c \otimes z^j \otimes p(z)$. Then the concavity condition holds for $p(z)$ if and only if it holds for $p'(z)$.*

Proof Let $p(z)$ and $p'(z)$ be two maxpolynomials such that

$$p'(z) = c \otimes p(z)$$

for some $c \in \mathbb{R}$. Then

$$c_s' - c_{s+1}' = c_s + c - c_{s+1} - c = c_s - c_{s+1}.$$

If $p'(z) = z^j \otimes p(z)$ for some $j \in \mathbb{R}$ then

$$j_{s+1}' - j_s' = j_{s+1} + j - j_s - j = j_{s+1} - j_s$$

and the statement follows. $\qquad \square$

Theorem 5.1.10 *A maxpolynomial has no inessential terms if and only if it satisfies the concavity condition.*

Proof Due to Lemma 5.0.1 we only need to prove the "if" part.

By Lemma 5.1.9 we may assume without loss of generality that $p(z)$ is standard. By applying RESOLUTION and then EVOLUTION the result now follows by Theorems 5.1.8, 5.1.2 and 5.1.7. $\qquad \square$

It follows from Theorem 5.1.8 that if a standard maxpolynomial $p(z)$ satisfies the concavity condition then the algorithm RESOLUTION applied to $p(z)$ will produce a standard product form equal to $p(z)$ as a function. If $p(z)$ does not satisfy the concavity condition then it contains an inessential term (Theorem 5.1.10). By removing an inessential term, $p(z)$ as a function does not change. Hence by a repeated removal of inessential terms we can find a standard maxpolynomial $p'(z)$ from $p(z)$ such that $p'(z)$ satisfies the concavity condition and $p(z) = p'(z)$ for all $z \in \mathbb{R}$. Formally, this process can be described by the following algorithm:

Algorithm 5.1.11 RECTIFICATION
Input: Standard maxpolynomial $p(z) = \sum_{r=0,\dots,p}^{\oplus} c_r \otimes z^{j_r}$.
Output: Standard maxpolynomial $p'(z)$ with no inessential terms and $p'(z) = p(z)$ for all $z \in \mathbb{R}$.
$p'(z) := c_{p-1} \otimes z^{j_{p-1}} \oplus c_p \otimes z^{j_p}$
$s := p - 1, t := p$
For $r = p - 2, p - 3, \dots, 0$ do
begin
    Until $\frac{c_s - c_t}{j_t - j_s} > \frac{c_r - c_s}{j_s - j_r}$ do
    begin
        Remove $c_s \otimes z^{j_s}$ from $p'(z)$, let $c_s \otimes z^{j_s}$ and $c_t \otimes z^{j_t}$ be the
        lowest and second-lowest order term in $p'(z)$, respectively.
    end
    $p'(z) := c_r \otimes z^{j_r} \oplus p'(z), t := s, s := r$
end

Clearly, RECTIFICATION runs in $O(p)$ time since every term enters and leaves $p(z)$ at most once.

We summarize the results of this section:

Theorem 5.1.12 [71] (*Max-algebraic Fundamental Theorem of Algebra*) *For every maxpolynomial $p(z)$ of length p it is possible to find using $O(p)$ operations a product form $q(z)$ such that $p(z) = q(z)$ for all $z \in \mathbb{R}$. This product form is unique up to the order of its factors.*

Proof Let $p(z)$ be the maxpolynomial $\sum_{r=0,\dots,p}^{\oplus} c_r \otimes z^{j_r}$. By taking out $c_p \otimes z^{j_0}$ it is transformed to a standard maxpolynomial, say $p'(z)$, which in turn is transformed using RECTIFICATION into a standard maxpolynomial $p''(z)$ with no inessential terms. The algorithm RESOLUTION then finds a standard product form $q(z)$ such that $q(z) = p''(z)$ for all $z \in \mathbb{R}$. By Theorems 5.1.8 and 5.1.10 we have $p''(z) = p'(z) = p(z)$ for all $z \in \mathbb{R}$ and the statement follows. $\qquad\square$

We may now extend the term "corner" to any maxpolynomial: Corners of a maxpolynomial $p(z)$ are corners of the product form that is equal to $p(z)$ as a function.

It will be important in the next section that it is possible to explicitly describe the greatest corner of a maxpolynomial:

Theorem 5.1.13 *The greatest corner of* $p(z) = \sum_{r=0,\dots,p}^{\oplus} c_r \otimes z^{j_r}$, $p > 0$, *is*

$$\max_{r=0,\dots,p-1} \frac{c_r - c_p}{j_p - j_r}.$$

Proof A corner exists since $p > 0$. Let γ be the greatest corner of $p(z)$. Then

$$c_p \otimes z^{j_p} \geq c_r \otimes z^{j_r}$$

for all $z \geq \gamma$ and for all $r = 0, 1, \dots, p$. At the same time there is an $r < p$ such that

$$c_p \otimes z^{j_p} < c_r \otimes z^{j_r}$$

for all $z < \gamma$. Hence $\gamma = \max_{r=0,1,\dots,p-1} \gamma_r$ where γ_r is the intersection point of $c_p \otimes z^{j_p}$ and $c_r \otimes z^{j_r}$, that is

$$\gamma_r = \frac{c_r - c_p}{j_p - j_r}$$

and the statement follows.   □

Note that an alternative treatment of maxpolynomials can be found in [8] and in [2] in terms of convex analysis and (in particular) Legendre–Fenchel transform.

5.2 Maxpolynomial Equations

Maxpolynomial equations are of the form

$$p(z) = q(z), \tag{5.9}$$

where $p(z)$ and $q(z)$ are maxpolynomials. Since both $p(z)$ and $q(z)$ are piecewise linear convex functions, it is clear geometrically that the solution S set to (5.9) is the union of a finite number of closed intervals in $\mathbb{R}$, including possibly one-element sets, and unbounded intervals (see Fig. 5.1, where S consists of one closed interval and two isolated points). Let us denote the set of boundary points of S (that is, the set of extreme points of the intervals) by S^*. The set S^* can easily be characterized:

Theorem 5.2.1 [64] *Every boundary point of S is a corner of $p(z) \oplus q(z)$.*

Proof Let $z \in S^*$. If z is not a corner of $p(z) \oplus q(z)$ then $p(z) \oplus q(z)$ does not change the slope in a neighborhood of z. By the convexity of $p(z)$ and $q(z)$ then neither $p(z)$ nor $q(z)$ can change slope in a neighborhood of z. But then z is an interior point to S, a contradiction.   □

Theorem 5.2.1 provides a simple solution method for maxpolynomial equations (5.9). After finding all corners of $p(z) \oplus q(z)$, say $\beta_1 < \dots < \beta_r$, it remains

Fig. 5.1 Solving
maxpolynomial equations

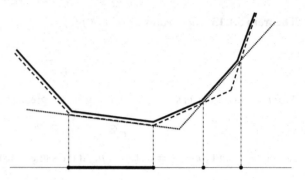

(1) to check which of them are in S, and
(2) if $\gamma_1 < \cdots < \gamma_t$ are the corners in S then by selecting arbitrary interleaving points $\alpha_0, \ldots, \alpha_t$ so that

$$\alpha_0 < \gamma_1 < \alpha_1 < \cdots < \gamma_t < \alpha_t$$

and checking whether $\alpha_j \in S$ for $j = 0, \ldots, t$, it is decided about each of the intervals $[\gamma_{j-1}, \gamma_j]$ ($j = 1, \ldots, t+1$) whether it is a subset of S. Here $\gamma_0 = -\infty$ and $\gamma_{t+1} = +\infty$.

Example 5.2.2 [64] Find all solutions to the equation

$$9 \oplus 8 \otimes z \oplus 4 \otimes z^2 \oplus z^3 = 10 \oplus 8 \otimes z \oplus 5 \otimes z^2.$$

If $p(z) = 9 \oplus 8 \otimes z \oplus 4 \otimes z^2 \oplus z^3$ and $q(z) = 10 \oplus 8 \otimes z \oplus 5 \otimes z^2$ then

$$p(z) \oplus q(z) = 10 \oplus 8 \otimes z \oplus 5 \otimes z^2 \oplus z^3$$
$$= (z \oplus 2) \otimes (z \oplus 3) \otimes (z \oplus 5).$$

All corners are solutions and by checking the interleaving points (say) $1, 2.5, 4, 6$ one can find $S = [2, 3] \cup \{5\}$.

5.3 Characteristic Maxpolynomial

5.3.1 Definition and Basic Properties

There are various ways of defining a characteristic polynomial in max-algebra, briefly characteristic maxpolynomial [62, 99]. We will study the concept defined in [62].

Let $A = (a_{ij}) \in \overline{\mathbb{R}}^{n \times n}$. Then the *characteristic maxpolynomial* of A is

$$\chi_A(x) = \text{maper}(A \oplus x \otimes I) = \text{maper} \begin{pmatrix} a_{11} \oplus x & a_{12} & \cdots & a_{1n} \\ a_{21} & a_{22} \oplus x & \cdots & a_{2n} \\ \vdots & \vdots & & \vdots \\ a_{n1} & a_{n2} & \cdots & a_{nn} \oplus x \end{pmatrix}.$$

It immediately follows from this definition that $\chi_A(x)$ is of the form

$$x^n \oplus \delta_1 \otimes x^{n-1} \oplus \cdots \oplus \delta_{n-1} \otimes x \oplus \delta_n,$$

or briefly, $\sum_{k=0,\dots,n}^{\oplus} \delta_{n-k} \otimes x^k$, where $\delta_0 = 0$. Hence the characteristic maxpolynomial of an $n \times n$ matrix is a standard maxpolynomial with exponents $0, 1, \dots, n$, degree n and length $n + 1$ or less.

Example 5.3.1 If

$$A = \begin{pmatrix} 1 & 3 & 2 \\ 0 & 4 & 1 \\ 2 & 5 & 0 \end{pmatrix}$$

then

$$\chi_A(x) = \text{maper} \begin{pmatrix} 1 \oplus x & 3 & 2 \\ 0 & 4 \oplus x & 1 \\ 2 & 5 & 0 \oplus x \end{pmatrix}$$

$$= (1 \oplus x) \otimes (4 \oplus x) \otimes (0 \oplus x) \oplus 3 \otimes 1 \otimes 2$$

$$\oplus 2 \otimes 0 \otimes 5 \oplus 2 \otimes (4 \oplus x) \otimes 2 \oplus (1 \oplus x) \otimes 1 \otimes 5 \oplus 3 \otimes 0 \otimes (0 \oplus x)$$

$$= x^3 \oplus 4 \otimes x^2 \oplus 6 \otimes x \oplus 8.$$

Theorem 5.3.2 [62] *If $A = (a_{ij}) \in \overline{\mathbb{R}}^{n \times n}$ then*

$$\delta_k = \sum_{B \in P_k(A)}^{\oplus} \text{maper}(B), \tag{5.10}$$

for $k = 1, \dots, n$, where $P_k(A)$ is the set of all principal submatrices of A of order k.

Proof The coefficient δ_k is associated with x^{n-k} in $\chi_A(x)$ and therefore is the maximum of the weights of all permutations that select $n - k$ symbols of x and k constants from different rows and columns of a submatrix of A obtained by removing the rows and columns of selected x. Since x only appear on the diagonal the corresponding submatrices are principal. □

Hence we can readily find $\delta_n = \text{maper}(A)$ and $\delta_1 = \max(a_{11}, a_{22}, \dots, a_{nn})$, but other coefficients cannot be found easily from (5.10) as the number of matrices in $P_k(A)$ is $\binom{n}{k}$.

If considered as a function, the characteristic maxpolynomial is a piecewise linear convex function in which the slopes of the linear pieces are n and some (possibly none) of the numbers $0, 1, \ldots, n - 1$. Note that it may happen that $\delta_k = \varepsilon$ for all $k = 1, \ldots, n$ and then $\chi_A(x)$ is just x^n. We can easily characterize such cases:

Proposition 5.3.3 *If $A = (a_{ij}) \in \overline{\mathbb{R}}^{n \times n}$ then $\chi_A(x) = x^n$ if and only if D_A is acyclic.*

Proof If D_A is acyclic then the weights of all permutations with respect to any principal submatrix of A are ε and thus all $\delta_k = \varepsilon$. If D_A contains a cycle, say $(i_1, \ldots, i_k, i_1)$ for some $k \in N$ then

$$\mathrm{maper}\,(A\,(i_1, \ldots, i_k)) > \varepsilon,$$

thus $\delta_k > \varepsilon$ by Theorem 5.3.2. $\qquad\square$

Note that the coefficients δ_k are closely related to the best submatrix problem and to the job rotation problem, see Sect. 2.2.3.

5.3.2 The Greatest Corner Is the Principal Eigenvalue

By Theorem 5.1.13 we know that the greatest corner of a maxpolynomial $p(z) = \sum_{r=0,\ldots,p}^{\oplus} c_r \otimes z^{j_r}$, $p > 0$, is

$$\max_{r=0,\ldots,p-1} \frac{c_r - c_p}{j_p - j_r}.$$

If $p(x) = \chi_A(x)$ where $A = (a_{ij}) \in \overline{\mathbb{R}}^{n \times n}$ then $p = n$, $j_r = r$ and $c_r = \delta_{n-r}$ for $r = 0, 1, \ldots, n$ with $c_n = \delta_0 = 0$. Hence the greatest corner of $\chi_A(x)$ is

$$\max_{r=0,\ldots,n-1} \frac{\delta_{n-r}}{n - r}$$

or, equivalently

$$\max_{k=1,\ldots,n} \frac{\delta_k}{k}. \tag{5.11}$$

We are ready to prove a remarkable property of characteristic maxpolynomials resembling the one in conventional linear algebra. As a convention, the greatest corner of a maxpolynomial with no corners (that is, $\lambda(A) = \varepsilon$, see Proposition 5.3.3) is by definition ε.

Theorem 5.3.4 [62] *If $A = (a_{ij}) \in \overline{\mathbb{R}}^{n \times n}$ then the greatest corner of $\chi_A(x)$ is $\lambda(A)$.*

Proof The statement is evidently true if $\lambda(A) = \varepsilon$. Thus assume now that $\lambda(A) > \varepsilon$, hence at least one corner exists. Let β be the greatest corner of $\chi_A(x)$ and $k \in \{1, \ldots, n\}$, then $\delta_k = \text{maper}(B)$, where $B \in P_k(A)$. We have

$$\text{maper}(B) = w(\pi, B) = w(\pi_1, B) \otimes \cdots \otimes w(\pi_s, B)$$

for some $\pi \in ap(B)$ and its constituent cycles $\pi_1, \ldots, \pi_s$. We also have

$$w(\pi_j, B) \leq (\lambda(A))^{l(\pi_j)}$$

for all $j = 1, \ldots, s$. Hence

$$\delta_k = \text{maper}(B) \leq (\lambda(A))^{l(\pi_1) + \cdots + l(\pi_s)} = (\lambda(A))^k$$

and so

$$\frac{\delta_k}{k} \leq \lambda(A),$$

yielding using (5.11):

$$\beta \leq \lambda(A).$$

Suppose now $\lambda(A) = \frac{w(\sigma, A)}{l(\sigma)}$, $\sigma = (i_1, \ldots, i_r)$, $r \in \{1, \ldots, n\}$. Let $\overline{B} = A(i_1, \ldots, i_r)$. Then

$$\delta_r \geq \text{maper}(\overline{B}) \geq w(\sigma, A) = (\lambda(A))^{l(\sigma)} = (\lambda(A))^r.$$

Therefore

$$\frac{\delta_r}{r} \geq \lambda(A),$$

yielding by (5.11):

$$\beta \geq \lambda(A),$$

which completes the proof. $\qquad\square$

Example 5.3.5 The principal eigenvalue of

$$A = \begin{pmatrix} 2 & 1 & 4 \\ 1 & 0 & 1 \\ 2 & 2 & 1 \end{pmatrix}$$

is $\lambda(A) = 3$. The characteristic maxpolynomial is

$$\chi_A(x) = x^3 \oplus 2 \otimes x^2 \oplus 6 \otimes x \oplus 7 = (x \oplus 1) \otimes (x \oplus 3)^2$$

and the greatest corner is 3.

5.3.3 Finding All Essential Terms of a Characteristic Maxpolynomial

As already mentioned in Sect. 2.2.3, no polynomial method is known for finding all coefficients of a characteristic maxpolynomial or, equivalently, to solve the job rotation problem. Recall (see Sect. 2.2.3) that this question is equivalent to the best principal submatrix problem (BPSM), which is the task to find the greatest optimal values δ_k for the assignment problem of all $k \times k$ principal submatrices of A, $k = 1, \ldots, n$. It will be convenient now to denote by BPSM(k) the task of finding this value for a particular integer k.

We will use the functional interpretation of a characteristic maxpolynomial to derive a method for finding coefficients of this maxpolynomial corresponding to all essential terms. Recall that as every maxpolynomial, the characteristic maxpolynomial is a piecewise linear and convex function which can be written using conventional notation as

$$\chi_A(x) = \max(\delta_n, \delta_{n-1} + x, \delta_{n-2} + 2x, \ldots, \delta_1 + (n-1)x, nx).$$

If for some $k \in \{0, \ldots, n\}$ the term $\delta_{n-k} \otimes x^k$ is inessential, then

$$\chi_A(x) = \sum_{i \neq k}^{\oplus} \delta_{n-i} \otimes x^i$$

holds for all $x \in \mathbb{R}$, and therefore all inessential terms may be ignored if $\chi_A(x)$ is considered as a function. We now present an $O(n^2(m + n \log n))$ method for finding all essential terms of a characteristic maxpolynomial for a matrix with m finite entries. It then follows that this method solves BPSM(k) for those $k \in \{1, \ldots, n\}$, for which $\delta_{n-k} \otimes x^k$ is essential and, in particular, when all terms are essential then this method solves BPSM(k) for all $k = 1, \ldots, n$.

We will first discuss the case of finite matrices. Let $A = (a_{ij}) \in \mathbb{R}^{n \times n}$ be given. For convenience we will denote $\chi_A(x)$ by $z(x)$ and $A \oplus x \otimes I$ by $A(x) = (a(x)_{ij})$. Hence

$$z(x) := \max_{\pi} \sum_{i=1}^{n} a(x)_{i,\pi(i)}$$

and

$$a(x)_{ij} := \begin{cases} \max(x, a_{ii}), & \text{for } i = j, \\ a_{ij}, & \text{for } i \neq j. \end{cases}$$

Since $z(x)$ is piecewise linear and convex and all its linear pieces are of the form $z_k(x) := kx + \delta_{n-k}$ for $k = 0, 1, \ldots, n$ and constants δ_{n-k}, the maxpolynomial $z(x)$ has at most n corners. Recall that $z_n(x) := nx$, that is, $\delta_0 = 0$. The main idea of the method for finding all linear pieces of $z(x)$ is based on the fact that it is easy to evaluate $z(x)$ for any real value of x as this is simply maper$(A \oplus x \otimes I)$, that is, the optimal value for the assignment problem for $A \oplus x \otimes I$. By a suitable choice of $O(n)$ values of x we will be able to identify all linear pieces of $z(x)$.

Let x be fixed and $\pi \in \mathrm{ap}(A(x)) = \mathrm{ap}(a(x)_{ij})$ (recall that $\mathrm{ap}(A)$ denotes the set of optimal permutations to the assignment problem for a square matrix A, see Sect. 1.6.4). We call a diagonal entry $a(x)_{ii}$ of the matrix $A(x)$ *active*, if $x \geq a_{ii}$ and if this diagonal position is selected by π, that is, $\pi(i) = i$. All other entries will be called *inactive*. If there are exactly k active values for a certain x and permutation π then this means that $z(x) = kx + \delta_{n-k} = x^k \otimes \delta_{n-k}$, that is, the value of $z(x)$ is determined by the linear piece with the slope k. Here δ_{n-k} is the sum of $n - k$ inactive entries of $A(x)$ selected by π. No two of these inactive entries can be from the same row or column and they are all in the submatrix, say B, obtained by removing the rows and columns of all active elements. Since all active elements are on the diagonal, B is principal and the $n - k$ inactive elements form a feasible solution to the assignment problem for B. This solution is also optimal by optimality of π. This yields the following:

Proposition 5.3.6 [20] *Let $x \in \mathbb{R}$ and $\pi \in P_n$. If $z(x) = \mathrm{maper}(A(x)) = \sum_{i=1}^{n} a(x)_{i,\pi(i)}$, $i_1, \ldots, i_k$ are indices of all active entries and $\{j_1, \ldots, j_{n-k}\} = N - \{i_1, \ldots, i_k\}$ then $A(j_1, \ldots, j_{n-k})$ is a solution to $\mathrm{BPSM}(n - k)$ for A and $\delta_{n-k} = \mathrm{maper}(A(j_1, \ldots, j_{n-k}))$.*

There may, of course, be several optimal permutations for the same value of x selecting different numbers of active elements which means that the value of $z(x)$ may be equal to the function value of several linear pieces with different slopes at x. We will pay special attention to this question in Proposition 5.3.14 below.

Proposition 5.3.7 [20] *If $z(\overline{x}) = z_r(\overline{x}) = z_s(\overline{x})$ for some $\overline{x} \in \mathbb{R}$ and integers $r < s$, then there are no essential terms with the slope $k \in (r, s)$ and $\overline{x}$ is a corner of $z(x)$.*

Proof Since $z_r(\overline{x}) = \delta_{n-r} + r\overline{x} = z(\overline{x}) \geq \delta_{n-k} + k\overline{x}$ for every k, we have $z_r(x) = \delta_{n-r} + rx \geq \delta_{n-k} + kx = z_k(x)$ for every $x < \overline{x}$ and $k > r$, thus $z(x) \geq z_r(x) \geq z_k(x)$ for every $x < \overline{x}$ and for every $k > r$.

Similarly, $z(x) \geq z_s(x) \geq z_k(x)$ for every $x > \overline{x}$ and for every $k < s$. Hence, $z(x) \geq z_k(x)$ for every x and for every integer slope k with $r + 1 \leq k \leq s - 1$. □

For $x \leq \widetilde{a} = \min(a_{11}, a_{22}, \ldots, a_{nn})$, $z(x)$ is given by $\max_\pi \sum_{i=1}^{n} a_{i,\pi(i)} = \mathrm{maper}(A) = \delta_n$. Then obviously, $z(x) = z_0(x) = \delta_n$ for $x \leq \widetilde{a}$.

Now, let $\alpha^* := \max_{ij} a_{ij}$ and let E be the matrix whose entries are all equal to 1. For $x \geq \alpha^*$ the matrix $A(x) - \alpha^* \cdot E$ (in conventional notation) has only nonnegative elements on its main diagonal. All off-diagonal elements are negative. Therefore we get $z(x) = nx = z_n(x)$ for $x \geq \alpha^*$. Note that for finding $z(x)$ there is no need to compute α^*.

The intersection point of $z_0(x)$ with $z_n(x)$ is $x_1 = \frac{\delta_n}{n}$. We find $z(x_1)$ by solving the assignment problem $\max_\pi \sum_{i=1}^{n} a(x_1)_{i,\pi(i)}$.

Corollary 5.3.8 *If $z(x_1) = z_0(x_1)$ then $z(x) = \max(z_0(x), z_n(x))$.*

Thus, if $z(x_1) = z_0(x_1)$, we are done and the function $z(x)$ has the form

$$z(x) = \begin{cases} z_0(x), & \text{for } x \leq x_1, \\ z_n(x), & \text{for } x \geq x_1. \end{cases} \qquad (5.12)$$

Otherwise we have found a new linear piece of $z(x)$. Let us call it $z_k(x) := kx + \delta_{n-k}$, where k is the number of active elements in the corresponding optimal solution and δ_{n-k} is given by $\delta_{n-k} := z(x_1) - kx_1$. We remove x_1 from the list.

Next we intersect $z_k(x)$ with $z_0(x)$ and with $z_n(x)$. Let x_2 and x_3, respectively, be the corresponding intersection points. We generate a list $L := (x_2, x_3)$. Let us choose an element from the list, say x_2, and determine $z(x_2)$. If $z(x_2) = z_0(x_2)$, then x_2 is a corner of $z(x)$. By Proposition 5.3.7 this means that there are no essential terms of the characteristic maxpolynomial with slopes between 0 and k. We delete x_2 from L and process a next point from L. Otherwise we have found a new linear piece of $z(x)$ and can proceed as above. Thus, for every point in the list we either find a new slope which leads to two new points in the list or we detect that the currently investigated point is a corner of L. In such a case this point will be deleted and no new points are generated. If the list L is empty, we are done and we have already found the function $z(x)$. Every point of the list either leads to a new slope (and therefore to two new points in L) or it is a corner of $z(x)$, in which case this point is deleted from L. Therefore only $O(n)$ entries will enter and leave the list. This means the procedure stops after investigating at most $O(n)$ linear assignment problems. Thus we have shown:

Theorem 5.3.9 [20] *All essential terms of the characteristic maxpolynomial of $A \in \mathbb{R}^{n \times n}$ can be found in $O(n^4)$ steps.*

The proof of the following statement is straightforward.

Proposition 5.3.10 *Let $A = (a_{ij})$, $B = (b_{ij}) \in \mathbb{R}^{n \times n}$, $r, s \in N$, $a_{rs} \leq b_{rs}$, $a_{ij} = b_{ij}$ for all $i, j \in N$, $i \neq r$, $j \neq s$. If $\pi \in \text{ap}(A)$ satisfies $\pi(r) = s$ then $\pi \in \text{ap}(B)$.*

Corollary 5.3.11 *If $\text{id} \in \text{ap}(A(\overline{x}))$ then $\text{id} \in \text{ap}(A(x))$ for all $x \geq \overline{x}$.*

Remarks

1. A diagonal element of $A(y)$ may not be active for some y with $y > x$ even if it is active in $A(x)$. For instance, consider the following 4×4 matrix A:

$$\begin{pmatrix} 0 & 0 & 0 & 29 \\ 0 & 8 & 20 & 0 \\ 0 & 0 & 12 & 28 \\ 29 & 28 & 0 & 16 \end{pmatrix}.$$

For $x = 4$ the unique optimal permutation is $\pi = (1)(2, 3, 4)$ of value 80, for which the first diagonal element is active. For $y = 20$ the unique optimal permutation is $\pi = (1, 4)(2)(3)$ of value 98, in which the second and third, but not the first, diagonal elements of the matrix are active.

2. If an intersection point x is found by intersecting two linear functions with the slopes k and $k + 1$ respectively, this point is immediately deleted from the list L since it cannot lead to a new essential term (as there is no slope strictly between k and $k + 1$).

3. If at an intersection point y the slope of $z(x)$ changes from k to l with $l - k \geq 2$, then an upper bound for δ_{n-r} related to an inessential term $rx + \delta_{n-r}, k < r < l$, can be obtained by $z(y) - ry$. Due to the convexity of the function $z(x)$ this is the least upper bound on δ_{n-r} which can be obtained by using the values of $z(x)$.

Taking into account our previous discussion, we arrive at the following algorithm. The values x which have to be investigated are stored as triples $(x, k(l), k(r))$ in a list L. The interpretation of such a triple is that x has been found as the intersection point of two linear functions with the slopes $k(l)$ and $k(r)$, $k(l) < k(r)$.

Algorithm 5.3.12 ESSENTIAL TERMS
Input: $A = (a_{ij}) \in \mathbb{R}^{n \times n}$.
Output: All essential terms of the characteristic maxpolynomial of A, in the form $kx + \delta_{n-k}$.

1. Solve the assignment problem with the cost matrix A and set $\delta_n := \mathrm{maper}(A)$ and $z_0(x) := \delta_n$.
2. Determine x_1 as the intersection point of $z_0(x)$ and $z_n(x) := nx$.
3. Let $L := \{(x_1, 0, n)\}$.
4. If $L = \emptyset$, stop. The function $z(x)$ has been found. Otherwise choose an arbitrary element $(x_i, k_i(l), k_i(r))$ from L and remove it from L.
5. If $k_i(r) = k_i(l) + 1$, then (see Remark 2 above) go to step 4. (x_i is a corner of $z(x)$; for x close to x_i the function $z(x)$ has slope $k_i(l)$ for $x < x_i$, and $k_i(r)$ for $x > x_i$.)
6. Find $z(x_i) = \mathrm{maper}(A(x_i))$. Take an arbitrary optimal permutation to the assignment problem for the matrix $A(x_i)$ and let k_i be the number of active elements in this solution. Set $\delta_{n-k_i} := z(x_i) - k_i x_i$.
7. Set $z_i(x) := k_i x + \delta_{n-k_i}$.
8. Intersect $z_i(x)$ with the lines having slopes $k_i(l)$ and $k_i(r)$. Let y_1 and y_2 be the intersection points, respectively. Add the triples $(y_1, k_i(l), k_i)$ and $(y_2, k_i, k_i(r))$ to the list L and go to step 4. [See a refinement of this step after Proposition 5.3.14.]

Example 5.3.13 Let

$$A := \begin{pmatrix} 0 & 4 & -2 & 3 \\ 2 & 1 & 3 & -1 \\ -2 & -3 & 1 & 0 \\ 7 & -2 & 8 & 4 \end{pmatrix}.$$

We solve the assignment problem for A by the Hungarian method and transform A to a normal form. The asterisks indicate entries selected by an optimal permutation:

$$
\begin{pmatrix}
-4 & 0 & -6 & -1 \\
-1 & -2 & 0 & -4 \\
-3 & -4 & 0 & -1 \\
-1 & -10 & 0 & -4
\end{pmatrix},
$$

$$
\begin{pmatrix}
-3 & 0^* & -6 & 0 \\
0^* & -2 & 0 & -3 \\
-2 & -4 & 0 & 0^* \\
0 & -10 & 0^* & -3
\end{pmatrix}.
$$

Thus $z_0(x) = 14$.

Now we solve $14 = 4x$ and we get $x_1 = 3.5$. By solving the assignment problem for $x_1 = 3.5$ we get:

$$
\begin{pmatrix}
3.5 & 4 & -2 & 3 \\
2 & 3.5 & 3 & -1 \\
-2 & -3 & 3.5 & 0 \\
7 & -2 & 8 & 4
\end{pmatrix},
$$

$$
\begin{pmatrix}
-0.5 & 0 & -6 & -1 \\
-1.5 & 0 & -0.5 & -4.5 \\
-5.5 & -6.5 & 0 & -3.5 \\
-1 & -10 & 0 & -4
\end{pmatrix},
$$

$$
\begin{pmatrix}
0 & 0 & -6 & 0 \\
-1 & 0 & -0.5 & -3.5 \\
-5 & -6.5 & 0 & -2.5 \\
-0.5 & -10 & 0 & -3
\end{pmatrix},
$$

$$
\begin{pmatrix}
0 & -0.5 & -6.5 & 0^* \\
-0.5 & 0^* & -0.5 & -3 \\
-4.5 & -6.5 & 0^* & -2 \\
0^* & -10 & 0 & -2.5
\end{pmatrix}.
$$

Thus $z_2(3.5) = 17$ and we get $z_2(x) := 2x + 10$. Intersecting this function with $z_0(x)$ and $z_4(x)$ yields the two new points $x_2 := 2$ (solving $14 = 2x + 10$) and $x_3 := 5$ (solving $2x + 10 = 4x$). Investigating $x = 2$ shows that the slope changes at this point from 0 to 2. Thus we have here a corner of $z(x)$. Finding the value $z(5)$ amounts to solving the assignment problem with the cost matrix

$$
\begin{pmatrix}
5 & 4 & -2 & 3 \\
2 & 5 & 3 & -1 \\
-2 & -3 & 5 & 0 \\
7 & -2 & 8 & 5
\end{pmatrix}.
$$

This assignment problem yields the solution $z(5) = 20 = z_4(5)$. Thus no new essential term has been found and we have $z(x)$ completely determined as

$$z(x) = \begin{cases} 14 & \text{for } 0 \le x \le 2 \\ 2x + 10 & \text{for } 2 \le x \le 5 \\ 4x & \text{for } x \ge 5. \end{cases}$$

In max-algebraic terms $z(x) = 14 \oplus 10 \otimes x^2 \oplus x^4$.

The following proposition enables us to make a computational refinement of the algorithm ESSENTIAL TERMS. We refer to the assignment problem terminology introduced in Sect. 1.6.4.

Proposition 5.3.14 *Let $\overline{x} \in \mathbb{R}$ and let $B = (b_{ij})$ be a normal form of $A(\overline{x})$. Let $C = (c_{ij})$ be the matrix obtained from B as follows:*

$$c_{ij} = \begin{cases} 0, & \text{if } b_{ij} = 0 \text{ and } (i, j) \text{ is inactive,} \\ 1, & \text{if } (i, j) \text{ is active,} \\ \varepsilon, & \text{otherwise.} \end{cases}$$

Then every $\pi \in ap(C)$ $[\pi \in ap(-C)]$ is an optimal solution to the assignment problem for $A(\overline{x})$ with maximal [minimal] number of active elements.

Proof The statement immediately follows from the definitions of C and of a normal form of a matrix. ☐

If for some value of $\overline{x}$ there are two or more optimal solutions to the assignment problem for $A(\overline{x})$ with different numbers of active elements then using Proposition 5.3.14 we can find an optimal solution with the smallest number and another one with the greatest number of active elements. This enables us to find two new lines (rather than one) in step 6 of Algorithm 5.3.12:

(a) $z_k(x) := kx + \delta_{n-k}$, where k is the minimal number of active elements of an optimal solution to the assignment problem for $A(\overline{x})$ and δ_{n-k} is given by $\delta_{n-k} := z(\overline{x}) - k\overline{x}$;

(b) $z_{k'}(x) := k'x + \delta_{n-k'}$, where k' is the maximal number of active elements of an optimal solution to the assignment problem for $A(\overline{x})$ and $\delta_{n-k'}$ is given by $\delta_{n-k'} := z(\overline{x}) - k'\overline{x}$.

In step 8 of Algorithm 5.3.12 we then intersect $z_i(x)$ with the line having the slope $k_i(l)$ and $z_{k'}(x)$ with the line having slope $k_i(r)$.

So far we have assumed in this subsection that all entries of the matrix are finite. If some (but not all) entries of A are ε, the same algorithm as in the finite case can be used except that the lowest order finite term has to be found since a number of the coefficients of the characteristic maxpolynomial may be ε. The following theorem is useful here. In this theorem we denote

$$\underline{\delta} = \min(0, nA_{\min}), \qquad \overline{\delta} = \max(0, nA_{\max}),$$

where $A_{\min}$ [$A_{\max}$] is the least [greatest] finite entry of A. We will also denote in this and the next subsection

$$K = \{k; \delta_k \text{ finite}\}$$

and

$$k_0 = \max K. \tag{5.13}$$

Clearly, the lowest-order finite term of the characteristic maxpolynomial is $z_{k_0}(x) = \delta_{k_0} \otimes x^{n-k_0}$.

Theorem 5.3.15 [38] *If $A \in \overline{\mathbb{R}}^{n \times n}$ then $n - k_0$ is the number of active elements in $A(\overline{x})$, where $\overline{x}$ is any real number satisfying*

$$\overline{x} < \underline{\delta} - \overline{\delta}$$

and $\delta_{k_0} = z(\overline{x}) - (n - k_0)\overline{x}$.

Proof It is sufficient to prove that if x_0 is a point of intersection of two different linear pieces of $\chi_A(x)$ then

$$x_0 \geq \underline{\delta} - \overline{\delta}.$$

Suppose that

$$\delta_r + (n - r)x_0 = \delta_s + (n - s)x_0$$

for some $r, s \in \{0, 1, \ldots, n\}, r > s$. Then

$$(r - s)x_0 = \delta_r - \delta_s.$$

If $A_{\min} \leq 0$ then $\delta_r \geq s A_{\min} \geq n A_{\min} = \underline{\delta}$. If $A_{\min} \geq 0$ then $\delta_r \geq s A_{\min} \geq 0 = \underline{\delta}$. Hence $\delta_r \geq \underline{\delta}$.

If $A_{\max} \leq 0$ then $\delta_s \leq r A_{\max} \leq 0 = \overline{\delta}$. If $A_{\max} \geq 0$ then $\delta_s \leq r A_{\max} \leq n A_{\max} = \overline{\delta}$. Hence $\delta_s \leq \overline{\delta}$.

We deduce that $\delta_r - \delta_s \geq \underline{\delta} - \overline{\delta}$ and the rest follows from the fact that $r - s \geq 1$ and $\underline{\delta} - \overline{\delta} \leq 0$.   □

It follows from this result that for a general matrix, k_0 can be found using $O(n^3)$ operations. Note that for symmetric matrices this problem can be converted to the maximum cardinality bipartite matching problem and thus solved in $O(n^{2.5}/\sqrt{\log n})$ time [37].

Theorem 5.3.15 enables us to modify the beginning of the algorithm ESSENTIAL TERMS for $A \in \overline{\mathbb{R}}^{n \times n}$ by finding the intersection of the lowest order finite term $z_{k_0}(x)$ (rather than $z_0(x)$) with x^n. Moreover, instead of considering the classical assignment problem we rather formulate the problem in step 6 of the algorithm as the maximum weight perfect matching problem in a bipartite graph $(N, N; E)$. This graph has an arc $(i, j) \in E$ if and only if a_{ij} is finite. It is known [1] that the maximum weight perfect matching problem in a graph with m arcs can be solved by a shortest augmenting path method using Fibonacci heaps in $O(n(m + n \log n))$ time. Since in the worst case $O(n)$ such maximum weight perfect matching problems must be solved, we get the following result.

Theorem 5.3.16 [20] *If $A \in \overline{\mathbb{R}}^{n \times n}$ has m finite entries, then all essential terms of $\chi_A(x)$ can be found in $O(n^2(m + n \log n))$ time.*

5.3.4 Special Matrices

Although no polynomial method seems to exist for finding all coefficients of a characteristic maxpolynomial for general matrices or even for matrices over $\{0, -\infty\}$, there are a number of special cases for which this problem can be solved efficiently. These include permutation, pyramidal, Hankel and Monge matrices and special matrices over $\{0, -\infty\}$ [28, 37, 116].

We briefly discuss two special types: diagonally dominant matrices and matrices over $\{0, -\infty\}$.

Proposition 5.3.17 *If $A = (a_{ij}) \in \overline{\mathbb{R}}^{n \times n}$ is diagonally dominant then so are all principal submatrices of A and all coefficients of the characteristic maxpolynomial can be found by the formula*

$$\delta_k = a_{i_1 i_1} + a_{i_2 i_2} + \cdots + a_{i_k i_k},$$

for $k = 1, \ldots, n$, where $a_{i_1 i_1} \geq a_{i_2 i_2} \geq \cdots \geq a_{i_n i_n}$.

Proof Let A be a diagonally dominant matrix, $B = A(i_1, i_2, \ldots, i_k)$ for some indices $i_1, i_2, \ldots, i_k$ and suppose that id $\notin$ ap(B). Take any $\pi \in$ ap(B) and extend π to a permutation σ of the set N by setting $\sigma(i) = i$ for every $i \notin \{i_1, i_2, \ldots, i_k\}$. Then obviously σ is a permutation of a weight greater than that of id $\in P_n$, a contradiction. The formula follows. □

Matrices over $T = \{0, -\infty\}$ have implications for problems outside max-algebra and in particular for the conventional permanent, which for a real matrix $A = (a_{ij})$ we denote as usual by per(A), that is

$$\mathrm{per}(A) = \sum_{\pi \in P_n} \prod_{i \in N} a_{i, \pi(i)}.$$

If $A = (a_{ij}) \in T^{n \times n}$ then $\delta_k = 0$ or $\delta_k = -\infty$ for every $k = 1, \ldots, n$. Clearly, $\delta_k = 0$ if and only if there is a $k \times k$ principal submatrix of A with k independent zeros, that is, with k zeros selected by a permutation or, equivalently, k zeros no two of which are either from the same row or from the same column.

It is easy to see that if $A = (a_{ij}) \in T^{n \times n}$ then $B = 2^A = (2^{a_{ij}}) = (b_{ij})$ is a zero-one matrix. If $\pi \in P_n$ then

$$\prod_{i \in N} b_{i,\pi(i)} = \prod_{i \in N} 2^{a_{i,\pi(i)}} = 2^{\sum_{i \in N} a_{i,\pi(i)}}.$$

Hence $\mathrm{per}(B) > 0$ is equivalent to

$$(\exists \pi \in P_n) \, (\forall i \in N) \, b_{i,\pi(i)} = 1.$$

But this is equivalent to

$$(\exists \pi \in P_n) \, (\forall i \in N) \, a_{i,\pi(i)} = 0.$$

Thus, the task of finding the coefficient δ_k of the characteristic maxpolynomial of a square matrix over T is equivalent to the following problem expressed in terms of the classical permanents:

PRINCIPAL SUBMATRIX WITH POSITIVE PERMANENT: *Given an $n \times n$ zero-one matrix A and a positive integer k ($k \leq n$), is there a $k \times k$ principal submatrix B of A with positive (conventional) permanent?*

Another equivalent version for matrices over T is graph-theoretical: Since every permutation is a product of cycles, $\delta_k = 0$ means that in D_A (and F_A) there is a set of pairwise node-disjoint cycles covering exactly k nodes. Hence deciding whether $\delta_k = 0$ is equivalent to the following:

EXACT CYCLE COVER: *Given a digraph D with n nodes and a positive integer k ($k \leq n$), is there a set of pairwise node-disjoint cycles covering exactly k nodes of D?*

Finally, it may be useful to see that the value of k_0 defined by (5.13) can explicitly be described for matrices over $\{0, -\infty\}$:

Theorem 5.3.18 [28] *If $A \in T^{n \times n}$ then $k_0 = n + \mathrm{maper}(A \oplus (-1) \otimes I)$.*

Proof Since all finite δ_k are 0 in conventional notation we have:

$$\chi_A(x) = \max_{k \in K} (n - k) x.$$

Therefore, for $x < 0$:

$$\chi_A(x) = x . \min_k (n - k) = x . (n - k_0),$$

from which the result follows by setting $x = -1$. $\qquad \square$

5.3.5 Cayley–Hamilton in Max-algebra

A max-algebraic analogue of the Cayley–Hamilton Theorem was proved in [119] and [140], see also [8]. Some notation used here has been introduced in Sect. 1.6.4.

Let $A = (a_{ij}) \in \mathbb{R}^{n \times n}$ and $v \in \mathbb{R}$. Let us denote

$$p^+(A, v) = \left| \{\pi \in P_n^+; w(\pi, A) = v\} \right|$$

and

$$p^-(A, v) = \left| \{\pi \in P_n^-; w(\pi, A) = v\} \right|.$$

The following equation is called the (*max-algebraic*) *characteristic equation* for A (recall that $\max \emptyset = \varepsilon$):

$$\lambda^n \oplus \sum_{k \in J}^{\oplus} c_{n-k} \otimes \lambda^k = c_1 \otimes \lambda^{n-1} \oplus \sum_{k \in \bar{J}}^{\oplus} c_{n-k} \otimes \lambda^k,$$

where

$$c_k = \max \left\{ v; \sum_{B \in P_k(A)} p^+(B, v) \neq \sum_{B \in P_k(A)} p^-(B, v) \right\}, \quad k = 1, \ldots, n,$$

$$d_k = (-1)^k \left(\sum_{B \in P_k(A)} p^+(B, c_k) - \sum_{B \in P_k(A)} p^-(B, c_k) \right), \quad k = 1, \ldots, n$$

and

$$J = \{j; d_j > 0\}, \qquad \bar{J} = \{j; d_j < 0\}.$$

Theorem 5.3.19 (Cayley–Hamilton in max-algebra) *Every real square matrix A satisfies its max-algebraic characteristic equation.*

An application of this result in the theory of discrete-event dynamic systems can be found in Sect. 6.4.

In general it is not easy to find a max-algebraic characteristic equation for a matrix. However, as the next theorem shows, unlike for characteristic maxpolynomials it is relatively easy to do so for matrices over $T = \{0, -\infty\}$. Given a matrix $A = (a_{ij})$, the symbol 2^A will stand for the matrix $(2^{a_{ij}})$.

Theorem 5.3.20 [28] *If $A \in T^{n \times n}$ then the coefficients d_k in the max-algebraic characteristic equation for A are the coefficients at λ^{n-k} of the conventional characteristic polynomial for the matrix 2^A.*

Proof If $A \in T^{n \times n}$ then all finite c_k are 0. Note that if $k \in N$ and $\mathrm{maper}(B) = \varepsilon$ for all $B \in P_k(A)$ then the term $c_k \otimes \lambda^{n-k}$ does not appear on either side of the equation. If $B = (b_{ij}) \in T^{k \times k}$ then $p^+(B, 0)$ is the number of even permutations that select only zeros from B. The matrix 2^B is zero-one, zeros corresponding to $-\infty$ in B and ones corresponding to zeros in B. Thus $p^+(B, 0)$ is the number of even permutations that select only ones from 2^B. Similarly for $p^-(B, 0)$.

Since 2^B is zero-one, all terms in the standard determinant expansion of 2^B are either 1 (if the corresponding permutation is even and selects only ones), or -1 (if the corresponding permutation is odd and selects only ones), or 0 (otherwise). Hence $\det 2^B = p^+(B, 0) - p^-(B, 0)$. Since

$$d_k = (-1)^k \sum_{B \in P_k(A)} \left(p^+(B, 0) - p^-(B, 0) \right),$$

it follows that

$$d_k = (-1)^k \sum_{B \in P_k(A)} \det 2^B,$$

which is the coefficient at λ^{n-k} of the conventional characteristic polynomial of the matrix 2^A.   □

5.4 Exercises

Exercise 5.4.1 Find the standard form of

$$p(z) = 3 \otimes z^{2.5} \oplus 2 \otimes z^{4.7} \oplus 4 \otimes z^{6.2} \oplus 1 \otimes z^{8.3}$$

and then factorize it using RECTIFICATION and RESOLUTION. $[1 \otimes z^{2.5} \otimes (2 \oplus 1 \otimes z^{2.2} \oplus 3 \otimes z^{3.7} \oplus z^{5.8}); 1 \otimes z^{2.5} \otimes (-\frac{1}{3.7} \oplus x)^{3.7} \otimes (\frac{3}{2.1} \oplus z)^{2.1}]$

Exercise 5.4.2 Find the characteristic maxpolynomial and characteristic equation for the following matrices; factorize the maxpolynomial and check whether $\chi_A(x) = \text{LHS} \oplus \text{RHS}$ of the maxpolynomial equation:

(a) $A = \begin{pmatrix} 3 & -2 & 1 \\ 4 & 0 & 5 \\ 3 & 1 & 2 \end{pmatrix}$. $[\chi_A(x) = 9 \oplus 6 \otimes x \oplus 3 \otimes x^2 \oplus x^3 = (3 \oplus x)^3; \lambda^3 \oplus 9 = 3 \otimes \lambda^2 \oplus 6 \otimes \lambda]$

(b) $A = \begin{pmatrix} 1 & 0 & -3 \\ 2 & 3 & 1 \\ 4 & -2 & 0 \end{pmatrix}$. $[\chi_A(x) = 5 \oplus 4 \otimes x \oplus 3 \otimes x^2 \oplus x^3 = (1 \oplus x)^2 \otimes (3 \oplus x); \lambda^3 \oplus 4 \otimes \lambda = 3 \otimes \lambda^2 \oplus 5]$

(c) $A = \begin{pmatrix} 1 & 2 & 5 \\ -1 & 0 & 3 \\ 1 & 1 & 1 \end{pmatrix}$. $[\chi_A(x) = 6 \oplus 6 \otimes x \oplus 1 \otimes x^2 \oplus x^3 = (3 \oplus x)^2 \otimes (0 \oplus x); \lambda^3 = 1 \otimes \lambda^2 \oplus 6 \otimes \lambda]$

Exercise 5.4.3 A square matrix A is called strictly diagonally dominant if $\text{ap}(A) = \{\text{id}\}$. Find a formula for the characteristic equation of strictly diagonally dominant matrices. $[\lambda^n \oplus \delta_2 \otimes \lambda^{n-2} \oplus \delta_4 \otimes \lambda^{n-4} \oplus \cdots = \delta_1 \otimes \lambda^{n-1} \oplus \delta_3 \otimes \lambda^{n-3} \oplus \delta_5 \otimes \lambda^{n-5} \oplus \cdots$ where $\delta_k = $ the sum of k greatest diagonal values]

Chapter 6
Linear Independence and Rank. The Simple Image Set

We introduced a concept of linear independence in Sect. 3.3 in geometric terms. For finite systems of vectors (such as columns of a matrix) this definition reads:

Vectors $a_1, \ldots, a_n \in \overline{\mathbb{R}}^m$ are called *linearly dependent* (LD) if

$$a_k = \sum_{i \in N - \{k\}}^{\oplus} \alpha_i \otimes a_i$$

for some $k \in N$ and $\alpha_i \in \overline{\mathbb{R}}$, $i \in N - \{k\}$. The vectors $a_1, \ldots, a_n$ are *linearly independent* (LI) if they are not linearly dependent. We presented efficient methods for checking linear independence and for finding the coefficients of linear dependence in Sect. 3.3. That section also contains results on anomalies of linear independence. For these and other reasons various alternative concepts of linear independence have been studied. In most cases they would be equivalent to the above mentioned definition if formulated in linear algebra; however, in max-algebra they are nonequivalent.

We will discuss and compare two other concepts of independence in this chapter: strong linear independence [60] and Gondran–Minoux independence [98]. It will be of particular interest to see and compare these concepts in the setting of square matrices, that is, to compare regularity, strong regularity and Gondran–Minoux regularity.

6.1 Strong Linear Independence

In Chap. 3 we introduced the notation

$$S(A, b) = \{x \in \overline{\mathbb{R}}^n; A \otimes x = b\}$$

for $A \in \overline{\mathbb{R}}^{m \times n}$ and $b \in \overline{\mathbb{R}}^m$. Now we also denote for $A \in \overline{\mathbb{R}}^{m \times n}$:

$$T(A) = \{|S(A, b)| ; b \in \mathbb{R}^m\}.$$

P. Butkovič, *Max-linear Systems: Theory and Algorithms*,
Springer Monographs in Mathematics 151,
DOI 10.1007/978-1-84996-299-5_6, © Springer-Verlag London Limited 2010

The set $T(A)$ will be called the *type of* A. Note that the definition of $T(A)$ uses **finite** vectors b (see Sect. 3.1).

A set $C \subseteq \overline{\mathbb{R}}^n$ is said to be *max-convex* if $\alpha \otimes x \oplus \beta \otimes y \in C$ for every $x, y \in C, \alpha, \beta \in \overline{\mathbb{R}}$ with $\alpha \oplus \beta = 0$.

Lemma 6.1.1 *The set $S(A, b)$ is max-convex for every $A \in \overline{\mathbb{R}}^{m \times n}$ and $b \in \overline{\mathbb{R}}^m$.*

Proof $A \otimes (\alpha \otimes x \oplus \beta \otimes y) = \alpha \otimes A \otimes x \oplus \beta \otimes B \otimes y = (\alpha \oplus \beta) \otimes b = b.$ □

Regularity and linear independence are closely related to the number of solutions of max-linear systems. Similarly as in conventional linear algebra the number of solutions to a max-linear system can only be 0, 1 or ∞:

Theorem 6.1.2 [24] $|S(A, b)| \in \{0, 1, \infty\}$ *for any $A \in \overline{\mathbb{R}}^{m \times n}$ and $b \in \overline{\mathbb{R}}^m$.*

Proof We only need to prove that if a system $A \otimes x = b$ has more than one solution then it has an infinite number of solutions. Suppose $A \otimes x = b$, $A \otimes y = b$ and $x \neq y$ for some $x, y \in \overline{\mathbb{R}}^n$. Then by Lemma 6.1.1 $\alpha \otimes x \oplus \beta \otimes y$ is also a solution for any $\alpha, \beta \in \overline{\mathbb{R}}$ such that $\alpha \oplus \beta = 0$. Let without loss of generality $x_k < y_k$ and take $\alpha = 0$ and β between $x_k \otimes y_k^{-1}$ and 0. Then $\alpha \oplus \beta = 0, \alpha \otimes x \oplus \beta \otimes y$ is different from both x and y and there is an infinite number of such vectors.   □

For reasons explained at the beginning of Chap. 3, we will concentrate on doubly $\mathbb{R}$-astic matrices.

Lemma 6.1.3 *If $A \in \overline{\mathbb{R}}^{m \times n}$ is doubly $\mathbb{R}$-astic and $\{1, \infty\} \subseteq T(A)$ then $T(A) = \{0, 1, \infty\}$.*

Proof Let $A \in \overline{\mathbb{R}}^{m \times n}$ be doubly $\mathbb{R}$-astic and $\{1, \infty\} \subseteq T(A)$. Suppose that every column of A has only one finite entry. Then $m \leq n$ because there are no ε rows. If $m = n$ then A is a generalized permutation matrix for which $T(A) = \{1\}$. If $m < n$ then A contains an $m \times m$ generalized permutation submatrix. By choosing the remaining $n - m$ variables sufficiently, small we get $T(A) = \{\infty\}$.

Thus A has a column with at least two finite entries. Therefore the number of finite entries in A is more than n.

Suppose that in every row there is a finite entry unique in its column. Then $m \leq n$. If $m < n$ then a contradiction is obtained as above, hence $m = n$, there are only m finite entries, a contradiction. Therefore there is a row, say the kth, whose every finite entry is nonunique in its column. Hence there is a value, say c, such that $c \otimes a_{kj} < \sum_{i \neq k}^{\oplus} a_{ij}$ for all $j \in N$. Then by Corollary 3.1.2 the system $A \otimes x = b$ has no solution where $b_k = c^{-1}$ and $b_i = 0$ for $i \neq k$.   □

We can deduce a full list of types for doubly $\mathbb{R}$-astic matrices.

Proposition 6.1.4 *Each of the sets* $\{1\}, \{\infty\}, \{0, 1\}, \{0, \infty\}, \{0, 1, \infty\}$ *is the type of a doubly* $\mathbb{R}$*-astic matrix and there are no other types of doubly* $\mathbb{R}$*-astic matrices.*

Proof If $A \in \overline{\mathbb{R}}^{m \times n}$ is doubly $\mathbb{R}$-astic and $x \in \mathbb{R}^n$ then $A \otimes x \in \mathbb{R}^m$ and $x \in S(A, A \otimes x)$, thus $T(A) = \{0\}$ is obviously impossible. Due to Theorem 6.1.2 and Lemma 6.1.3, which includes the type $\{1, \infty\}$, it remains to show that the remaining five cases are all possible. The examples of doubly $\mathbb{R}$-astic matrices are (in the order stated in the proposition):

$$I, \begin{pmatrix} 0 & \varepsilon & 0 \\ \varepsilon & 0 & 0 \end{pmatrix}, \begin{pmatrix} 0 & \varepsilon \\ \varepsilon & 0 \\ 0 & 0 \end{pmatrix}, \begin{pmatrix} 0 & 0 \\ 0 & 0 \end{pmatrix}, \begin{pmatrix} 1 & 0 \\ 0 & 0 \end{pmatrix}. \qquad \square$$

The types $T(A)$ of matrices in conventional linear algebra are $\{1\}, \{\infty\}, \{0, 1\}, \{0, \infty\}$. They correspond to the following cases expressed using the linear-algebraic rank: $r(A) = m = n$, $r(A) = m < n$, $r(A) = n < m$, $r(A) < \min(m, n)$.

This comparison is even more striking if we consider finite matrices A in max-algebra with $m \geq 2$. For such matrices we can always find a b such that $A \otimes x = b$ has no solution and another b for which the system has an infinite number of solutions. More precisely:

Theorem 6.1.5 [24] $T(A)$ *is either* $\{0, \infty\}$ *or* $\{0, 1, \infty\}$ *for any* $A \in \mathbb{R}^{m \times n}, m \geq 2$.

Proof If $b = a_1$ then $A \otimes x = b$ has an infinite number of solutions x, where $x_1 = 0$ and $x_2, \ldots, x_n$ are sufficiently small.

If $b_1 < \min\{a_{1j} \otimes a_{ij}^{-1}; i \in M, i \neq 1, j \in N\}$ and $b_i = 0$ for all $i > 1$ then $A \otimes x = b$ has no solution since then $b_1^{-1} \otimes a_{1j} > a_{ij}$ for every $i \in M, i \neq 1$ and $j \in N$, thus $M_j = \{1\}$ for all $j \in N$, implying that there is no solution by Corollary 3.1.2. $\qquad \square$

We say that the columns of $A \in \overline{\mathbb{R}}^{m \times n}$ are *strongly linearly independent* (SLI) if $1 \in T(A)$, that is, the system $A \otimes x = b$ has a unique solution for at least one $b \in \mathbb{R}^m$; otherwise they are called *strongly linearly dependent*. A square matrix with strongly linearly independent columns is called *strongly regular*.

The next three statements are indicating some similarity between max-algebra and conventional linear algebra.

Lemma 6.1.6 *If* $A \in \overline{\mathbb{R}}^{m \times n}$ *has SLI columns then* A *is doubly* $\mathbb{R}$*-astic and if* y *is the unique solution to* $A \otimes x = b$ *for some* b *then* y *is finite.*

Proof The statement follows straightforwardly from the definitions. $\qquad \square$

Theorem 6.1.7 *A doubly* $\mathbb{R}$*-astic matrix* $A \in \overline{\mathbb{R}}^{m \times n}$ *has strongly linearly independent columns if and only if it contains a strongly regular* $n \times n$ *submatrix.*

Proof Suppose that A is doubly $\mathbb{R}$-astic and the unique solution to $A \otimes x = b$ is $\bar{x} \in \mathbb{R}^n$. It also follows from Corollary 3.1.3 that for every $j \in N$ there is at least one $i \in M_j$ such that $i \notin M_k$ for all $k \neq j$. Let us denote this index i by i_j (take any in the case of a tie). Consider the subsystem with row indices $i_1, i_2, \ldots, i_n$ (and with all column indices). This is an $n \times n$ system with a unique column maximum in every column and in every row. Hence again by Corollary 3.1.3 this system has a unique solution and so A contains an $n \times n$ strongly regular submatrix.

Suppose now that $A' \in \overline{\mathbb{R}}^{n \times n}$ is a strongly regular submatrix of A. Then there exists a $b' \in \mathbb{R}^n$ such that $A' \otimes x = b'$ has a unique solution, say z. Take $b = A \otimes z$. Then $b \in \mathbb{R}^m$ and the system $A \otimes x = b$ has a solution. If it had more than one solution then the subsystem $A' \otimes x = b'$ would also have more than one solution. Hence $A \otimes x = b$ has a unique solution, which completes the proof. $\qquad\square$

Corollary 6.1.8 *If a matrix $A \in \overline{\mathbb{R}}^{m \times n}$ has strongly linearly independent columns then $m \geq n$.*

The question of checking whether the columns of a given matrix are SLI may be of interest. It seems that currently no polynomial method for answering this question exists, see Chap. 11. On the other hand it is possible to check strong regularity of an $n \times n$ matrix in $O(n^3)$ time. This is presented in the next section and it enables us, using Theorem 6.1.7, to decide SLI (with nonpolynomial complexity) by checking $n \times n$ submatrices for strong regularity.

6.2 Strong Regularity of Matrices

6.2.1 A Criterion of Strong Regularity

Recall that $A \in \overline{\mathbb{R}}^{n \times n}$ is called strongly regular if the system $A \otimes x = b$ has a unique solution for some $b \in \mathbb{R}^n$. Our aim now is to characterize strongly regular matrices and the sets of vectors b for which $A \otimes x = b$ has a unique solution. Recall that a strongly regular matrix is doubly $\mathbb{R}$-astic (Lemma 6.1.6) and we will therefore assume throughout that A has this property. By the same lemma we need to consider $f(x) = A \otimes x$ as a mapping $\mathbb{R}^n \longrightarrow \mathbb{R}^n$. The set $\{A \otimes x; x \in \mathbb{R}^n\}$ is the set of images of this mapping. We will therefore call this set the *image set* and denote it by $\mathrm{Im}(A)$. Clearly, $\mathrm{Im}(A) \subseteq \mathrm{Col}(A)$ and $\mathrm{Col}(A) = \mathrm{Im}(A) \cup \{\varepsilon\}$ if A is finite.

We also define

$$S_A = \{b \in \mathbb{R}^n; A \otimes x = b \text{ has a unique solution}\}.$$

The set S_A is called the *simple image set* of A. The elements of $\mathrm{Im}(A)$ and S_A will be called *images* and *simple images*, respectively. Observe that $S_A \subseteq \mathrm{Im}(A)$ for every A and $S_A \neq \emptyset$ if and only if A is strongly regular.

A unique column maximum in every column and in every row is a feature that characterizes every uniquely solvable square system. To see this just realize that (see Corollary 3.1.3) $A \otimes x = b$ has a unique solution for $b \in \mathbb{R}^n$ if and only if the sets $M_1(A, b), \ldots, M_n(A, b)$ form a minimal covering of the set $N = \{1, \ldots, n\}$. It is easily seen that this is only possible if all the sets $M_1(A, b), \ldots, M_n(A, b)$ are one-element and pairwise-disjoint.

If a square matrix has a unique column maximum in every column and in every row then the column maxima determine a permutation of the set N whose weight is strictly greater than the weight of any other permutation and thus this matrix has strong permanent (see Sect. 1.6.4). In other words, if A is a square matrix and $A \otimes x = 0$ has a unique solution then A has strong permanent. However, a normalization of a system $A \otimes x = b$ means to multiply A by a diagonal matrix from the left. Lemma 1.6.32 states that this does not affect $\mathrm{ap}(A)$ and so we have proved:

Proposition 6.2.1 *If $A \in \overline{\mathbb{R}}^{n \times n}$ is strongly regular then A has strong permanent.*

The converse is also true (see the next theorem) and therefore verifying that a matrix is strongly regular is converted to the checking that it has strong permanent. This can be done by checking that a digraph is acyclic (see Sect. 1.6.4) and therefore is solvable using $O(n^3)$ operations.

Theorem 6.2.2 (Criterion of strong regularity) *A square matrix over $\overline{\mathbb{R}}$ is strongly regular if and only if it has strong permanent.*

This result has originally been proved in [36] for finite matrices over linearly ordered commutative groups; however, it is the aim of this subsection to present a simpler proof, which is similar to that in [26]. We refer to terminology introduced in Sect. 1.6.4 and start with a few lemmas and a theorem that may be of interest also on their own:

Lemma 6.2.3 *If $A \in \overline{\mathbb{R}}^{n \times n}$ is strongly regular then $\mathrm{maper}(A)$ is finite.*

Proof The statement immediately follows from Proposition 6.2.1.   □

Lemma 6.2.4 *If $A \approx B$ then A is strongly regular if and only if B is strongly regular.*

Proof Permutations of the rows and columns of A as well as $\otimes$ multiplying them by finite constants does not affect the existence of a unique solution to a max-linear system.   □

Due to Corollary 1.6.38 and (1.29) we may now assume without loss of generality that the doubly $\mathbb{R}$-astic matrix whose strong regularity we wish to check is strongly definite.

Lemma 6.2.5 *If $A = (a_{ij}) \in \overline{\mathbb{R}}^{n \times n}$ is strongly definite and $b \in S_A$ then $B = (b_i^{-1} \otimes a_{ij})$ has column maxima only on the diagonal.*

Proof $A \otimes x = b$ has a unique solution, thus by Corollary 3.1.3 the column maxima are unique and determine a permutation, say π. Hence, if $\pi \neq \mathrm{id}$ then $w(\pi, A) > w(\mathrm{id}, A)$, which is a contradiction since A is strongly definite and therefore diagonally dominant by (1.29). □

If A is a square matrix then $\widetilde{A}$ will stand for the matrix obtained from A after replacing all diagonal entries by ε.

Theorem 6.2.6 *If $A \in \overline{\mathbb{R}}^{n \times n}$ is strongly definite then*

$$S_A = \left\{ b \in \mathbb{R}^n ; \widetilde{A} \otimes b \le g \otimes b \text{ for some } g < 0 \right\}.$$

Proof Let $A = (a_{ij}) \in \overline{\mathbb{R}}^{n \times n}$ be strongly definite and $\widetilde{A} = (\widetilde{a}_{ij})$. Then we have:

b is a simple image of A

$\Longleftrightarrow$ $A \otimes x = b$ has a unique solution

$\Longleftrightarrow$ $B = \left(b_i^{-1} \otimes a_{ij} \right)$ has column maxima only on the diagonal

(Lemma 6.2.5)

$\Longleftrightarrow$ $\left(b_i^{-1} \otimes a_{ij} \otimes b_j \right)$ is strictly normal

$\Longleftrightarrow$ $(\forall i \neq j) \left(b_i^{-1} \otimes a_{ij} \otimes b_j < 0 \right)$

$\Longleftrightarrow$ $(\exists g < 0)\, (\forall i, j) \left(b_i^{-1} \otimes \widetilde{a}_{ij} \otimes b_j \le g \right)$

$\Longleftrightarrow$ $(\exists g < 0)\, (\forall i, j) \left(\widetilde{a}_{ij} \otimes b_j \le g \otimes b_i \right)$

$\Longleftrightarrow$ $(\exists g < 0)\, (\forall i) \left(\max_j \left(\widetilde{a}_{ij} \otimes b_j \right) \le g \otimes b_i \right)$

$\Longleftrightarrow$ $(\exists g < 0) \left(\widetilde{A} \otimes b \le g \otimes b \right).$ □

Corollary 6.2.7 *If $A \in \overline{\mathbb{R}}^{n \times n}$ is strongly definite then A is strongly regular if and only if the set*

$$\left\{ b \in \mathbb{R}^n ; \widetilde{A} \otimes b \le g \otimes b \right\}$$

is nonempty for some $g < 0$.

For the proof of our principal result, Theorem 6.2.2, we need to prove a few more properties:

Lemma 6.2.8 *If $A \in \overline{\mathbb{R}}^{n \times n}$ is strongly definite then A has strong permanent if and only if every cycle in $D_{\widetilde{A}}$ is negative.*

Proof If A is strongly definite then $w(\pi, A) \leq 0 = w(\mathrm{id}, A)$ for every $\pi \in P_n$. If $\pi \in P_n$ then $w(\pi, A) = w(\pi_1, A) \otimes \cdots \otimes w(\pi_k, A)$, where $\pi_1, \ldots, \pi_k$ are the constituent cycles. Hence A has strong permanent if and only if all cycles of length two or more in D_A are negative. This is equivalent to saying that all cycles in $D_{\widetilde{A}}$ are negative. $\qquad \square$

Lemma 6.2.9 *If $A \in \overline{\mathbb{R}}^{n \times n}$ and $B = A \otimes Q$ where Q is a generalized permutation matrix then $S_A = S_B$. That is the simple image set of a matrix is unaffected by adding constants to its columns.*

Proof $(A \otimes Q) \otimes x = A \otimes (Q \otimes x)$ hence

$$(A \otimes Q) \otimes x = b$$

has a unique solution if and only if

$$A \otimes z = b$$

has a unique solution and $x = Q^{-1} \otimes z$. $\qquad \square$

We are ready to prove Theorem 6.2.2.

Proof By Proposition 1.6.40 and Lemma 6.2.9 we may assume without loss of generality that A is strongly definite.

Due to Proposition 6.2.1 it remains to prove the "if" part.

By Lemma 6.2.8 every cycle in $D_{\widetilde{A}}$ is negative. Hence $\lambda(\widetilde{A}) < 0$.

If $\widetilde{A} \neq \varepsilon$ then by Theorem 1.6.18 for any $g \geq \lambda(\widetilde{A})$ and $g > \varepsilon$ there is a solution $x \in \mathbb{R}^n$ to $\widetilde{A} \otimes x \leq g \otimes x$. If $\widetilde{A} = \varepsilon$ then $\widetilde{A} \otimes x \leq g \otimes x$ is satisfied by any g and $x \in \mathbb{R}^n$. Hence the statement now follows by Corollary 6.2.7 by taking $g = \lambda(\widetilde{A})$ if $\lambda(\widetilde{A}) > \varepsilon$ and any $g \in \mathbb{R}$, $g < 0$, otherwise. $\qquad \square$

Corollary 6.2.10 *If $A \in \overline{\mathbb{R}}^{n \times n}$ is strongly definite then A is strongly regular if and only if $\lambda(\widetilde{A}) < 0$.*

Due to Theorems 1.6.18 and 6.2.6 we also have:

Corollary 6.2.11 *Let $A \in \overline{\mathbb{R}}^{n \times n}$ be strongly definite and strongly regular. If $\widetilde{A} \neq \varepsilon$ then*

$$S_A = \left\{ \Delta \left(g^{-1} \otimes \widetilde{A} \right) \otimes x; x \in \mathbb{R}^n, \lambda(\widetilde{A}) \leq g < 0, g \neq \varepsilon \right\}.$$

If $\widetilde{A} = \varepsilon$ then $S_A = \mathbb{R}^n$.

Example 6.2.12 Let

$$A = \begin{pmatrix} 1 & 2 & 3 \\ 1 & 0 & 5 \\ 5 & 6 & 3 \end{pmatrix}.$$

Then the weights of all permutations are 4, 12, 10, 8, 12, 6; $|ap(A)| = 2$, A does not have strong permanent and hence is not strongly regular. For matrices of higher orders this would be decided algorithmically by checking whether the associated digraph is acyclic after a transformation to a normal form. We illustrate on the current matrix how this might be done:

$$A \longrightarrow B = \begin{pmatrix} 2 & 3 & 1 \\ 0 & 5 & 1 \\ 6 & 3 & 5 \end{pmatrix}$$

$$\longrightarrow C = \begin{pmatrix} 0 & -2 & -4 \\ -2 & 0 & -4 \\ 4 & -2 & 0 \end{pmatrix}$$

$$\longrightarrow G = \begin{pmatrix} 0 & 0 & 0 \\ -4 & 0 & -2 \\ 0 & -4 & 0 \end{pmatrix}.$$

Here B is a diagonally dominant matrix obtained from A by moving the first column to become the last; C is strongly definite, obtained from B by subtracting the diagonal elements from their columns and G is normal, obtained from C using the Hungarian method by adding 4 and 2 to the first and second rows, respectively and subtracting 4 and 2 from the first and second columns, respectively. The digraph $Z_{\tilde{G}}$ contains arcs $(1, 2)$, $(1, 3)$ and $(3, 1)$ and thus also the cycle $(1, 3, 1)$. This confirms that A does not have strong permanent (Theorem 1.6.39).

Example 6.2.13 Let

$$A = \begin{pmatrix} 1 & 2 & 3 \\ 1 & 0 & 5 \\ 5 & 4 & 3 \end{pmatrix}.$$

Then the weights of all permutations are 4, 12, 8, 8, 10, 6; $|ap(A)| = 1$, A has strong permanent and hence is strongly regular. The unique optimal permutation is $\pi = (1, 2, 3)$. As in the previous example we now illustrate how this would be done algorithmically by transforming A to a normal form:

$$A \longrightarrow B = \begin{pmatrix} 2 & 3 & 1 \\ 0 & 5 & 1 \\ 4 & 3 & 5 \end{pmatrix}$$

$$\longrightarrow C = \begin{pmatrix} 0 & -2 & -4 \\ -2 & 0 & -4 \\ 2 & -2 & 0 \end{pmatrix}$$

$$\longrightarrow G = \begin{pmatrix} 0 & 0 & -2 \\ -4 & 0 & -4 \\ 0 & -2 & 0 \end{pmatrix}.$$

Here $B = A \otimes Q_1$, where

$$Q_1 = \begin{pmatrix} \varepsilon & \varepsilon & 0 \\ 0 & \varepsilon & \varepsilon \\ \varepsilon & 0 & \varepsilon \end{pmatrix},$$

$C = B \otimes Q_2$ where $Q_2 = \mathrm{diag}(-2, -5, -5)$,
$G = P \otimes C \otimes Q_3$, where $P = \mathrm{diag}(2, 0, 0)$ and $Q_3 = \mathrm{diag}(-2, 0, 0)$.

The digraph $Z_{\widetilde{G}}$ contains arcs $(1, 2)$ and $(3, 1)$ and is therefore acyclic which confirms that A is strongly regular.

Since $C = A \otimes Q$, where Q is the generalized permutation matrix $Q_1 \otimes Q_2$, by Lemma 6.2.9 we have $S_A = S_C$ and we can find a simple image of A as suggested by Corollary 6.2.11: Set $g = \lambda(\widetilde{C}) = -1$ (say) and calculate

$$\Delta\left(g^{-1} \otimes \widetilde{C}\right) = \Delta \begin{pmatrix} \varepsilon & -1 & -3 \\ -1 & \varepsilon & -3 \\ 3 & -1 & \varepsilon \end{pmatrix} = \begin{pmatrix} 0 & -1 & -3 \\ 0 & -1 & -3 \\ 3 & 2 & 0 \end{pmatrix}.$$

The column space of $\Delta(g^{-1} \otimes \widetilde{C})$ is one-dimensional and every multiple of the vector $(0, 0, 3)^T$ is a simple image of both C and A.

6.2.2 The Simple Image Set

The simple image set (SIS) of strongly definite matrices is fully described by Theorem 6.2.6 and Corollary 6.2.11. We will now present some further properties of simple image sets [26], in particular their relation to eigenspaces.

Let us first consider strongly definite matrices. Since

$$A^{k+1} \otimes x = A^k \otimes (A \otimes x),$$

we have $\mathrm{Im}(A^{k+1}) \subseteq \mathrm{Im}(A^k)$ for every k natural. It follows then from Proposition 1.6.12 that for a strongly definite matrix A

$$\mathrm{Im}(A) \supseteq \mathrm{Im}\left(A^2\right) \supseteq \mathrm{Im}\left(A^3\right) \supseteq \cdots$$
$$\supseteq \mathrm{Im}\left(A^{n-1}\right) = \mathrm{Im}\left(A^n\right) = \mathrm{Im}\left(A^{n+1}\right) = \cdots = V(A),$$

see Fig. 6.1. It turns out that if A is strongly definite then A is strongly regular if and only if $V(A)$ has nonempty interior, which is then S_A. More precisely, we have the following result, whose proof is omitted here.

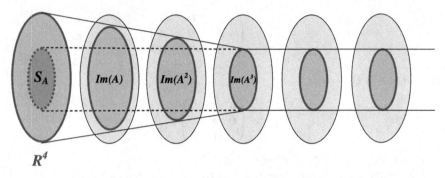

R^4

Fig. 6.1 $S_A = \text{int}(V(A))$

Theorem 6.2.14 [26] *If $A \in \mathbb{R}^{n \times n}$ is strongly definite and strongly regular then $S_A = \text{int}(V(A))$ or, equivalently, $S_A = \text{int}(\text{Im}(A^k))$ for every $k \geq n - 1$.*

Recall that by Proposition 1.6.40 for any square matrix $A \in \mathbb{R}^{n \times n}$ there is a generalized permutation matrix Q so that $A \otimes Q$ is strongly definite. Using Lemma 6.2.9 we can remove the assumption that A is strongly definite and deduce the following:

Corollary 6.2.15 *If $A \in \mathbb{R}^{n \times n}$ is strongly regular then*

$$S_A = \text{int}(V(A \otimes Q))$$

or, equivalently, $S_A = \text{int}(\text{Im}((A \otimes Q)^k))$ for every $k \geq n - 1$, where Q is any permutation matrix such that $A \otimes Q$ is strongly definite.

Example 6.2.16 Consider the matrix A of Example 6.2.13. By Corollary 6.2.15 we deduce that $S_A = \text{int}(V(C))$.

Lemma 6.2.17 *If $A \in \overline{\mathbb{R}}^{n \times n}$ is strongly definite then the set of finite eigenvectors of A is convex.*

Proof If A is strongly definite then $\Lambda(A) = \{0\}$. By Lemma 1.6.14 we then have $V(A) = V_0^*(A)$. The latter is convex by Lemma 1.6.20.   □

Theorem 6.2.18 *Let $A \in \mathbb{R}^{n \times n}$ be strongly definite. Then A is strongly regular if and only if the topological dimension of $V(A)$ is n.*

Proof Let $A \in \mathbb{R}^{n \times n}$ be strongly definite. It follows from Theorem 6.2.14 that A is strongly regular if and only if $\text{int}(V(A)) \neq \emptyset$. But $V(A)$ for strongly definite matrices is convex (Lemma 6.2.17) and thus this property is equivalent to the topological dimension of $V(A)$ being n [125].   □

Recall that pd(A) stands for the (max-algebraic) dimension of the principal eigenspace of A, that is, the maximal number of nonequivalent fundamental eigenvectors of A or, equivalently, the number of critical components of $C(A)$.

Theorem 6.2.19 *If $A \in \mathbb{R}^{n \times n}$ be strongly definite. Then A is strongly regular if and only if* $\mathrm{pd}(A) = n$.

Proof If A is not strongly regular then D_A and therefore also $D_{\Gamma(A)}$ contains a zero cycle of length two or more. By Theorem 4.3.3 then at least two columns of $\Gamma(A)$ are multiples of each other. Hence $\mathrm{pd}(A) < n$.

If A is strongly regular then the critical cycles are exactly loops at all nodes and so $C(A)$ has n critical components. Hence $\mathrm{pd}(A) = n$.   □

For a strongly definite matrix $A \in \mathbb{R}^{n \times n}$ it follows from Theorems 6.2.18 and 6.2.19 that $\mathrm{pd}(A) = n$ if and only if the topological dimension of the principal eigenspace is n. In fact equality holds between these two types of dimension, see Exercise 6.6.5.

6.2.3 Strong Regularity in Linearly Ordered Groups

In this subsection we use terminology and notation introduced in Sect. 1.4.

A generalization of Theorem 6.2.2 to linearly ordered commutative groups is straightforward provided that the group is radicable. It is less straightforward but still possible to prove this theorem when the underlying group is dense, but it is not true for sparse groups such as $\mathcal{G}_3$. For instance the matrix $A = \left(\begin{smallmatrix} 1 & 0 \\ 0 & 0 \end{smallmatrix}\right)$ over the additive group of integers has strong permanent but it is not possible to add integer constants to its rows so that both column maxima would be strict and in different rows.

An alternative criterion for sparse groups is based on the existence of the smallest positive element, which we denote here as α. Observe that if we denote for $A \in \overline{\mathbb{R}}^{n \times n}$ and $g < 0$:

$$U_g(A) = \{b \in \mathbb{R}^n; A \otimes b \leq g \otimes b\}$$

then

$$g_1 \leq g_2 \Longrightarrow U_{g_1}(A) \subseteq U_{g_2}(A).$$

Using Corollary 6.2.7 we deduce:

Theorem 6.2.20 *If $\mathcal{G}$ is a sparse linearly ordered commutative group, α is the smallest positive element of this group and A is an $n \times n$ strongly definite matrix over $\mathcal{G}$ then A is strongly regular if and only if $U_{\alpha^{-1}}(\widetilde{A}) \neq \emptyset$ or, equivalently, there is no positive cycle in $D_{\alpha \otimes \widetilde{A}}$.*

Example 6.2.21 Consider the matrix $A = \left(\begin{smallmatrix} 0 & 0 \\ -2 & 0 \end{smallmatrix}\right)$. In $\mathcal{G}_3$ (see Sect. 1.4) $\alpha = 1$ and $\alpha \otimes \widetilde{A} = \left(\begin{smallmatrix} \varepsilon & 1 \\ -1 & \varepsilon \end{smallmatrix}\right)$, hence A is strongly regular. In $\mathcal{G}_4$ we have $\alpha = 2$ and $\alpha \otimes \widetilde{A} = \left(\begin{smallmatrix} \varepsilon & 2 \\ 0 & \varepsilon \end{smallmatrix}\right)$, hence there is a positive cycle in $D_{\alpha \otimes \widetilde{A}}$ and thus A is not strongly regular.

6.2.4 Matrices Similar to Strictly Normal Matrices

We continue our analysis in the principal interpretation and discuss the question raised in Sect. 1.6.4: Which matrices are similar to strictly normal ones? Recall that every matrix is similar to a normal matrix (Theorem 1.6.37) and normal matrices are strongly definite.

So, assume that $A = (a_{ij}) \in \overline{\mathbb{R}}^{n \times n}$ is strongly definite and that $b \in \mathbb{R}^n$ is a vector for which the system $A \otimes x = b$ has a unique solution, thus $\mathrm{ap}(A) = \{\mathrm{id}\}$. Let

$$B = \mathrm{diag}(b_1^{-1}, b_2^{-1}, \ldots, b_n^{-1}) \otimes A \otimes \mathrm{diag}(b_1, b_2, \ldots, b_n).$$

Then $\mathrm{ap}(B) = \mathrm{ap}(A) = \{\mathrm{id}\}$ and B has a unique column maximum in every row and column and it also has zero diagonal. Hence B is strictly normal. We deduce that strong regularity is a sufficient condition for a matrix to be similar to a strictly normal one.

Conversely, if A is strongly definite and

$$\mathrm{diag}(c_1, \ldots, c_n) \otimes A \otimes \mathrm{diag}(b_1, b_2, \ldots, b_n)$$

is strictly normal then $c_i \otimes b_i = 0$ for all $i \in N$, yielding $c_i = b_i^{-1}$ for all $i \in N$. Therefore in

$$\mathrm{diag}(b_1^{-1}, b_2^{-1}, \ldots, b_n^{-1}) \otimes A$$

all column maxima are on the diagonal only and thus $A \otimes x = b$ has a unique solution. We have proved:

Theorem 6.2.22 *Let $A \in \overline{\mathbb{R}}^{n \times n}$ be strongly definite. Then*

$$\mathrm{diag}(c_1, \ldots, c_n) \otimes A \otimes \mathrm{diag}(b_1, b_2, \ldots, b_n)$$

is strictly normal if and only if $c_i = b_i^{-1}$ for all $i \in N$ and the system $A \otimes x = b$ has a unique solution.

Corollary 6.2.23 *Let $A \in \overline{\mathbb{R}}^{n \times n}$ be a matrix with finite $\mathrm{maper}(A)$. Then A is similar to a strictly normal matrix if and only if A has strong permanent or, equivalently, if and only if A is strongly regular.*

6.3 Gondran–Minoux Independence and Regularity

Another concept of linear independence in max-algebra is Gondran–Minoux independence. In this section we restrict our attention to finite matrices. We say that the vectors $a_1, \ldots, a_n \in \mathbb{R}^m$ are *Gondran–Minoux dependent* (GMD) if

$$\sum_{j \in S}^{\oplus} \alpha_j \otimes a_j = \sum_{j \in T}^{\oplus} \alpha_j \otimes a_j \tag{6.1}$$

holds for some $\alpha_1, \ldots, \alpha_n \in \mathbb{R}$ and two nonempty, disjoint subsets S and T of the set N. If the vectors are not GMD then we call them *Gondran–Minoux independent (GMI)*. A square matrix with Gondran–Minoux independent columns is called *Gondran–Minoux regular*.

In the formulation of a Gondran–Minoux regularity criterion below we use the symbols introduced in Sect. 1.6.4. This result was first presented in [99]. It was later revisited in [25].

Theorem 6.3.1 (Gondran–Minoux) *Let $A \in \mathbb{R}^{n \times n}$. Then the following hold*:

(a) *A is Gondran–Minoux regular if and only if either* $\mathrm{ap}(A) \subseteq P_n^+$ *or* $\mathrm{ap}(A) \subseteq P_n^-$
    (equivalently, either $\mathrm{ap}^+(A) = \emptyset$ *or* $\mathrm{ap}^-(A) = \emptyset$*);*
(b) *If permutations* $\pi \in \mathrm{ap}^+(A), \sigma \in \mathrm{ap}^-(A)$ *are known then the sets S and T and all α_j in (6.1) can be found using $O(n^2)$ operations.*

Proof (a) Suppose that (6.1) holds for nonempty, disjoint subsets S and T of the set N and $\alpha_1, \ldots, \alpha_n \in \mathbb{R}$. We prove that $\mathrm{ap}^+(A) \neq \emptyset$ and $\mathrm{ap}^-(A) \neq \emptyset$. The converse will follow from part (b).

By Lemma 1.6.43 it is sufficient to prove that $\mathrm{ap}^+(B) \neq \emptyset$ and $\mathrm{ap}^-(B) \neq \emptyset$ for some matrix B, $A \approx B$. Let us permute the columns of the matrix $A \otimes \mathrm{diag}(\alpha_1, \ldots, \alpha_n)$ so that $S = \{1, \ldots, k\}$ for some k. Denote the obtained matrix by $A' = (a'_{ij})$ and its columns by $a'_1, \ldots, a'_n$. Then

$$\sum_{j \leq k}^{\oplus} a'_j = \sum_{j > k}^{\oplus} a'_j.$$

Let us denote this vector by $c = (c_1, \ldots, c_n)^T$. Let $B = (b_{ij})$ be any matrix obtained from the matrix $(c_i^{-1} \otimes a'_{ij})$ by permuting its rows so that $\mathrm{id} \in \mathrm{ap}(B)$. Then B has the following two properties:

$$b_{ij} \leq 0 \quad \text{for all } i, j \in N$$

and

$$(\forall i)\,(\exists j_1 \leq k)\,(\exists j_2 > k)\, b_{ij_1} = 0 = b_{ij_2}.$$

We construct a sequence of indices as follows: Let $i_1 = 1$; if i_r has already been defined and $i_r \leq k$ then i_{r+1} is any $j > k$ such that $b_{i_r, j} = 0$ and if $i_r > k$ then i_{r+1} is any $j \leq k$ such that $b_{i_r, j} = 0$. By finiteness of N, $i_r = i_s$ for some r, s and $s < r$. Let r, s be the first such indices and set

$$L = \{i_s, i_{s+1}, \ldots, i_{r-1}\}.$$

Clearly, if $i_s \leq k$ then $i_{s+1} > k, i_{s+2} \leq k, \ldots$ and hence (using a similar reason if $i_s > k$) the size of L is even. Set $\pi(i_t) = i_{t+1}$ for $t = s, s+1, \ldots, r-1$ and $\pi(i) = i$ for $i \in N - L$. Hence

$$w(\pi, B) = \prod_{i \notin L}^{\otimes} b_{ii} \otimes \prod_{i \in L}^{\otimes} b_{i, \pi(i)}$$

$$= \prod_{i \notin L}^{\otimes} b_{ii} \otimes \prod_{i \in L}^{\otimes} 0$$

$$\geq \prod_{i \in N}^{\otimes} b_{ii}$$

$$= w(\mathrm{id}, B)$$

$$\geq w(\pi, B).$$

Hence $\pi \in \mathrm{ap}(B)$ and $\pi \in P_n^-$, thus $\pi \in \mathrm{ap}^-(B)$ an $\mathrm{id} \in \mathrm{ap}^+(B)$.

(b) Suppose now that $\pi \in \mathrm{ap}^+(A), \sigma \in \mathrm{ap}^-(A)$ are known. By Theorem 1.6.35 a matrix $A' \approx A$ with $\mathrm{maper}(A') = 0$ and $A' \leq 0$ can be found in $O(n)$ time. We can then permute the columns of A' in $O(n^2)$ time so that for the obtained matrix A'' we have $\pi' = \mathrm{id} \in \mathrm{ap}(A'')$ and thus A'' is normal. A permutation $\sigma' \in \mathrm{ap}^-(A'')$ can be derived from σ in $O(n)$ time. At least one of the constituent cyclic permutations of σ' is of odd parity and thus of even length (see Sect. 1.6.4) and it can also be found in $O(n)$ time. By a simultaneous permutation of the rows and columns of A'' in $O(n^2)$ time we produce a matrix where this odd cycle is $(1, 2, \ldots, k)$ for some even integer $k \geq 2$. We denote the obtained normal matrix as $B = (b_{ij})$. The matrix B has the form:

	1	2	3	...	k	k+1	...	n
1	0	0						
2		0	0					
3			0					
⋮				⋱	0			
k	0				0			
k+1						0		
⋮							⋱	
n								0

Let us assign indices $1, 3, \ldots, k-1$ to S; $2, 4, \ldots, k$ to T and set $\alpha_1 = \cdots = \alpha_k = 0$. If $k = n$ then (6.1) is satisfied for B. Suppose now that $k < n$. We will set all $\alpha_{k+1}, \ldots, \alpha_n$ to certain nonpositive values and therefore (6.1) will hold for the first k equations independently of the choice of these values and of the assignment of the columns to S and T. To ensure equality in the rows $k + 1, \ldots, n$ we first compute for all $i = k+1, \ldots, n$:

$$L_i = \sum_{j \in S}^{\oplus} b_{ij} \otimes \alpha_j$$

and

$$R_i = \sum_{j \in T}^{\oplus} b_{ij} \otimes \alpha_j.$$

Let us denote

$$I = \{i > k; L_i \neq R_i\}.$$

If $I = \emptyset$ and $S \cup T = N$ then we have (6.1) for B. If $S \cup T \neq N$ we set for every $j \in N - S \cup T$:

$$\alpha_j = \min_i L_i$$

and assign j to S or T arbitrarily; the statement then follows for B.

If $I \neq \emptyset$ then let $s \in I$ be any index satisfying

$$L_s \oplus R_s = \max_{i \in I}(L_i \oplus R_i). \tag{6.2}$$

Set $S' = S \cup \{s\}$ and $T' = T$ if $L_s < R_s$ and set $S' = S$ and $T' = T \cup \{s\}$ if $L_s > R_s$. In both cases take $\alpha_s = L_s \oplus R_s$. Let us denote

$$L'_i = \sum_{j \in S'}^{\oplus} b_{ij} \otimes \alpha_j$$

and

$$R'_i = \sum_{j \in T'}^{\oplus} b_{ij} \otimes \alpha_j.$$

Since

$$b_{ss} \otimes \alpha_s = \alpha_s = L_s \oplus R_s$$

we then get $L'_s = R'_s$. At the same time

$$b_{is} \otimes \alpha_s \leq \alpha_s = L_s \oplus R_s \tag{6.3}$$

holds for all $i > k$ and therefore $L_i = R_i \geq L_s \oplus R_s$ implies $L'_i = R'_i$.

Let

$$I' = \left\{i > k; L'_i \neq R'_i\right\}.$$

As above we assume that $I' \neq \emptyset$. Let $q \in I'$ be defined by

$$L'_q \oplus R'_q = \max_{i \in I'} \left(L'_i \oplus R'_i\right).$$

Then

$$L'_q \oplus R'_q \leq L_s \oplus R_s \tag{6.4}$$

because either $L'_q \oplus R'_q = b_{qs} \otimes \alpha_s$ and then (6.4) follows from (6.3) or, $L'_q \oplus R'_q > b_{qs} \otimes \alpha_s$, implying $L'_i = L_i$ and $R'_i = R_i$, yielding $q \in I$ and thus (6.4) follows from (6.2). This also shows that if we continue in this way after resetting $S' \longrightarrow S, T' \longrightarrow T, L'_i \longrightarrow L_i, R'_i \longrightarrow R_i, I' \longrightarrow I, q \longrightarrow s$ then the process will be monotone ($L_s \oplus R_s$ will be nonincreasing) and therefore once $L_i = R_i \geq L_s \oplus R_s$, it will always imply $L_i' = R_i'$ (but note that if $L_i = R_i < L_s \oplus R_s$ then the ith equation may be violated at the end of the current iteration). Hence after at most $n - k$ repetitions we will have $I = \emptyset$ and we proceed as explained above for this case.

All computations necessary for assigning j and setting α_j are $O(n)$, hence the overall computational complexity is $O(n^2)$. In order to get (6.1) for A, we only need to carry out the inverse permutations of the rows and columns of B to get this identity for A'' and A' and similarly the inverse transformation of A', which will yield the result for A. All these operations are $O(n^2)$.   □

Example 6.3.2 We illustrate the method for finding the decomposition (6.1) presented in the previous proof on the following 9×9 matrix where the transformation to B has already been made (with an even cycle of length $k = 4$). Note that the entries in the last row are α_j.

S	T	S	T					
0	0							
	0	0						
		0	0					
0			0					
−8	−3	−4	−4	0	−1	−1		
−5	−8	−6	−5	−1	0	−6		
−1	−3	−7	−4	0	−2	0		
0	0	−3	−5	−2	−4	−3	0	
−6	−7	−8	−7	−2	−1	−4		0
0	0	0	0	−2	−3	−1	−4	−4

By applying the procedure we obtain successively:

$$I = \{5, 7, 9\}, \quad s = 7, \quad T := T \cup \{7\}, \quad \alpha_7 = -1,$$

$$I = \{5, 9\}, \quad s = 5, \quad S := S \cup \{5\}, \quad \alpha_5 = -2,$$

$$I = \{6, 9\}, \quad s = 6, \quad T := T \cup \{6\}, \quad \alpha_6 = -3,$$

$$I = \emptyset, \quad S := S \cup \{8, 9\}, \quad \alpha_8 = \alpha_9 = -4 \quad \text{(say)}.$$

Hence $S = \{1, 3, 5, 8, 9\}$, $T = \{2, 4, 6, 7\}$. The sets L_i, R_i develop in individual iterations as follows:

L_i	R_i	L_i	R_i	L_i	R_i	L_i	R_i
0	0	0	0	0	0	0	0
0	0	0	0	0	0	0	0
0	0	0	0	0	0	0	0
0	0	0	0	0	0	0	0
−4	−3	−4	−2	−2	−2	−2	−2
−5	−5	−5	−5	−3	−5	−3	−3
−1	−3	−1	−1	−1	−1	−1	−1
0	0	0	0	0	0	0	0
−6	−7	−6	−5	−4	−5	−4	−4

As a consequence of Theorems 6.3.1 and 1.6.44 we have:

Corollary 6.3.3 [25] *Let $A \in \mathbb{R}^{n \times n}$ and let B be any normal form of A. Then A is Gondran–Minoux regular if and only if Z_B does not contain an even cycle.*

Using Theorem 1.6.44 and the subsequent Remark 1.6.45 we deduce:

Corollary 6.3.4 *The problem of deciding whether a given matrix $A \in \mathbb{R}^{n \times n}$ is Gondran–Minoux regular can be solved using $O(n^3)$ operations.*

Corollary 6.3.5 *Every strongly regular matrix is Gondran–Minoux regular.*

The analogue of Theorem 6.1.7 is not true for Gondran–Minoux independence; this is demonstrated by a counterexample in [4]. In this example a 6×7 matrix is presented whose rows are Gondran–Minoux independent but none of the 6×6 submatrices is Gondran–Minoux regular.

Nevertheless we can prove an analogue of Corollary 6.1.8:

Theorem 6.3.6 *If a matrix $A \in \mathbb{R}^{m \times n}$ has Gondran–Minoux independent columns then $m \geq n$.*

Proof Let $A = (a_{ij}) \in \mathbb{R}^{m \times n}$ and $m < n$. We shall show that A has Gondran–Minoux dependent columns.

Since the Gondran–Minoux independence of columns is not affected by $\otimes$ multiplying the columns by constants, we may assume without loss of generality that the last row of A is zero. Let B be an $m \times m$ submatrix of A with greatest value of maper(B). We may assume that B consists of the first m columns of A and that id $\in$ ap(B) (if necessary, we appropriately permute the columns of A). Let C be the $n \times n$ matrix obtained by adding $n - m$ zero rows to A. Then clearly maper(C) = maper(B) and ap(C) contains any permutation that is an extension of id from ap(B) to a permutation of N. As A already had one zero row and we have added at least another one, C has at least two zero rows, thus ap(C) contains at least one pair of permutations of different parities (see Fig. 6.2).

Fig. 6.2 To Theorem 6.3.6

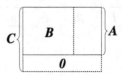

Hence, by Theorem 6.3.1 C is not Gondran–Minoux regular and if we denote the columns of C by $c_1, \ldots, c_n$ then

$$\sum_{j \in S}^{\oplus} \alpha_j \otimes c_j = \sum_{j \in T}^{\oplus} \alpha_j \otimes c_j$$

holds for some real numbers α_j and two nonempty, disjoint subsets S and T of the set N. This vector equality restricted to the first m components then yields the Gondran–Minoux dependence of the columns of A.   □

6.4 An Application to Discrete-event Dynamic Systems

In this section we present an application of the max-algebraic Cayley–Hamilton Theorem (Theorem 5.3.19) and Gondran–Minoux Theorem (Theorem 6.3.1) in the theory of discrete-event dynamic systems.

Given $A \in \overline{\mathbb{R}}^{n \times n}$ and $b, c \in \overline{\mathbb{R}}^n$, the sequence $\{g_j\}_{j=0}^{\infty}$, where

$$g_j = c^T \otimes A^j \otimes b$$

for all $j = 0, 1, 2, \ldots$, is called a *discrete-event dynamic system* (DEDS) with *starting vector b* and *observation vector c*. The scalars g_j are called *Markov parameters* of the system and the triple (A, b, c) is called a *realization* of the DEDS of dimension n.

Suppose that (A, b, c) is a realization of a DEDS $\{g_j\}_{j=0}^{\infty}$, and consider the Hankel matrices

$$H_r = \begin{pmatrix} g_0 & g_1 & \cdots & g_r \\ g_1 & g_2 & \cdots & g_{r+1} \\ \cdots & \cdots & \cdots & \cdots \\ g_r & g_{r+1} & \cdots & g_{2r} \end{pmatrix}$$

for $r = 0, 1, \ldots$. By Theorem 5.3.19 there exist $\alpha_0, \alpha_1, \ldots, \alpha_n \in \overline{\mathbb{R}}$ and disjoint sets $S, T \subseteq \{0, 1, \ldots, n\}$ such that

$$\sum_{j \in S}^{\oplus} \alpha_j \otimes A^j = \sum_{j \in T}^{\oplus} \alpha_j \otimes A^j.$$

If we multiply this equation by A^k (k positive integer) and then by c^T from the left and by b from the right we obtain

$$\sum_{j\in S}^{\oplus} \alpha_j \otimes g_{j+k} = \sum_{j\in T}^{\oplus} \alpha_j \otimes g_{j+k}$$

for any positive integer k. Hence

$$\sum_{j\in S}^{\oplus} \alpha_j \otimes h_j = \sum_{j\in T}^{\oplus} \alpha_j \otimes h_j,$$

where $h_0, h_1, \ldots, h_r$ are the columns of H_r. Using Theorem 6.3.1 we deduce:

Theorem 6.4.1 *Let $G = \{g_j\}_{j=0}^{\infty}$ be a real sequence and $r > 0$ an integer. If either $\mathrm{ap}^+(H_r) = \emptyset$ or $\mathrm{ap}^-(H_r) = \emptyset$ then no realization of G of dimension r or less exists.*

The *minimal-dimensional realization problem* (that is, the task of finding a realization of a given sequence of Markov parameters of minimal dimension) seems to be unresolved and hard for general sequences; however, using Theorem 6.4.1 it is possible to solve this question for some types of DEDS, such as for convex sequences. Let us recall that a sequence $\{g_j\}_{j=0}^{\infty}$ is called *convex* if

$$g_j + g_{j-2} \geq 2g_{j-1}$$

for every natural number $j \geq 2$. The following is providing a useful tool:

Proposition 6.4.2 [67] *If $\{g_j\}_{j=0}^{\infty}$ is convex then*

(a) $\mathrm{id} \in \mathrm{ap}(H_r)$ *for every* $r = 0, 1, \ldots$;
(b) *If* $\mathrm{ap}(H_0) = \mathrm{ap}(H_1) = \cdots = \mathrm{ap}(H_{r-1}) = \{\mathrm{id}\}$ *and* $\mathrm{ap}(H_r) \neq \{\mathrm{id}\}$ *then*

$$\mathrm{ap}(H_r) = \{\mathrm{id}, (1)(2) \ldots (r-1)(r, r+1)\}.$$

Corollary 6.4.3 *If $\{g_j\}_{j=0}^{\infty}$ is convex then $n = \min\{r; \mathrm{ap}(H_r) \neq \{\mathrm{id}\}\}$ if and only if n is the least integer satisfying*

$$g_{2n} + g_{2n-2} = 2g_{n-1}.$$

It is easily seen that for $A = \mathrm{diag}(d_1, \ldots, d_n), b = 0, c = (c_1, \ldots, c_n)^T$ the DEDS is $\{g_j\}_{j=0}^{\infty}$, where

$$g_j = \sum_{i\in N}^{\oplus} c_i \otimes d_i^j$$

or, in conventional notation

$$g_j = \max_{i\in N}(c_i + jd_i).$$

Hence the sequence $\{g_j\}_{j=0}^{\infty}$ is convex and has a constant slope starting from some $j = j_0$. This indicates that for a convex sequence of Markov parameters which ultimately has a constant slope and the transient (that is, the beginning of the sequence before the slope becomes constant) is strictly convex, a realization of dimension $\lceil j_0/2 \rceil + 1$ can be found [67]. For such sequences, in conjunction with Corollary 6.4.3 this provides a minimal-dimensional realization.

The minimal-dimensional realization problem for general convex sequences can also be efficiently solved [90]. The basic principles are the same but the proof of minimality is more evolved and requires different methodology.

6.5 Conclusions

In Sect. 3.3 and this chapter we have studied three concepts of independence in max-algebra: linear independence, strong linear independence and Gondran–Minoux independence. From the presented theory it follows that linear independence implies strong linear independence and strong linear independence implies Gondran–Minoux independence. For square matrices these three concepts turn to regularity, strong regularity and Gondran–Minoux regularity.

Following the resolution of the even cycle problem now all three types of regularity for an $n \times n$ matrix can be checked in $O(n^3)$ time; however, checking SLI and GMI in polynomial time seems to be an unresolved problem (see Chap. 11).

Note that further theory of strong regularity can be found in [113].

6.6 Exercises

Exercise 6.6.1 For each of the following matrices decide whether they are strongly regular and whether they are Gondran–Minoux regular:

(a) $A = \begin{pmatrix} 1 & 2 & 4 \\ -4 & 0 & 2 \\ 1 & 3 & 1 \end{pmatrix}$ [strongly regular, hence also Gondran–Minoux regular]

(b) $A = \begin{pmatrix} 1 & 2 & 5 \\ -4 & 0 & 2 \\ 1 & 3 & 1 \end{pmatrix}$ [Gondran–Minoux regular but not strongly regular]

(c) $A = \begin{pmatrix} 1 & 2 & 5 \\ -1 & 0 & 3 \\ 1 & 1 & 1 \end{pmatrix}$ [Not Gondran–Minoux regular, hence also not strongly regular]

Exercise 6.6.2 Decide whether the matrix below has strongly linearly independent columns:

$$A = \begin{pmatrix} 1 & 2 & 5 \\ -4 & 0 & 2 \\ 1 & 3 & 1 \\ 1 & 3 & 0 \end{pmatrix}$$

[It has, consider the 3×3 submatrix consisting of rows $1, 3, 4$]

Exercise 6.6.3 In the following matrix A find a 3×3 submatrix whose max-algebraic permanent is greatest without checking all 3×3 submatrices (a solution to this question can be found by inspection):

$$A = \begin{pmatrix} 1 & 3 & 0 \\ 2 & -1 & 4 \\ 5 & 6 & 3 \\ 1 & 2 & 2 \\ 4 & 1 & 3 \end{pmatrix}.$$

(Hint: Subtract the column maximum from each column.)

Exercise 6.6.4 Prove the statement: Let $A \in \overline{\mathbb{R}}^{n \times n}$ and M be any maximal set of nonequivalent eigennodes of A. Then the submatrix

$$\Gamma((\lambda(A))^{-1} \otimes A)[M]$$

is strongly regular.

Exercise 6.6.5 Let $A \in \overline{\mathbb{R}}^{n \times n}$ be strongly definite. Prove that pd(A) is equal to the topological dimension of the principal eigenspace. (Hint: Show that the topological dimension is equal to the number of strongly connected components of the critical digraph. See [139])

Exercise 6.6.6 A real square matrix is called typical if no two entries have the same fractional part. Prove the statement: If A is typical and Im(A) contains an integer vector then A is strongly regular.

Exercise 6.6.7 Consider systems $Ax = b$ in nonnegative linear algebra (A and B are nonnegative, with conventional addition and multiplication). Show that

(a) $T(A)$ are the same as in max-algebra;
(b) If A is positive, the possible types of $T(A)$ are $\{0, \infty\}$, $\{0, 1\}$ and $\{0, 1, \infty\}$.

(Hint: Consider convex sets in the plane.)

Exercise 6.6.8 (Izhakian linear dependence) Let $A \in \overline{\mathbb{R}}^{n \times n}$. Show that A is not strongly regular if and only if the following condition is satisfied: there exists x such that the maximum in each expression $(A \otimes x)_i$ is attained twice.

(Hint: Use the method used in the proof of Theorem 6.3.1.)

Chapter 7
Two-sided Max-linear Systems

Unlike in conventional linear algebra, moving from the task of finding a solution to a one-sided max-linear system of the form

$$A \otimes x = b$$

to finding a solution to a two-sided system

$$A \otimes x \oplus c = B \otimes x \oplus d, \tag{7.1}$$

where $A, B \in \overline{\mathbb{R}}^{m \times n}$ and $c, d \in \overline{\mathbb{R}}^m$, means a significant change of difficulty of the problem. Instead of finding a pre-image of a max-linear mapping we now have to find a vector in the intersection of two column spaces without the possibility of converting this task to the first one. The good news is that the solution set to (7.1) is finitely generated (Theorem 7.6.1), and that we feel reasonably confident in being able to solve such systems, see the pseudopolynomial Alternating Method of Sect. 7.3 and the corresponding Matlab codes downloadable from

http://web.mat.bham.ac.uk/P.Butkovic/software/index.htm.

Yet, a basic question remains open to date: are two-sided systems polynomially solvable? It follows from the results in [14] that two-sided systems are polynomially equivalent to mean payoff games, a well-known hard problem in $NP \cap Co - NP$. Thus, there is good reason to hope that the answer to the question is affirmative.

If $c = d = \varepsilon$ in (7.1) then this system has the form $A \otimes x = B \otimes x$ and is called *homogeneous*, otherwise it is *nonhomogeneous*. A system of the form

$$A \otimes x = B \otimes y, \tag{7.2}$$

where $A \in \overline{\mathbb{R}}^{m \times n}$ *and* $B \in \overline{\mathbb{R}}^{m \times k}$ is a special homogeneous system and will be called a *system with separated variables*. In fact we can transform nonhomogeneous systems into homogeneous and these in turn into systems with separated variables (see Sect. 7.4). We will, of course, be interested in *nontrivial solutions*, that is, when

P. Butkovič, *Max-linear Systems: Theory and Algorithms*,   149
Springer Monographs in Mathematics 151,
DOI 10.1007/978-1-84996-299-5_7, © Springer-Verlag London Limited 2010

$x \neq \varepsilon$ for homogeneous systems and when $\binom{x}{y} \neq \varepsilon$ for systems with separated variables. In some cases (such as the Alternating Method) we will restrict our attention to finite solutions.

We start by presenting a few easily solvable special cases, then we continue with the Alternating Method for solving the systems with separated variables with a proof of pseudopolynomial computational complexity and then we show how to convert general systems to systems with separated variables. A proof of finite generation of the solution set concludes this chapter. Nonhomogeneous systems are also studied in Chap. 10 in connection with max-linear programs.

Note that the generalized eigenproblem

$$A \otimes x = \lambda \otimes B \otimes x,$$

which will be studied in Chap. 9, may be seen as a generalization of both the eigenproblem and two-sided linear systems. It is providing, among other benefits, useful information about the two-sided systems. For instance, it follows that compared to the general case a randomly considered system $A \otimes x = B \otimes x$ is less likely to have a nontrivial solution if both A and B are symmetric.

An alternative approach to solving two-sided systems can be found in [148].

7.1 Basic Properties

For $A, B \in \overline{\mathbb{R}}^{m \times n}$ and $c, d \in \overline{\mathbb{R}}^{m}$ we denote

$$S(A, B, c, d) = \left\{ x \in \overline{\mathbb{R}}^{n}; A \otimes x \oplus c = B \otimes x \oplus d \right\}$$

and

$$S(A, B) = \left\{ x \in \overline{\mathbb{R}}^{n}; A \otimes x = B \otimes x \right\}.$$

Proposition 7.1.1 *For any $A, B \in \overline{\mathbb{R}}^{m \times n}$ and $c, d \in \overline{\mathbb{R}}^{m}$ the set $S(A, B, c, d)$ is max-convex and the set $S(A, B)$ is a subspace.*

Proof Let $\alpha, \beta \in \mathbb{R}$ and $\alpha \oplus \beta = 0$. Then

$$A \otimes (\alpha \otimes x \oplus \beta \otimes y) \oplus c$$

$$= A \otimes (\alpha \otimes x \oplus \beta \otimes y) \oplus \alpha \otimes c \oplus \beta \otimes c$$

$$= \alpha \otimes (A \otimes x \oplus c) \oplus \beta \otimes (A \otimes y \oplus c)$$

$$= \alpha \otimes (B \otimes x \oplus d) \oplus \beta \otimes (B \otimes y \oplus d)$$

$$= B \otimes (\alpha \otimes x \oplus \beta \otimes y) \oplus \alpha \otimes d \oplus \beta \otimes d$$

$$= B \otimes (\alpha \otimes x \oplus \beta \otimes y) \oplus d.$$

Hence $S(A, B, c, d)$ is max-convex; the second statement is proved similarly. □

Corollary 7.1.2 *The solution set of a homogeneous system with separated variables is a subspace.*

If A has an ε row, say the kth then in a solution $\binom{x}{y}$ to (7.2) $y_j = \varepsilon$ if $b_{kj} > \varepsilon$. All such variables y_j and the kth equation may removed from the system. Similarly, if B has an ε row. If one of A and B has an ε column then any such column may be removed from the system with no affect on the solution set. We may therefore assume without loss of generality that A and B are doubly $\mathbb{R}$-astic.

7.2 Easily Solvable Special Cases

In some situations it is not difficult to solve two-sided systems. For instance all solutions (if any) to the systems of the form

$$A \otimes x = \alpha \otimes x,$$

where $A \in \overline{\mathbb{R}}^{n \times n}$ and $\alpha \in \overline{\mathbb{R}}$ is given, can easily be found using the techniques of Chap. 4. This readily generalizes to the systems

$$A \otimes x = P \otimes x,$$

where $A \in \overline{\mathbb{R}}^{n \times n}$ and P is a generalized permutation matrix since the inverse to P exists. Let us discuss now a few other, less trivial, yet simple cases.

7.2.1 A Classical One

Special two-sided systems have been studied already in early works on max-algebra [97, 100] see also [8]. The best known example perhaps is the system

$$x = A \otimes x \oplus b. \tag{7.3}$$

If $\lambda(A) \leq 0$ then $\Delta(A) = I \oplus A \oplus A^2 \oplus \cdots \oplus A^{n-1}$ by Proposition 1.6.10 and hence

$$A \otimes \Delta(A) \otimes b \oplus b = \left(A \oplus A^2 \oplus \cdots \oplus A^n \right) \otimes b \oplus b$$

$$= \left(I \oplus A \oplus A^2 \oplus \cdots \oplus A^n \right) \otimes b$$

$$= (I \oplus \Gamma(A)) \otimes b = \Delta(A) \otimes b,$$

proving that $\Delta(A) \otimes b$ is a solution to (7.3). This solution is unique when $\lambda(A) < 0$ [8, 102].

7.2.2 Idempotent Matrices

Another special case is related to *idempotent* matrices, that is, square matrices $A \in \overline{\mathbb{R}}^{n \times n}$ such that

$$A \otimes A = A.$$

If A is idempotent then $\Gamma(A) = A$ and so $\lambda(A) \leq 0$ by Proposition 1.6.10. Also, A is definite if $A \neq \varepsilon$ since then $A \otimes v = v$ for some column $v \neq \varepsilon$, which means that $0 \in \Lambda(A)$.

In the next statement we consider finite matrices.

Theorem 7.2.1 [23] *If $A, B \in \mathbb{R}^{n \times n}$ are increasing and idempotent then the following are equivalent*:

(a) $A \otimes x = B \otimes y$ *is satisfied by some* $x, y \in \overline{\mathbb{R}}^n, x, y \neq \varepsilon$.
(b) $A \oplus B$ *is definite*.
(c) $A \otimes x = B \otimes x$ *is satisfied by some* $x \in \overline{\mathbb{R}}^n, x \neq \varepsilon$.

Proof (a) $\Longrightarrow$ (b) The vector $z = A \otimes x = B \otimes y$ is finite and

$$(A \oplus B) \otimes z = A \otimes z \oplus B \otimes z = A \otimes (A \otimes x) \oplus B \otimes (B \otimes y)$$

$$= A \otimes x \oplus B \otimes y = z \oplus z = z.$$

(b) $\Longrightarrow$ (c) If $A \oplus B$ is definite then for some $z \in \mathbb{R}^n$ we have

$$z = (A \oplus B) \otimes z = A \otimes z \oplus B \otimes z,$$

hence $A \otimes z \leq z, B \otimes z \leq z$ but these inequalities are satisfied with equality because A, B are increasing. Thus (c) follows.

(c) $\Longrightarrow$ (a) Trivial. $\square$

7.2.3 Commuting Matrices

We also briefly discuss the case of commuting matrices.

Theorem 7.2.2 *If $A, B \in \overline{\mathbb{R}}^{n \times n}$ and $A \otimes B = B \otimes A$ then the two-sided max-linear system with separated variables*

$$A \otimes x = B \otimes y$$

has a nontrivial solution and this solution can be found by solving the eigenproblem for one of A and B.

Proof If $A_k = \varepsilon$ then set $y = \varepsilon$, $x_k = 0$ and $x_j = \varepsilon$ for $j \neq k$. Similarly, if B has an ε column.

Suppose now that both A and B are column $\mathbb{R}$-astic. Let $z \in V(A, \lambda)$, $\lambda \in \overline{\mathbb{R}}$, $z \neq \varepsilon$, then by Lemma 4.7.1 $B \otimes z \in V(A, \lambda)$ and $B \otimes z \neq \varepsilon$ since B is column $\mathbb{R}$-astic. Also, $\lambda > \varepsilon$ because A is column $\mathbb{R}$-astic. Therefore we have $\lambda \otimes z \neq \varepsilon$ and

$$A \otimes (B \otimes z) = \lambda \otimes (B \otimes z) = B \otimes (\lambda \otimes z).$$

It remains to set $x = B \otimes z$ and $y = \lambda \otimes z$. □

Note that it follows from the proof of Theorem 7.2.2 that a solution (x, y) with $x \neq \varepsilon$ and $y \neq \varepsilon$ exists, provided that both A and B are commuting column $\mathbb{R}$-astic matrices.

Corollary 7.2.3 *If* $A, B \in \overline{\mathbb{R}}^{n \times n}$, $A \otimes B = B \otimes A$ *and* $\varphi(t), \psi(t)$ *are max-polynomials then the two-sided max-linear system*

$$\varphi(A) \otimes x = \psi(B) \otimes y$$

has a nontrivial solution, and this solution can be found by solving the eigenproblem for one of $\varphi(A)$ *and* $\psi(B)$.

Proof If $A \otimes B = B \otimes A$ then also $\varphi(A) \otimes \psi(B) = \psi(B) \otimes \varphi(A)$. □

7.2.4 Essentially One-sided Systems

If in a system (7.2), where A and B are doubly $\mathbb{R}$-astic, one of the vectors x, y is one-dimensional, then we have an essentially one-sided system and solution methods from Chap. 3 can be applied immediately. However, in this case we can describe the unique scaled basis of this set. This will be useful in the context of attraction spaces (Sect. 8.5).

Let us assume without loss of generality that y is one-dimensional. Thus B is a one-column matrix. Since B is assumed to be doubly $\mathbb{R}$-astic, it is finite. We may then assume that $B = 0$ and the system is

$$A \otimes x = y.$$

Note that by eliminating the variable y and equating all left-hand sides we can write this system equivalently as a chain of equations

$$\sum_{j \in N}^{\oplus} a_{1j} x_j = \sum_{j \in N}^{\oplus} a_{2j} \otimes x_j = \cdots = \sum_{j \in N}^{\oplus} a_{mj} \otimes x_j. \tag{7.4}$$

Since A is doubly $\mathbb{R}$-astic, we have $\max_{i \in M} a_{ij} > \varepsilon$ for every $j \in N$. Using the substitution

$$z_j = \left(\max_{i \in M} a_{ij} \right)^{-1} \otimes x_j, \quad j \in N, \tag{7.5}$$

we can now assume that the system is

$$A \otimes z = y, \tag{7.6}$$

where all column maxima in A are 0 (and y is a single variable). By Theorem 3.1.1 $\binom{z}{y} \in \mathbb{R}^{n+1}$ is a scaled solution to (7.6) if and only if $y = 0, z \le 0$ and for the sets

$$M_j = \{ i \in M; a_{ij} = 0 \}, \quad j \in N,$$

we have

$$\bigcup_{j:x_j=0} M_j = M. \tag{7.7}$$

Let us denote the solution set to (7.6) by S. It turns out that zero is the only possible value of any finite component of a scaled extremal in S:

Proposition 7.2.4 [134] *Let $\binom{z}{0} \in S$ be a scaled vector and $\varepsilon < z_j < 0$ for some $j \in N$. Then $\binom{z}{0}$ is not an extremal of S.*

Proof Let $K^< := \{ j \in N; \varepsilon < z_j < 0 \}$ and $K^0 := \{ j \in N; z_j = 0 \}$, and define vectors $\binom{v^0}{0} \in \overline{\mathbb{R}}^{n+1}$ and $\binom{v(k)}{0} \in \overline{\mathbb{R}}^{n+1}$ for each $k \in K^<$ by

$$v_j^0 = \begin{cases} 0, & \text{if } j \in K^0 \\ \varepsilon, & \text{otherwise} \end{cases}, \quad v_j(k) = \begin{cases} 0, & \text{if } j \in K^0 \cup \{k\} \\ \varepsilon, & \text{otherwise} \end{cases}.$$

Observe that both $\binom{v^0}{0}$ and $\binom{v(k)}{0}$ for any $k \in K^<$, are (at least two) solutions to (7.4), different from $\binom{z}{0}$. We have:

$$\binom{z}{0} = \binom{v^0}{0} \oplus \sum_{k \in K^<}^{\oplus} z_k \otimes \binom{v(k)}{0},$$

hence $\binom{z}{0}$ is not an extremal. $\qquad \square$

We have seen above that $\binom{z}{y} \in \mathbb{R}^{n+1}$ is a scaled solution to (7.6) if and only if $y = 0, z \le 0$ and (7.7) holds, that is, the sets $M_j, j \in K$ form a covering of M, where

$$K = \{ j \in N; z_j = 0 \}.$$

Recall that a covering is called minimal if it does not contain any proper subcovering. We will now also say that a covering is *nearly minimal* if it contains no more than one proper subcovering. Hence, a covering $M_j, j \in K$ is nearly minimal if and only if there exists no more than one $r \in K$ such that $M_j, j \in K \setminus \{r\}$ is also a covering. Recall that by e^j ($j \in N$) we denote the vector that has the jth coordinate zero and all other are ε.

Proposition 7.2.5 [134] *The unique scaled basis of S consists of the vectors of the form $\binom{v^K}{0}$, where $v^K = \sum_{j \in K}^{\oplus} e^j$, and $M_j, j \in K$ is a nearly minimal covering of M.*

Proof By Corollary 3.3.11 we only need to prove that a vector is an extremal in S if and only if it is $\binom{v^K}{0}$ for a nearly minimal covering of M.

Let $\binom{v}{0}$ be an extremal of S. By Proposition 7.2.4, all its finite components are zero and thus $v = v^K$ for some $K \subseteq N$, such that $M_j, j \in K$ is a covering of M. If K is not nearly minimal, then there exist r and s such that $M_j, j \in K[r] := K \setminus \{r\}$ and $M_j, j \in K[s] := K \setminus \{s\}$ are both coverings of M. Then $\binom{v^{K[r]}}{0}$ and $\binom{v^{K[s]}}{0}$ are both solutions to (7.6) and $\binom{v^K}{0} = \binom{v^{K[r]}}{0} \oplus \binom{v^{K[s]}}{0}$, hence $\binom{v^K}{0}$ is not an extremal.

Conversely, if $\binom{v^K}{0}$ is a scaled solution but not an extremal, then there exist $\binom{u}{0} \neq \binom{v^K}{0}$ and $\binom{w}{0} \neq \binom{v^K}{0}$ such that $\binom{v^K}{0} = \binom{u}{0} \oplus \binom{w}{0}$. Evidently $\binom{u}{0} \leq \binom{v^K}{0}$ and $\binom{w}{0} \leq \binom{v^K}{0}$. By Proposition 7.2.4 we can represent $\binom{u}{0}$ and $\binom{w}{0}$ as combinations of solutions to (7.6) over $\{0, \varepsilon\}$. These solutions correspond to coverings, which are proper subcoverings of $M_j, j \in K$. At least two of these coverings are different from each other, hence K is not nearly minimal. $\square$

Thus, the problem of finding the unique scaled basis of system (7.6) is equivalent to the problem of finding all nearly minimal subcoverings of $M_j, j \in N$.

The following special case will also be useful. Here we denote for every $i \in M$:

$$L_i = \{j \in N; a_{ij} = 0\}.$$

Corollary 7.2.6 *If $L_1, \ldots, L_m$ are pairwise disjoint, then the unique scaled basis of S is the set of vectors $\binom{v^K}{0}$, where $v^K = \sum_{j \in K}^{\oplus} e^j$, and K is an index set which contains exactly one index from each set L_i ($i \in M$).*

Proof In this case there are no nearly minimal coverings of M other than minimal. If K is an index set which contains exactly one index from each set L_i ($i \in M$) then $M_j, j \in K$ is a minimal covering of M. $\square$

7.3 Systems with Separated Variables—The Alternating Method

Consider the problem of solving max-linear systems with separated variables:

Given $A \in \overline{\mathbb{R}}^{m \times n}$ and $B \in \overline{\mathbb{R}}^{m \times k}$ find $x \in \mathbb{R}^n$, $y \in \mathbb{R}^k$ such that

$$A \otimes x = B \otimes y. \qquad (7.8)$$

The method we will present finds a finite solution to (7.8) or decides that no such solution exists. We will therefore assume in this section that "solution" means finite solution. As explained at the beginning of this chapter, we may assume without loss of generality that A and B are doubly $\mathbb{R}$-astic. Note that the product of a doubly $\mathbb{R}$-astic matrix and a finite vector is a finite vector.

The algebraic method for solving one-sided systems (Sect. 3.2) will be helpful for solving (7.8). Recall that for any $A \in \overline{\mathbb{R}}^{m \times n}$ and $b \in \overline{\mathbb{R}}^m$ the vector $\bar{x} = A^* \otimes' b$ (the principal solution) is the greatest solution to $A \otimes x \le b$, and $A \otimes x = b$ has a solution if and only if $\bar{x}$ is a solution. So a rather natural idea is starting from some $x = x(0)$ to take for $y(0)$ the principal solution to $B \otimes y = A \otimes x(0)$, then for $x(1)$ the principal solution to $A \otimes x = B \otimes y(0)$, for $y(1)$ the principal solution to $B \otimes y = A \otimes x(1)$ and so on. It is probably not immediately obvious whether the sequences $\{x(k)\}_{k=0}^{\infty}$, $\{y(k)\}_{k=0}^{\infty}$ yield anything useful. We will show that under reasonable assumptions they either converge to a solution to (7.8) or we can deduce that there is no solution. But first we formally present the algorithm. This section is based on [68].

Algorithm 7.3.1 ALTERNATING METHOD
Input: $A \in \overline{\mathbb{R}}^{m \times n}$, $B \in \overline{\mathbb{R}}^{m \times k}$, doubly $\mathbb{R}$-astic.
Output: A solution (x, y) to (7.8) or an indication that no such solution exists.
Let $x(0) \in \mathbb{R}^n$ be any vector.
$r := 0$
again:
$y(r) := B^* \otimes' (A \otimes x(r))$
$x(r+1) := A^* \otimes' (B \otimes y(r))$
If $x_i(r+1) < x_i(0)$ for every $i \in N$ then stop (no solution)
If $A \otimes x(r+1) = B \otimes y(r)$ then stop $((x(r+1), y(r))$ is a solution)
Go to again

Example 7.3.2 Let

$$A = \begin{pmatrix} 3 & -\infty & 0 \\ 1 & 1 & 0 \\ -\infty & 1 & 2 \end{pmatrix}, \qquad B = \begin{pmatrix} 1 & 1 \\ 3 & 2 \\ 3 & 1 \end{pmatrix}.$$

Then

$$A^* = \begin{pmatrix} -3 & -1 & +\infty \\ +\infty & -1 & -1 \\ 0 & 0 & -2 \end{pmatrix}, \qquad B^* = \begin{pmatrix} -1 & -3 & -3 \\ -1 & -2 & -1 \end{pmatrix}.$$

Set (randomly) $x(0) = (5, 3, 1)^T$. The algorithm then finds

$$r = 0: \quad x(0) = \begin{pmatrix} 5 \\ 3 \\ 1 \end{pmatrix}, \quad A \otimes x(0) = \begin{pmatrix} 8 \\ 6 \\ 4 \end{pmatrix}, \quad y(0) = \begin{pmatrix} 1 \\ 3 \end{pmatrix},$$

$$B \otimes y(0) = \begin{pmatrix} 4 \\ 5 \\ 4 \end{pmatrix};$$

$$r = 1: \quad x(1) = \begin{pmatrix} 1 \\ 3 \\ 2 \end{pmatrix}, \quad A \otimes x(1) = \begin{pmatrix} 4 \\ 4 \\ 4 \end{pmatrix}, \quad y(1) = \begin{pmatrix} 1 \\ 2 \end{pmatrix},$$

$$B \otimes y(1) = \begin{pmatrix} 3 \\ 4 \\ 4 \end{pmatrix};$$

$$r = 2: \quad x(2) = \begin{pmatrix} 0 \\ 3 \\ 2 \end{pmatrix}, \quad A \otimes x(2) = \begin{pmatrix} 3 \\ 4 \\ 4 \end{pmatrix}.$$

Since $A \otimes x(2) = B \otimes y(1)$, the algorithm stops yielding the solution $(x(2), y(1))$.

In order to prove correctness of the Alternating Method, first recall that by Corollary 3.2.4 the following hold for any matrices U, V, W of compatible sizes:

$$U \otimes (U^* \otimes' W) \leq W, \tag{7.9}$$

$$U \otimes (U^* \otimes' (U \otimes W)) = U \otimes W. \tag{7.10}$$

The following operators will be useful:

$$\pi : y \longrightarrow A^* \otimes' (B \otimes y)$$

and

$$\psi : x \longrightarrow B^* \otimes' (A \otimes x).$$

Hence the Alternating Method generates the pair-sequence

$$\{(x(r), y(r))\}_{r=0,1,\dots}$$

satisfying

$$x(r+1) = \pi(y(r)) \tag{7.11}$$

and

$$y(r) = \psi(x(r)). \tag{7.12}$$

Let $x \in \mathbb{R}^n$, $y \in \mathbb{R}^k$. We shall say that (x, y) is *stable* if $(x, y) = (\pi(y), \psi(x))$.

Lemma 7.3.3 *Every stable pair (x, y) is a solution.*

Proof If (x, y) is stable then using (7.9) we have

$$A \otimes x = A \otimes \pi (y) = A \otimes \left(A^* \otimes' (B \otimes y)\right) \leq B \otimes y$$
$$= B \otimes \psi (x) = B \otimes \left(B^* \otimes' (A \otimes x)\right) \leq A \otimes x,$$

implying equality between all terms and hence also the lemma statement. $\qquad\square$

A solution that is stable will be called a *stable solution*.

Lemma 7.3.4 *If (x, y) is a solution then $(\pi(y), \psi(x))$ is a stable solution.*

Proof If (x, y) is a solution then using (7.10) we have

$$\psi (\pi (y)) = B^* \otimes' \left(A \otimes \left(A^* \otimes' (B \otimes y)\right)\right)$$
$$= B^* \otimes' \left(A \otimes \left(A^* \otimes' (A \otimes x)\right)\right)$$
$$= B^* \otimes' (A \otimes x) = \psi (x).$$

Similarly, $\pi (\psi (x)) = \pi (y)$, whence $(\pi (y), \psi (x))$ is stable and therefore a solution. $\qquad\square$

The next two lemmas present important monotonicity features of the Alternating Method, which will be crucial for the proof of performance.

Lemma 7.3.5 *The sequence $\{A(x(r))\}_{r=0,1,...}$ is nonincreasing.*

Proof Applying (7.9) to (7.11) and (7.12) we get

$$A \otimes x (r + 1) \leq B \otimes y (r) \leq A \otimes x (r).$$ $\qquad\square$

Lemma 7.3.6 *The sequence $\{x(r)\}_{r=0,1,...}$ is nonincreasing.*

Proof $x(r + 1) = \pi(y(r)) = \pi(B^* \otimes' (A \otimes x(r)))$. This implies that $x(r + 1)$ is an isotone function of the nonincreasing $A \otimes x(r)$. $\qquad\square$

The next lemma and theorem are a further preparation for the proof of correctness of the Alternating Method.

Lemma 7.3.7 *If a solution exists then the sequence $\{x(r)\}_{r=0,1,...}$ is lower-bounded for any $x(0)$.*

Proof For any stable solution (x, y) and $\alpha \in \mathbb{R}$ it is immediate that $\alpha \otimes (x, y)$ is also a stable solution, and α may be chosen small enough so that $\alpha \otimes x \leq x(0)$.

By Lemma 7.3.4 if a solution exists then a stable solution (u, v) exists such that $x(0) \geq u$. And if $x(r) \geq u$ for some r then by (7.9) and isotonicity we have

$$x(r+1) = (\pi \circ \psi)(x(r)) \geq (\pi \circ \psi)(u) = \pi(v) = u$$

and the result follows by induction. $\qquad\square$

Theorem 7.3.8 *If all components of $x(r)$ or $y(r)$ have properly decreased after a number of steps of the Alternating Method then (7.8) has no solution.*

Proof In the proof of Lemma 7.3.7 the value of α may be taken so that $\alpha \otimes x \leq x(0)$ but with equality in at least one component. Lemmas 7.3.6 and 7.3.7 then imply that component of $x(r)$ remains fixed in value for all $r \geq 0$. Moreover it is clear that analogues of Lemmas 7.3.5, 7.3.6 and 7.3.7 are provable for the sequence $\{y(r)\}_{r=0,1,\ldots}$. $\qquad\square$

We are ready to prove the correctness of the Alternating Method and deduce corollaries.

Theorem 7.3.9 *The pair-sequence $\{(x(r), y(r))\}_{r=0,1,\ldots}$ generated by the Alternating Method converges if and only if a solution exists. Convergence is then monotonic, to a stable solution, for any choice of $x(0) \in \mathbb{R}^n$.*

Proof If $(x(r), y(r)) \longrightarrow (\xi, \eta)$ then (ξ, η) by Lemma 7.3.7 and by continuity

$$(\xi, \eta) = \lim(x(r+1), y(r)) = \lim(\pi(y(r)), \psi(x(r))) = (\pi(\eta), \psi(\xi)).$$

Hence (ξ, η) is stable, thus a stable solution by Lemma 7.3.3.

Conversely, if a solution exists the monotonic convergence of $\{x(r)\}$ follows from Lemmas 7.3.6 and 7.3.7, and that of $\{y(r)\}$ by isotonicity and continuity. $\qquad\square$

If all finite entries in (7.8) are integer, A, B are doubly $\mathbb{R}$-astic and $x(0)$ is an integer vector then the integrality is preserved throughout the work of the Alternating Method. Hence if a solution exists, it will be found in a finite number of steps. We may summarize these observations in the following.

Theorem 7.3.10 *If $A \in \overline{\mathbb{Z}}^{m \times n}$ and $B \in \overline{\mathbb{Z}}^{m \times k}$ are doubly $\mathbb{R}$-astic and a solution to (7.8) exists then the Alternating Method starting from an $x(0) \in \mathbb{Z}^n$ will find an integer solution in a finite number of steps.*

In the integer case we may estimate the computational complexity of the Alternating Method provided that $x(0)$ is an integer vector and at least one of A, B is finite (and as before, the other one is doubly $\mathbb{R}$-astic). We will now assume without loss of generality that A is finite.

Theorem 7.3.11 *If* $A \in \mathbb{Z}^{m \times n}$, $B \in \overline{\mathbb{Z}}^{m \times k}$ *and the Alternating Method starts with* $x(0) \in \mathbb{Z}^n$ *then it will terminate after at most*

$$(n-1)\left(1 + \gamma^* \otimes A^* \otimes A \otimes \gamma\right)$$

iterations where $\gamma = x(0)$.

Proof Suppose first that a solution exists. By Theorem 7.3.8, there is a component of γ, say γ_j, that will not change during the run of the Alternating Method; let us call such a component a *sleeper*. The algorithm will halt as soon as $A \otimes x$ does not change. This is guaranteed to happen at the latest when all components of x become so small compared to γ_j that they will not affect the value of $A \otimes x$, more precisely when for every k and i

$$a_{ik} + x_k \leq a_{ij} + \gamma_j,$$

that is, when $x_k \leq u_{kj}$, where

$$u_{kj} = \min_i \left(a_{ij} + \gamma_j - a_{ik}\right) = \left(A^* \otimes' A\right)_{kj} + \gamma_j.$$

This inequality means that the nonsleeper x_k has become dominated by the sleeper γ_j. Since x_k is nonincreasing, the domination will persist in subsequent iterations. Since the value of j is not known we guarantee domination for x_k by considering all components as potential sleepers, that x_k is certainly dominated if it falls in value below

$$\beta_k = \min_j u_{kj} = \min_j \left(\left(A^* \otimes' A\right)_{kj} + \gamma_j\right) = \left(A^* \otimes' A \otimes' \gamma\right)_k.$$

Hence the fall of x_k before domination is at most

$$w_k = \gamma_k - \beta_k + 1.$$

There at most $n - 1$ nonsleepers and at every iteration at least one nonsleeper falls by at least 1 (otherwise $A \otimes x$ does not change and the algorithm stops). Hence the total number of iterations before domination is not exceeding

$$(n-1)\max_k w_k = (n-1)\max_k (\gamma_k - \beta_k + 1) = (n-1)\left(1 + \beta^* \otimes \gamma\right)$$

$$= (n-1)\left(1 + \gamma^* \otimes A^* \otimes A \otimes \gamma\right).$$

If a solution does not exist then after at most $(n-1)(1 + \gamma^* \otimes A^* \otimes A \otimes \gamma)$ iterations all components of x fall (since otherwise $A \otimes x$ does not change, yielding a solution) and the algorithm stops indicating infeasibility. $\qquad \square$

If $C \in \overline{\mathbb{R}}^{n \times n}$, $\lambda(C) > \varepsilon$, then it follows from Lemma 1.6.28 that

$$\min_{x \in \mathbb{R}^n} x^* \otimes C \otimes x = z^* \otimes C \otimes z = \lambda(C),$$

where z is any finite subeigenvector of C. Therefore (by Theorem 7.3.11) a plausible vector to start the Alternating Method with is a finite subeigenvector of $A^* \otimes A$. Note that $A^* \otimes A$ is finite since A is finite and thus all eigenvectors of $A^* \otimes A$ are finite (subeigenvectors). Then the number of iterations is bounded by

$$(n-1)\left(1 + \lambda(A^* \otimes A)\right).$$

Let $K(A) = \max\{|a_{ij}|; i \in M, j \in N\}$ for any matrix $A \in \mathbb{R}^{m \times n}$. It is easily seen that

$$|\lambda(A)| \le K(A).$$

Also, let $C = (c_{ij}) = A^* \otimes A$. Then C is increasing and $K(C) \le 2K(A)$, thus $0 \le \lambda(C) \le 2K(A)$. At the same time the individual (four) lines in the main loop of the Alternating Method require

$$O\left((mn + mk) + (mn + mk) + n + (mn + mk + m)\right)$$

operations (including comparisons). Hence the computational complexity of the Alternating Method is

$$(n-1)(1 + 2K(A))\,O(m(n+k)) = O(mn(n+k)K(A)). \tag{7.13}$$

We conclude:

Theorem 7.3.12 *The Alternating Method is pseudopolynomial if applied to instances with integer entries where one of the matrices A, B is finite and the other one is doubly $\mathbb{R}$-astic.*

The Alternating Method as stated here is not polynomial [132]. To see this consider the system $A \otimes x = B \otimes y$ with

$$A = \begin{pmatrix} 1 & 1 \\ 0 & k \\ 0 & 0 \end{pmatrix}, \qquad B = \begin{pmatrix} 0 & 1 \\ 0 & k \\ 0 & 0 \end{pmatrix}$$

and starting vector $x_0 = (k/2, 0)$, which is an eigenvector of $A^* \otimes A$. It can be verified that the Alternating Method will produce a sequence of vectors starting from x_0 in which the first component will decrease in every iteration by 1 until it eventually reaches $(0, 0)^T$, a solution to $A \otimes x = B \otimes y$.

Remark 7.3.13 In [136] the concept of cyclic projectors is studied. It enabled the author to generalize the Alternating Method to the case of homogeneous multi-sided systems, and to prove using the cellular decomposition idea, that the Alternating Method converges in a finite number of iterations to a finite solution of a multi-sided system with real entries, if such a solution exists. The paper also present new bounds on the number of iterations of the Alternating Method, expressed in terms of the Hilbert projective distance.

7.4 General Two-sided Systems

Following the presentation of a pseudopolynomial method for finding a solution to systems with separated variables in the previous section, a question arises whether the general two-sided systems can be converted to those with separated variables. The answer is affirmative and will be given next.

Consider a general two-sided system (7.1). We start with a cancellation rule that in many cases significantly simplifies it.

Lemma 7.4.1 (Cancellation Law) *Let $v, w, a, b \in \mathbb{R}, a > b$. Then for any real x we have*

$$v \oplus a \otimes x = w \oplus b \otimes x \qquad (7.14)$$

if and only if

$$v \oplus a \otimes x = w. \qquad (7.15)$$

Proof If x satisfies (7.14) then LHS $\geq a \otimes x > b \otimes x$. Hence RHS $= w$ and (7.15) follows. If (7.15) holds then $w \geq a \otimes x > b \otimes x$ and thus $w = w \oplus b \otimes x$. $\square$

It follows from Lemma 7.4.1 that from a two-sided system we may always remove a term involving a variable without changing the solution set if a term with the same variable appears on the other side of the same equation with a greater coefficient. This is, of course, not possible if the coefficients of a variable on both sides of an equation are equal. Also conversely, if a variable appears on one side only we may "reinstate" it on the other side with any coefficient smaller than the existing one. Thus for instance when studying systems where every equation contains each variable on at least one side with a finite coefficient, we may assume without loss of generality that all coefficients of such a system are finite.

As another consequence we have that if a column (row) of A is $\mathbb{R}$-astic then we may assume without loss of generality that so is the corresponding column (row) of B and vice versa.

If for a variable the corresponding columns in both A and B are ε then these columns and variable may be removed without affecting the solution set of (7.1). Similarly, if for some i both the ith rows of A and B are ε then either this system has no solution (when $c_i \neq d_i$) or is satisfied by any x (when $c_i = d_i$). In the latter case the ith equation may be removed. Hence we may assume without loss of generality that both A and B are doubly $\mathbb{R}$-astic.

By introducing an extra variable, say x_{n+1}, (7.1) can be converted to a homogeneous system

$$\tilde{A} \otimes z = \tilde{B} \otimes z \qquad (7.16)$$

where $\tilde{A} = (A|c)$, $\tilde{B} = (B|d)$ and $z = (z_1, \ldots, z_{n+1})^T$. This conversion is supported by the following:

Lemma 7.4.2 *Let* $A, B \in \overline{\mathbb{R}}^{m \times n}$ *and* $c, d \in \overline{\mathbb{R}}^m$. *Then* (7.1) *has a solution if and only if* (7.16) *has a solution with* $z_{n+1} = 0$.

Proof It follows immediately from the definitions. □

It is easily seen that if all entries in a homogeneous system are finite then a nontrivial solution exists if and only if a finite solution exists. Hence we have a slight modification of Lemma 7.4.2:

Lemma 7.4.3 *Let* $A, B \in \mathbb{R}^{m \times n}$ *and* $c, d \in \mathbb{R}^m$. *Then* (7.1) *has a solution if and only if* (7.16) *has a nontrivial solution.*

Consider now homogeneous systems of the form

$$A \otimes x = B \otimes x \tag{7.17}$$

where $A, B \in \overline{\mathbb{R}}^{m \times n}$ are (without loss of generality) doubly $\mathbb{R}$-astic. System (7.17) is equivalent to

$$A \otimes x = y$$
$$B \otimes x = y$$

or, in compact form

$$\begin{pmatrix} A \\ B \end{pmatrix} \otimes x = \begin{pmatrix} I \\ I \end{pmatrix} \otimes y. \tag{7.18}$$

This is a system with separated variables and both $\begin{pmatrix} A \\ B \end{pmatrix}$ and $\begin{pmatrix} I \\ I \end{pmatrix}$ are doubly $\mathbb{R}$-astic. Hence the Alternating Method may immediately be applied to this system with guaranteed convergence as specified in Theorem 7.3.9.

To achieve a complexity result based on Theorem 7.3.12 and (7.13) we will assume that $A, B \in \mathbb{Z}^{m \times n}$. As discussed above, this case actually covers all systems with entries from $\overline{\mathbb{Z}}$ with every variable appearing on at least one side of each equation with a finite coefficient. Then $\begin{pmatrix} A \\ B \end{pmatrix}$ is finite and $\begin{pmatrix} I \\ I \end{pmatrix}$ is doubly $\mathbb{R}$-astic. Hence if we denote $K(A|B)$ for convenience by K and we use the fact that (7.18) has $2m$ equations and $n + m$ variables we deduce by (7.13) that the Alternating Method applied to this system will terminate in a finite number of steps and its computational complexity is

$$O\left(2mn\left(n + m\right) K\right) = O\left(mn\left(m + n\right) K\right). \tag{7.19}$$

We conclude:

Theorem 7.4.4 *Homogeneous system* (7.17) *with finite, integer matrices* A, B *can be solved using the Alternating Method in pseudopolynomial time.*

7.5 The Square Case: An Application of Symmetrized Semirings

Symmetrized semirings [8, 86] are sometimes useful to study two-sided systems of
equations in max-algebra. We now give a brief account of this theory and its application to two-sided systems, although their practical use for solving the two-sided
systems is rather limited, since in general they only provide a necessary solvability condition. Another application of this idea is to the generalized eigenproblem
(Chap. 9).

Denote $\mathbb{S} = \overline{\mathbb{R}} \times \overline{\mathbb{R}}$ and extend $\oplus$ and $\otimes$ to $\mathbb{S}$ as follows:

$$(a, a') \oplus (b, b') = (a \oplus b, a' \oplus b'),$$

$$(a, a') \otimes (b, b') = (a \otimes b \oplus a' \otimes b', a \otimes b' \oplus a' \otimes b).$$

It is easy to check that $\varepsilon = (-\infty, -\infty)$ is the neutral element of $\mathbb{S}$ with respect
to $\oplus$ and $(0, -\infty)$ is the neutral element with respect to $\otimes$.

If $x = (a, a')$ then $\ominus x$ stands for (a', a), $x \ominus y$ means $x \oplus (\ominus y)$, the *modulus*
of $x \in \mathbb{S}$ is $|x| = a \oplus a'$, the *balance operator* is $x^{\bullet} = x \ominus x = (|x|, |x|)$. Note that
we are using the symbol $| \cdot |$ for both the modulus of an element of a symmetrized
semiring and for the absolute value of a real number since no confusion should arise.
The following identities are easily verified from the definitions:

$$\ominus(\ominus x) = x$$

$$\ominus(x \oplus y) = (\ominus x) \oplus (\ominus y)$$

$$\ominus(x \otimes y) = (\ominus x) \otimes y.$$

Lemma 7.5.1 *Let* $x, y \in \mathbb{S}$. *Then the following hold*:

(a) $|x \oplus y| = |x| \oplus |y|$,
(b) $|x \otimes y| = |x| \otimes |y|$,
(c) $|\ominus x| = |x|$.

Proof Let $x = (a, b)$, $y = (c, d)$. Then $|x \oplus y| = a \oplus c \oplus b \oplus d$ and $|x| \oplus |y| = a \oplus b \oplus c \oplus d$, hence the first identity. Also, we have

$$|x \otimes y| = (a \otimes c \oplus b \otimes d) \oplus (a \otimes d \oplus b \otimes c)$$

$$= (a \oplus b) \otimes (c \oplus d) = |x| \otimes |y|.$$

Part (c) is trivial.   □

Let $x = (a, a')$, $y = (b, b')$. We say that x *balances* y (notation $x \, \triangledown \, y$) if
$a \oplus b' = a' \oplus b$. Note that although $\triangledown$ is reflexive and symmetric, it is not transitive.

If $x = (a, b)$ then x is called *sign-positive (sign-negative)*, if $a > b$ $(a < b)$ or
$x = \varepsilon$; x is called *signed* if it is either sign-positive or sign-negative; x is called
balanced if $a = b$, otherwise it is called *unbalanced*. Thus, ε is the only element of
$\mathbb{S}$ that is both signed and balanced.

Proposition 7.5.2 *Let* $x, y \in \mathbb{S}$. *Then* $x \otimes y$ *is balanced if either of* x, y *is balanced;* $x \oplus y$ *is balanced if both* x *and* y *are balanced.*

Proof Straightforwardly from the definitions. □

Due to the bijective semiring morphism $t \longrightarrow (t, -\infty)$ we will identify, when appropriate, the elements of $\overline{\mathbb{R}}$ and the sign-positive elements of $\mathbb{S}$ of the form $(t, -\infty)$. Conversely, a sign-positive element (a, b) may be identified with $a \in \overline{\mathbb{R}}$. So for instance 3 may denote the real number as well as the element $(3, -\infty)$ of $\mathbb{S}$. By these conventions we may write $3 \ominus 2 = 3, 3 \ominus 7 = \ominus 7, 3 \ominus 3 = 3^{\bullet}$.

The following are easily proved (see Exercise 7.7.6) for $x, y, u, v \in \mathbb{S}$:

$$x \triangledown y, u \triangledown v \Longrightarrow x \oplus u \triangledown y \oplus v, \tag{7.20}$$

$$x \triangledown y \Longrightarrow x \otimes u \triangledown y \otimes u, \tag{7.21}$$

$$x \triangledown y \text{ and } x = (a, a'), y = (b, b') \text{ are sign-positive} \Longrightarrow a = b. \tag{7.22}$$

The operations $\oplus$ and $\otimes$ are extended to matrices and vectors over $\mathbb{S}$ in the same way as in linear algebra; $\triangledown$ is extended componentwise. A vector is called *sign-positive (sign-negative, signed)*, if all its components are sign-positive (sign-negative, signed). The properties mentioned above hold if they are appropriately modified for vectors. For more details see [123].

Proposition 7.5.3 [123] *Let* $A, B \in \overline{\mathbb{R}}^{m \times n}$. *To every solution* $x \in \overline{\mathbb{R}}^n$ *of the system* $A \otimes x = B \otimes x$ *there exists a sign-positive solution to the system of linear balances* $(A \ominus B) \otimes x \triangledown \varepsilon$, *and conversely.*

Proof Let $A, B \in \overline{\mathbb{R}}^{m \times n}$. Then the following are equivalent:

$$A \otimes x = B \otimes x, \quad x \in \overline{\mathbb{R}}^n,$$

$$A \otimes x \triangledown B \otimes x, \quad x \text{ sign-positive,}$$

$$A \otimes x \ominus B \otimes x \triangledown \varepsilon, \quad x \text{ sign-positive,}$$

$$(A \ominus B) \otimes x \triangledown \varepsilon, \quad x \text{ sign-positive.} \qquad \square$$

We now define the determinant of matrices in symmetrized semirings. The (symmetrized) sign of a permutation σ is $\text{sgn}(\sigma) = 0$ if σ is even and it is $\ominus 0$ if σ is odd, see Sect. 1.6.4. The *determinant* of $A = (a_{ij}) \in \mathbb{S}^{n \times n}$ is

$$\det(A) = \sum_{\sigma \in P_n}^{\oplus} \left(\text{sgn}(\sigma) \otimes \prod_{i \in N}^{\otimes} a_{i,\sigma(i)} \right).$$

The following is an analogue of the classical result in conventional linear algebra and is proved essentially in the same way.

Theorem 7.5.4 [123] *Let $A \in \mathbb{S}^{n \times n}$. Then the system of balances $A \otimes x \; \triangledown \; \varepsilon$ has a signed nontrivial (i.e. $\neq \varepsilon$) solution if and only if A has balanced determinant.*

Since a signed vector may or may not be sign-positive, it is not true in general, that the system $A \otimes x = B \otimes x$ has a nontrivial solution if and only if $A \ominus B$ has a balanced determinant (see Proposition 7.5.3). But the necessary condition obviously follows:

Corollary 7.5.5 *Let $A, B \in \overline{\mathbb{R}}^{n \times n}$ and $C = A \ominus B$. Then a necessary condition that the system $A \otimes x = B \otimes x$ have a nontrivial solution is that C has balanced determinant.*

We therefore need a method for deciding whether a given square matrix has balanced determinant. In principle this is, of course, possible by calculating the determinant. However, such a computation is only practical for matrices of small sizes (see Examples 7.5.8, 7.5.9 and 7.5.10), since unlike in conventional linear algebra there is no obvious way to avoid considering all $n!$ permutations. We will show that this task can be converted using the max-algebraic permanent (or, in conventional terms, the assignment problem) to the question of sign-nonsingularity of matrices.

Recall that the max-algebraic permanent of $A = (a_{ij}) \in \overline{\mathbb{R}}^{n \times n}$ is

$$\text{maper}(A) = \sum_{\sigma \in P_n}^{\oplus} \prod_{i \in N}^{\otimes} a_{i,\sigma(i)}.$$

Clearly, since $\text{maper}(A) = \max_{\sigma \in P_n} \sum_{i \in N} a_{i,\sigma(i)}$, the value of $\text{maper}(A)$ can be found by solving the linear assignment problem for A (see Sect. 1.6.4). Recall that we denoted

$$\text{ap}(A) = \left\{ \sigma \in P_n; \; \text{maper}(A) = \sum_{i \in N} a_{i,\sigma(i)} \right\}.$$

We refer the reader to Sect. 1.6.4 for definitions and more details on the relation between the max-algebraic permanent and the assignment problem. We only recall that perhaps the best known solution method for the assignment problem is the Hungarian method of computational complexity $O(n^3)$. This algorithm transforms A to a nonpositive matrix $B = (b_{ij})$ with $\text{ap}(A) = \text{ap}(B)$ and $\text{maper}(B) = 0$. Thus for $\pi \in \text{ap}(B)$ we have $b_{i,\pi(i)} = 0$ for all $i \in N$. If $b_{ij} = 0$ for some $i, j \in N$ then a $\pi \in \text{ap}(B)$ with $j = \pi(i)$ may or may not exist. But this can easily be decided by checking that $\text{maper}(B_{ij}) = 0$ where B_{ij} is the matrix obtained from B by removing row i and column j.

If $C = (c_{ij}) \in \mathbb{S}^{n \times n}$ then we denote $|C| = (|c_{ij}|) \in \overline{\mathbb{R}}^{n \times n}$. We also have $\det(C) = (d^+(C), d^-(C))$ or, for simplicity just (d^+, d^-), and so $|\det(C)| = d^+ \oplus d^-$.

Proposition 7.5.6 *For every $C = (c_{ij}) \in \mathbb{S}^{n \times n}$ we have:*

$$|\det(C)| = \text{maper} \, |C|. \tag{7.23}$$

Proof By a repeated use of Lemma 7.5.1 we have

$$|\det(C)| = \left| \sum_{\sigma \in P_n}^{\oplus} \left(\text{sgn}(\sigma) \otimes \prod_{i \in N}^{\otimes} c_{i,\sigma(i)} \right) \right| = \sum_{\sigma \in P_n}^{\oplus} \left| \text{sgn}(\sigma) \otimes \prod_{i \in N}^{\otimes} c_{i,\sigma(i)} \right|$$

$$= \sum_{\sigma \in P_n}^{\oplus} \left| \prod_{i \in N}^{\otimes} c_{i,\sigma(i)} \right| = \sum_{\sigma \in P_n}^{\oplus} \prod_{i \in N}^{\otimes} |c_{i,\sigma(i)}| = \text{maper} |C|. \qquad \square$$

A square $(0, 1, -1)$ matrix is called *sign-nonsingular* (SNS) [18] if at least one term of its standard determinant expansion is nonzero and all nonzero terms have the same sign.

Given $C = (c_{ij}) \in \mathbb{S}^{n \times n}$ we define $\widetilde{C} = (\widetilde{c}_{ij})$ to be the $n \times n$ $(0, 1, -1)$ matrix satisfying

$\widetilde{c}_{ij} = 1$ if $j = \sigma(i)$ for some $\sigma \in \text{ap}|C|$ and c_{ij} is sign-positive,

$\widetilde{c}_{ij} = -1$ if $j = \sigma(i)$ for some $\sigma \in \text{ap}|C|$ and c_{ij} is sign-negative,

$\widetilde{c}_{ij} = 0$ else.

The matrix $\widetilde{C}$ can easily be constructed since, as mentioned above, it is straightforward to check whether $j = \sigma(i)$ for some $\sigma \in \text{ap}|C|$.

Theorem 7.5.7 [86] *Let $C \in \mathbb{S}^{n \times n}$. A sufficient condition that C have balanced determinant is that $\widetilde{C}$ is not SNS. If C has no balanced entry then this condition is also necessary.*

Proof If $\widetilde{C}$ is not SNS then either all terms of the standard determinant expansion of $\widetilde{C}$ are zero or there are two nonzero terms of opposite signs. In the first case every permutation $\sigma \in \text{ap}|C|$ selects a balanced element, thus by Proposition 7.5.2 every permutation has balanced weight and so $\det(C)$ is balanced. In the second case there are $\sigma, \sigma' \in \text{ap}|C|$ such that $\text{sgn}(\sigma)w(\sigma, \widetilde{C}) = 1$ and $\text{sgn}(\sigma)w(\sigma', \widetilde{C}) = -1$. Hence $\det(C)$ contains two maximal terms, one sign-positive and the other one sign-negative. Therefore $\det(C)$ is balanced.

Suppose now that $\det(C)$ is balanced. Since C has no balanced entry, $\det(C)$ contains a sign-positive and a sign-negative entry of maximal value. For the corresponding permutations $\sigma, \sigma' \in \text{ap}|C|$ we then have that they contribute to standard determinant expansion of $\widetilde{C}$ with $+1$ and -1 and so $\widetilde{C}$ is not SNS. $\qquad \square$

The problem of checking whether a $(0, 1, -1)$ matrix is SNS or not is equivalent to the even cycle problem in digraphs [18, 143] and therefore polynomially solvable (Remark 1.6.45). Therefore the necessary solvability condition in Corollary 7.5.5 can be checked in polynomial time and enables us to prove for some systems that no nontrivial solution to $A \otimes x = B \otimes x$ exists. Yet, it does not provide a solution method for solving the two-sided systems as this condition is not sufficient in general.

Note that in the examples below the question whether the determinant is balanced is decided directly using the definition and the sign-nonsingularity is not used. This would not be practical for matrices of bigger sizes.

Example 7.5.8 Let

$$A = \begin{pmatrix} 3 & 8 & 2 \\ 7 & 1 & 4 \\ 0 & 6 & 3 \end{pmatrix}, \qquad B = \begin{pmatrix} 4 & 4 & 3 \\ 2 & 3 & 4 \\ 3 & 2 & 1 \end{pmatrix}.$$

Then

$$C = \begin{pmatrix} \ominus 4 & 8 & \ominus 3 \\ 7 & \ominus 3 & 4^\bullet \\ \ominus 3 & 6 & 3 \end{pmatrix},$$

$$d^+ = \max\left(10, 15^\bullet, 9\right),$$

$$d^- = \max\left(16, 14^\bullet, 18\right),$$

$$\text{maper } |C| = 18.$$

Since $d^+ \neq d^-$, the determinant of $C = A \ominus B$ is unbalanced and so the system $A \otimes x = B \otimes x$ has no nontrivial solution.

Example 7.5.9 Let

$$A = \begin{pmatrix} 3 & 8 & 2 \\ 7 & 1 & 4 \\ 0 & 5 & 3 \end{pmatrix}, \qquad B = \begin{pmatrix} 5 & 5 & 5 \\ 3 & 4 & 5 \\ 5 & 3 & 2 \end{pmatrix}.$$

Then

$$C = \begin{pmatrix} \ominus 5 & 8 & \ominus 5 \\ 7 & \ominus 4 & \ominus 5 \\ \ominus 5 & 5 & 3 \end{pmatrix},$$

$$d^+ = \max(12, 18, 14),$$

$$d^- = \max(17, 15, 18),$$

$$\text{maper } |C| = 18 = d^+ = d^-.$$

Hence $\det(A \ominus B)$ is balanced, and indeed $x = (2, 1, 4)^T$ is a solution to $A \otimes x = B \otimes x$.

Example 7.5.10 Let

$$A = \begin{pmatrix} 4 & 6 \\ 7 & 9 \end{pmatrix}, \qquad B = \begin{pmatrix} 0 & 1 \\ 3 & 1 \end{pmatrix}.$$

$$C = \begin{pmatrix} 4 & 6 \\ 7 & 9 \end{pmatrix}$$

and

$$\text{maper}\,|C| = 13 = d^+ = d^-.$$

Hence the determinant is balanced but no nontrivial solution to $A \otimes x = B \otimes x$ exists as (by the cancellation law) B is effectively ε.

7.6 Solution Set is Finitely Generated

In this section the set $S(A, B) = \{x \in \overline{\mathbb{R}}^n; A \otimes x = B \otimes x\}$ will be denoted shortly by S. Also, in this section only, the letter I denotes an index set (not the unit matrix). The aim of this section is to prove the following fundamental result:

Theorem 7.6.1 [35] *If $A, B \in \overline{\mathbb{R}}^{m \times n}$ then S is finitely generated, that is, there is an integer $w \geq 1$ and a matrix $T \in \overline{\mathbb{R}}^{n \times w}$ such that*

$$S = \left\{ T \otimes z; z \in \overline{\mathbb{R}}^w \right\}.$$

Lemma 7.6.2 [35] *If $A, B \in \overline{\mathbb{R}}^{1 \times n}$ then there is an integer $w \geq 1$ and a matrix $T \in \overline{\mathbb{R}}^{n \times w}$ such that*

$$S = \left\{ T \otimes z; z \in \overline{\mathbb{R}}^w \right\}.$$

We postpone the proof of the lemma for a while and first prove the theorem.

Proof of Theorem 7.6.1 Let us denote (in this proof only) the rows of A and B by $A_1, \ldots, A_m$ and $B_1, \ldots, B_m$, respectively. By Lemma 7.6.2 there is a matrix $T_1 \in \overline{\mathbb{R}}^{n \times w_1}$ for some integer $w_1 \geq 1$ such that

$$\left\{ x \in \overline{\mathbb{R}}^n; A_1 \otimes x = B_1 \otimes x \right\} = \left\{ T_1 \otimes z^{(1)}; z^{(1)} \in \overline{\mathbb{R}}^{w_1} \right\}.$$

Similarly, there is an integer $w_2 \geq 1$ and a matrix $T_2 \in \overline{\mathbb{R}}^{w_1 \times w_2}$ such that

$$\left\{ z^{(1)} \in \overline{\mathbb{R}}^{w_1}; A_2 \otimes T_1 \otimes z^{(1)} = B_2 \otimes T_1 \otimes z^{(1)} \right\} = \left\{ T_2 \otimes z^{(2)}; z^{(2)} \in \overline{\mathbb{R}}^{w_2} \right\}.$$

This process continues until at the end we have that there is an integer $w_m \geq 1$ and a matrix $T_m \in \overline{\mathbb{R}}^{w_{m-1} \times w_m}$ such that

$$\left\{ z^{(m-1)} \in \overline{\mathbb{R}}^{w_{m-1}}; A_m \otimes T_1 \otimes \cdots \otimes T_{m-1} \otimes z^{(m-1)} \right.$$

$$= B_m \otimes T_1 \otimes \cdots \otimes T_{m-1} \otimes z^{(m-1)} \Big\}$$

$$= \Big\{ T_m \otimes z^{(m)}; z^{(m)} \in \overline{\mathbb{R}}^{w_m} \Big\}.$$

We now show that for the wanted T we can take $T_1 \otimes \cdots \otimes T_m$ and $w = w_m$. Suppose first that $x = T \otimes z$ for some $z \in \overline{\mathbb{R}}^w$ and $k \in M$. Then

$$A_k \otimes (T \otimes z) = A_k \otimes T_1 \otimes \cdots \otimes T_m \otimes z$$

$$= (A_k \otimes T_1 \otimes \cdots \otimes T_{k-1}) \otimes T_k \otimes (T_{k+1} \otimes \cdots \otimes T_m \otimes z)$$

$$= (B_k \otimes T_1 \otimes \cdots \otimes T_{k-1}) \otimes T_k \otimes (T_{k+1} \otimes \cdots \otimes T_m \otimes z)$$

by the definition of T_k. Hence $A_k \otimes (T \otimes z) = B_k \otimes (T \otimes z)$ and thus $T \otimes z \in S$.

Suppose now that $x \in S$. Then $A_1 \otimes x = B_1 \otimes x$, thus

$$x = T_1 \otimes z^{(1)}, \quad z^{(1)} \in \overline{\mathbb{R}}^{w_1}.$$

At the same time $A_2 \otimes x = B_2 \otimes x$ and so

$$A_2 \otimes T_1 \otimes z^{(1)} = B_2 \otimes T_1 \otimes z^{(1)}$$

implying

$$z^{(1)} = T_2 \otimes z^{(2)}, \quad z^{(2)} \in \overline{\mathbb{R}}^{w_2}$$

and therefore $x = T_1 \otimes z^{(1)} = T_1 \otimes T_2 \otimes z^{(2)}$. By induction then

$$x = T_1 \otimes T_2 \otimes \cdots \otimes T_m \otimes z^{(m)}, \quad z^{(m)} \in \overline{\mathbb{R}}^{w_m}. \qquad \square$$

One equation of the form $A \otimes x = B \otimes x$ can be written as follows:

$$\sum_{j \in N}^{\oplus} a_j \otimes x_j = \sum_{j \in N}^{\oplus} b_j \otimes x_j. \qquad (7.24)$$

Due to Lemma 7.4.1 we may assume without loss of generality that

$$a_j \neq b_j \implies \min(a_j, b_j) = \varepsilon \qquad (7.25)$$

holds for every $j \in N$. Hence after a suitable renumbering of variables this equation can symbolically be written as

$$(\varepsilon, \ldots, \varepsilon, e, \ldots, e, a, \ldots, a, \varepsilon, \ldots \varepsilon) \otimes x = (\varepsilon, \ldots, \varepsilon, e, \ldots, e, \varepsilon, \ldots \varepsilon, b, \ldots, b) \otimes x.$$

This form corresponds to the partition of N into four subsets:

$$I = \big\{ j \in N; a_j = b_j = \varepsilon \big\},$$

$$J = \big\{ j \in N; a_j = b_j \neq \varepsilon \big\},$$

$$K = \{j \in N; a_j > b_j\},$$
$$L = \{j \in N; a_j < b_j\}.$$

We now define five sets of vectors:

$$e^i = \left(e^i_1, \ldots, e^i_n\right)^T, i \in I,$$

where (as before) $e^i_j = \varepsilon$, if $j \neq i$ and $e^i_j = 0$, if $j = i$;

$$r^i = \left(r^i_1, \ldots, r^i_n\right)^T, \quad i \in J,$$

where $r^i_j = \varepsilon$, if $j \neq i$ and $r^i_j = a_i^{-1} = b_i^{-1}$, if $j = i$;

$$s^{k,l} = \left(s^{k,l}_1, \ldots, s^{k,l}_n\right)^T, \quad k \in K, l \in L,$$

where $s^{k,l}_j = \varepsilon$, if $j \notin \{k, l\}$, $s^{k,l}_j = a_k^{-1}$, if $j = k$ and $s^{k,l}_j = b_l^{-1}$, if $j = l$;

$$r^{i,h} = \left(r^{i,h}_1, \ldots, r^{i,h}_n\right)^T, \quad i \in J, h \in K \cup L,$$

where $r^{i,h}_j = r^i_j$, if $j \neq h, r^{i,h}_j = a_h^{-1}$, if $j = h \in K$ and $r^{i,h}_j = b_h^{-1}$, if $j = h \in L$;

$$s^{k,l,h} = \left(s^{k,l,h}_1, \ldots, s^{k,l,h}_n\right)^T, \quad k \in K, l \in L, h \in K \cup L - \{k, l\},$$

where $s^{k,l,h}_j = s^{k,l}_j$, if $j \neq h$, $s^{k,l,h}_j = a_h^{-1}$, if $j = h \in K - \{k\}$ and $s^{k,l,h}_j = b_h^{-1}$, if $j = h \in L - \{l\}$.

Lemma 7.6.3 *Equation* (7.24) *has a nontrivial solution if and only if*

$$I \cup J \cup (K \times L) \neq \emptyset.$$

Proof If $I \cup J \cup (K \times L) \neq \emptyset$ then at least one of the vectors $e^i, i \in I$; $r^i, i \in J$; $s^{k,l}, k \in K, l \in L$ exists and each of these vectors is a nontrivial solution.

If $I \cup J \cup (K \times L) = \emptyset$ then $I = J = K \times L = \emptyset$. Since $I \cup J \cup K \cup L = N$ and $K \cap L = \emptyset$ we have either that $L = \emptyset$ and $K = N$ or $K = \emptyset$ and $L = N$. In the first case equation (7.24) reduces to

$$\max_{i \in N} a_i \otimes x_i = \varepsilon$$

and $a_i > \varepsilon$ for all $i \in N$ which implies that $x = \varepsilon$ is the unique solution. The second case can be dealt with in the same way. $\qquad \square$

We are now ready to present the proof of the key lemma.

Proof of Lemma 7.6.2 We prove that $y \in \overline{\mathbb{R}}^n$ is a solution to (7.24) if and only if it can be written in the form

$$y = \sum_{i \in I}^{\oplus} \pi^i \otimes e^i \oplus \sum_{i \in J}^{\oplus} \rho^i \otimes r^i \oplus \sum_{k \in K, l \in L}^{\oplus} \sigma^{k,l} \otimes s^{k,l}$$

$$\oplus \sum_{i \in J, h \in K \cup L}^{\oplus} \rho^{i,h} \otimes r^{i,h} \oplus \sum_{k \in K, l \in L, h \in K \cup L - \{k,l\}}^{\oplus} \sigma^{k,l,h} \otimes s^{k,l,h} \qquad (7.26)$$

where $\pi^i, \rho^i, \sigma^{k,l}, \rho^{i,h}, \sigma^{k,l,h} \in \mathbb{R}$. Note that if the index sets in this summation are empty then by Lemma 7.6.3 $S = \{\varepsilon\}$ and we may take any $w \geq 1$ and set $T = \varepsilon$.

It is easily seen that each of the vectors $e^i, i \in I; r^i, i \in J; s^{k,l}, k \in K, l \in L;$ $r^{i,h}, i \in J, h \in K \cup L; s^{k,l,h}, k \in K, l \in L, h \in K \cup L - \{k,l\}$ is a solution to (7.24) and thus by Proposition 7.1.1 also their max-algebraic linear combination is in S.

It remains to prove that every solution can be expressed as in (7.26). Let $x = (x_1, \ldots, x_n)^T \in S$ and let

$$v = \sum_{j \in N}^{\oplus} a_j \otimes x_j = \sum_{j \in N}^{\oplus} b_j \otimes x_j.$$

At least one of the following will occur:

Case 1: $v = \varepsilon$.
Case 2: $v \neq \varepsilon$ and $v = a_j \otimes x_j = b_j \otimes x_j$ for some $j \in J$.
Case 3: $v \neq \varepsilon$ and $v = a_f \otimes x_f = b_g \otimes x_g$ for some $f \in K$ and $g \in L$.

In Case 1 $x_i = \varepsilon$ for all $i \in J \cup K \cup L$ and thus it is sufficient to set $\pi^i = x_i$ for all $i \in I$ and all other coefficients set to ε.

In Case 2 we have $a_j = b_j > \varepsilon$ and $a_i \otimes x_i \leq v$, $b_i \otimes x_i \leq v$ for all $i \in N$, implying

$$\left. \begin{array}{l} a_j^{-1} \otimes a_i \otimes x_i \leq x_j, \\ b_j^{-1} \otimes b_i \otimes x_i \leq x_j. \end{array} \right\} \qquad (7.27)$$

Set

$$\pi^i = x_i, i \in I,$$

$$\rho^i = a_i \otimes x_i, i \in J,$$

$$\rho^{j,h} = a_h \otimes x_h, h \in K,$$

$$\rho^{j,h} = b_h \otimes x_h, h \in L,$$

and set $\rho^{i,h} = \varepsilon$ for all $i \in J - \{j\}, h \in K \cup L$ and also all other coefficients to ε. Let y be defined by (7.26), we show that $y = x$.

Let $t \in I$. Then

$$y_t = \sum_{i \in I}^{\oplus} \pi^i \otimes e_t^i = \pi^t \otimes e_t^t = x_t \otimes 0 = x_t.$$

Take $t \in J - \{j\}$. Then

$$y_t = \sum_{i \in J}^{\oplus} \rho^i \otimes r_t^i \oplus \sum_{h \in K \cup L}^{\oplus} \rho^{j,h} \otimes r_t^{j,h}$$

$$= \rho^t \otimes a_t^{-1} \oplus \varepsilon = a_t \otimes x_t \otimes a_t^{-1} = x_t$$

since here $t \notin K \cup L$ and $t \neq j$. Also, using (7.27) we have

$$y_j = \rho^j \otimes a_j^{-1} \oplus \sum_{h \in K}^{\oplus} \rho^{j,h} \otimes r_j^j \oplus \sum_{h \in L}^{\oplus} \rho^{j,h} \otimes r_j^j$$

$$= x_j \oplus \sum_{h \in K}^{\oplus} a_h \otimes x_h \otimes a_j^{-1} \oplus \sum_{h \in L}^{\oplus} b_h \otimes x_h \otimes b_j^{-1} = x_j.$$

Now take $t \in K$. Then

$$y_t = \sum_{i \in J, h \in K \cup L}^{\oplus} \rho^{i,h} \otimes r_t^{i,h} = \sum_{h \in K}^{\oplus} \rho^{j,h} \otimes r_t^{j,h}$$

$$= \rho^{j,t} \otimes r_t^{j,t} = a_t \otimes x_t \otimes a_t^{-1} = x_t.$$

Similarly it can be shown that $y_t = x_t$ for $t \in L$.

In Case 3 we have $a_f, b_g > \varepsilon$ and for all $i \in N$ there is

$$\left. \begin{array}{l} a_f^{-1} \otimes a_i \otimes x_i \leq x_f, \\ b_g^{-1} \otimes b_i \otimes x_i \leq x_g. \end{array} \right\} \tag{7.28}$$

Set

$$\pi^i = x_i, i \in I,$$

$$\rho^i = a_i \otimes x_i, i \in J,$$

$$\sigma^{f,g} = a_f \otimes x_f = b_g \otimes x_g,$$

$$\sigma^{f,g,h} = a_h \otimes x_h, \quad \text{if } h \in K,$$

$$= b_h \otimes x_h, \quad \text{if } h \in L,$$

and all other coefficients to ε. Let y be again defined by (7.26), and take any $t \in I$. Then

$$y_t = \sum_{i \in I}^{\oplus} \pi^i \otimes e_t^i = \pi^t \otimes e_t^t = x_t \otimes 0 = x_t.$$

Take $t \in J$. Then

$$y_t = \sum_{i \in J}^{\oplus} \rho^i \otimes r_t^i$$

$$= \rho^t \otimes a_t^{-1} = a_t \otimes x_t \otimes a_t^{-1} = x_t.$$

Now take $t \in K - \{f\}$. Then

$$y_t = \sigma^{f,g} \otimes s_t^{f,g} \oplus \sum_{h \in K \cup L - \{f,g\}}^{\oplus} \sigma^{f,g,h} \otimes s_t^{f,g,h}$$

$$= \varepsilon \oplus \sigma^{f,g,t} \otimes s_t^{f,g,t} = a_t \otimes x_t \otimes a_t^{-1} = x_t,$$

since $t \notin \{f, g\}$. Also, using (7.28) we have

$$y_f = \sigma^{f,g} \otimes s_f^{f,g} \oplus \sum_{h \in K \cup L - \{f,g\}}^{\oplus} \sigma^{f,g,h} \otimes s_f^{f,g,h}$$

$$= a_f \otimes x_f \otimes a_f^{-1} \oplus \sum_{h \in K \cup L - \{f,g\}}^{\oplus} a_h \otimes x_h \otimes a_f^{-1} = x_f.$$

The subcase $t \in L$ can be proved in a similar way.   □

Alongside the theoretical value, the constructive proofs of Theorem 7.6.1 and Lemma 7.6.2 show how to solve systems $A \otimes x = B \otimes x$. The number of variables is likely to grow rapidly during this process and so the method is unlikely to be useful except for the systems with a small number of variables and equations. Obviously, columns of a matrix T_i that are a max-combination of the others may be eliminated. We will illustrate this in the two examples below. Note that the $\mathcal{A}$-test specified in Theorem 3.4.2 may be used to find and eliminate the linearly dependent columns.

There are several improvements of this method; one of them can be found in [7].

Example 7.6.4 Let

$$A = \begin{pmatrix} 3 & 2 & \varepsilon \\ \varepsilon & \varepsilon & 2 \\ 2 & 0 & 3 \end{pmatrix}, \qquad B = \begin{pmatrix} 3 & \varepsilon & 0 \\ 0 & \varepsilon & 2 \\ \varepsilon & 0 & 3 \end{pmatrix}.$$

Then for the first equation we have $I = \emptyset$, $J = \{1\}$, $K = \{2\}$, $L = \{3\}$ and thus

$$r^1 = (-3, \varepsilon, \varepsilon)^T,$$

$$s^{2,3} = (\varepsilon, -2, 0)^T,$$

$$r^{1,2} = (-3, -2, \varepsilon)^T; \qquad r^{1,3} = (-3, \varepsilon, 0)^T.$$

Hence $w_1 = 4$ and

$$T_1 = \begin{pmatrix} -3 & \varepsilon & -3 & -3 \\ \varepsilon & -2 & -2 & \varepsilon \\ \varepsilon & 0 & \varepsilon & 0 \end{pmatrix}.$$

Therefore

$$A_2 \otimes T_1 = (\varepsilon, \varepsilon, 2) \otimes T_1 = (\varepsilon, 2, \varepsilon, 2)^T,$$

$$B_2 \otimes T_1 = (0, \varepsilon, 2) \otimes T_1 = (-3, 2, -3, 2)^T .$$

For the equation $A_2 \otimes T_1 \otimes z^{(1)} = B_2 \otimes T_1 \otimes z^{(1)}$ we then get $I = \emptyset$, $J = \{2, 4\}$, $K = \emptyset$, $L = \{1, 3\}$ and thus

$$r^2 = (\varepsilon, -2, \varepsilon, \varepsilon)^T , \qquad r^4 = (\varepsilon, \varepsilon, \varepsilon, -2)^T ;$$

$$r^{2,1} = (3, -2, \varepsilon, \varepsilon)^T , \qquad r^{2,3} = (\varepsilon, -2, 3, \varepsilon)^T ;$$

$$r^{4,1} = (3, \varepsilon, \varepsilon, -2)^T , \qquad r^{4,3} = (\varepsilon, \varepsilon, 3, -2)^T .$$

Hence $w_1 = 6$ and

$$T_2 = \begin{pmatrix} \varepsilon & \varepsilon & 3 & \varepsilon & 3 & \varepsilon \\ -2 & \varepsilon & -2 & -2 & \varepsilon & \varepsilon \\ \varepsilon & \varepsilon & \varepsilon & 3 & \varepsilon & 3 \\ \varepsilon & -2 & \varepsilon & \varepsilon & -2 & -2 \end{pmatrix} .$$

Therefore

$$T_1 \otimes T_2 = \begin{pmatrix} \varepsilon & -5 & 0 & 0 & 0 & 0 \\ -4 & \varepsilon & -4 & 1 & \varepsilon & 1 \\ -2 & -2 & -2 & -2 & -2 & -2 \end{pmatrix} .$$

By inspection we see that columns 4 and 6 are equal and column 3 is the max-algebraic sum of columns 1 and 5, thus we may remove redundant columns 3 and 6 and continue to work with the reduced matrix

$$T' = \begin{pmatrix} \varepsilon & -5 & 0 & 0 \\ -4 & \varepsilon & 1 & \varepsilon \\ -2 & -2 & -2 & -2 \end{pmatrix} .$$

Since

$$A_3 \otimes T' = (1, 1, 2, 2), \qquad B_3 \otimes T' = (1, 1, 1, 1),$$

for the equation $A_3 \otimes T' \otimes z^{(2)} = B_3 \otimes T' \otimes z^{(2)}$ we then get $I = \emptyset$, $J = \{1, 2\}$, $K = \{3, 4\}$, $L = \emptyset$ and thus

$$r^1 = (-1, \varepsilon, \varepsilon, \varepsilon)^T , \qquad r^2 = (\varepsilon, -1, \varepsilon, \varepsilon)^T ;$$

$$r^{1,3} = (-1, \varepsilon, -2, \varepsilon)^T , \qquad r^{1,4} = (-1, \varepsilon, \varepsilon, -2)^T ;$$

$$r^{2,3} = (\varepsilon, -1, -2, \varepsilon)^T , \qquad r^{2,4} = (\varepsilon, -1, \varepsilon, -2)^T .$$

Hence $w_3 = 6$ and

$$T = T' \otimes T_3$$

$$= T' \otimes \begin{pmatrix} -1 & \varepsilon & -1 & -1 & \varepsilon & \varepsilon \\ \varepsilon & -1 & \varepsilon & \varepsilon & -1 & -1 \\ \varepsilon & \varepsilon & -2 & \varepsilon & -2 & \varepsilon \\ \varepsilon & \varepsilon & \varepsilon & -2 & \varepsilon & -2 \end{pmatrix}$$

$$= \begin{pmatrix} \varepsilon & -6 & -2 & -2 & -2 & -2 \\ -5 & \varepsilon & -1 & -5 & -1 & \varepsilon \\ -3 & -3 & -3 & -3 & -3 & -3 \end{pmatrix}.$$

Columns 4 and 5 are dependent and so we conclude that x is a solution to $A \otimes x = B \otimes x$ if and only if

$$x = \begin{pmatrix} \varepsilon & -6 & -2 & -2 \\ -5 & \varepsilon & -1 & \varepsilon \\ -3 & -3 & -3 & -3 \end{pmatrix} \otimes z$$

where $z \in \overline{\mathbb{R}}^4$.

Example 7.6.5 Let $A = \left(\begin{smallmatrix} 1 & 0 \\ 0 & 1 \end{smallmatrix}\right)$, $B = \left(\begin{smallmatrix} \varepsilon & 0 \\ 0 & \varepsilon \end{smallmatrix}\right)$. Then for the first equation we have $I = \emptyset$, $J = \{2\}$, $K = \{1\}$, $L = \emptyset$ and thus

$$r^2 = (\varepsilon, 0)^T ,$$

$$r^{2,1} = (-1, 0)^T .$$

Hence $w_1 = 2$ and $T_1 = \left(\begin{smallmatrix} \varepsilon & -1 \\ 0 & 0 \end{smallmatrix}\right)$. Therefore

$$A_2 \otimes T_1 = (1, 1)^T$$

and

$$B_2 \otimes T_1 = (\varepsilon, -1)^T .$$

For the equation $A_2 \otimes T_1 \otimes z^{(1)} = B_2 \otimes T_1 \otimes z^{(1)}$ we then get $I = \emptyset$, $J = \emptyset$, $K = \{1, 2\}$, $L = \emptyset$ and thus by Lemma 7.6.3 the system $A \otimes x = B \otimes x$ has only the trivial solution.

7.7 Exercises

Exercise 7.7.1 Let

$$A = \begin{pmatrix} 4 & 6 \\ 1 & 2 \\ 3 & 0 \\ 6 & 6 \end{pmatrix}, \qquad B = \begin{pmatrix} 7 & 1 \\ 3 & 0 \\ 0 & 3 \\ 1 & 8 \end{pmatrix}.$$

Use the Gondran–Minoux Theorem to prove that the system $A \otimes x = B \otimes y$ has no nontrivial solution.

Exercise 7.7.2 Simplify each of the following systems using the cancellation law and then find a nontrivial solution or prove that there is none:

(a)

$$3 \otimes x_1 \oplus 4 \otimes x_2 \oplus 7 \otimes x_3 = 5 \otimes x_1 \oplus 1 \otimes x_2 \oplus 2 \otimes x_3$$
$$6 \otimes x_1 \oplus 3 \otimes x_2 \oplus 1 \otimes x_3 = 5 \otimes x_1 \oplus 2 \otimes x_2 \oplus 4 \otimes x_3.$$

[No solution]

(b)

$$1 \otimes x_1 \oplus 4 \otimes x_2 \oplus 2 \otimes x_3 = 0 \otimes x_1 \oplus 5 \otimes x_2 \oplus 3 \otimes x_3$$
$$2 \otimes x_1 \oplus 1 \otimes x_2 \oplus 6 \otimes x_3 = 1 \otimes x_1 \oplus 7 \otimes x_2 \oplus 0 \otimes x_3.$$

$[(4, 0, 2)^T]$

Exercise 7.7.3 Find a nontrivial solution to the system $A \otimes x = B \otimes x$, where

$$A = \begin{pmatrix} -4 & 3 & 0 & 2 \\ 5 & -1 & 6 & 3 \\ 7 & 3 & 0 & 4 \end{pmatrix}, \qquad B = \begin{pmatrix} 0 & 2 & 6 & 1 \\ 3 & 5 & 0 & 7 \\ 2 & 12 & 6 & 3 \end{pmatrix}.$$

$[x = (7, 2, 0, 5)^T]$

Exercise 7.7.4 Find a nontrivial solution to the system $A \otimes x = B \otimes y$, where

$$A = \begin{pmatrix} 5 & 8 & 1 \\ 3 & 6 & 2 \\ 5 & 0 & 3 \end{pmatrix}, \qquad B = \begin{pmatrix} 4 & 2 & 8 & 1 \\ 3 & 0 & 5 & 0 \\ 2 & -3 & 4 & 1 \end{pmatrix}.$$

$[x = (1, 2, 0)^T, y = (1, 8, 0, 5)^T]$

Exercise 7.7.5 Show that if A, B have all entries from $\{0, -\infty\}$ then the system $A \otimes x = B \otimes x$ can be solved in polynomial time. (Hint: Transform the system to an equivalent one where A and B have no ε rows.)

Exercise 7.7.6 Prove (7.20), (7.21) and (7.22).

Exercise 7.7.7 For each of the matrices below decide whether it is sign-nonsingular:

(a) $A = \begin{pmatrix} 1 & 0 & -1 \\ 1 & -1 & -1 \\ 0 & 1 & 1 \end{pmatrix}$. [Not SNS]

(b) $A = \begin{pmatrix} 1 & 1 & -1 \\ -1 & 0 & 1 \\ 1 & 1 & 1 \end{pmatrix}$. [Not SNS]

(c) $A = \begin{pmatrix} 1 & 1 & -1 \\ -1 & 0 & -1 \\ 0 & 1 & 1 \end{pmatrix}$. [SNS]

Exercise 7.7.8 For each pair of matrices A, B below consider the system $A \otimes x = B \otimes x$. Find $C = A \ominus B$ and decide whether $\tilde{C}$ is SNS. Then decide whether the system has a nontrivial solution and find one if applicable.

(a) $A = \begin{pmatrix} 3 & 1 & 7 \\ 2 & 4 & 0 \\ 6 & 3 & 5 \end{pmatrix}$, $B = \begin{pmatrix} 2 & 3 & 2 \\ 4 & 0 & 3 \\ 2 & 1 & 7 \end{pmatrix}$. [$\tilde{C}$ is SNS, no nontrivial solution]

(b) $A = \begin{pmatrix} 6 & 1 & 2 \\ 1 & 5 & 0 \\ 2 & 1 & 6 \end{pmatrix}$, $B = \begin{pmatrix} 4 & 5 & 5 \\ 2 & 3 & 1 \\ 7 & 3 & 1 \end{pmatrix}$. [$\tilde{C}$ is not SNS, a solution is $x = (3, 0, 4)^T$]

Exercise 7.7.9 Show by an example that ∇ is not transitive.

Chapter 8
Reachability of Eigenspaces

One of the aims of this book is analysis of multi-machine interactive production processes (see Sect. 1.3.3). Recall that in these processes machines $M_1, \ldots, M_n$ work
interactively and in stages. In each stage all machines simultaneously produce components necessary for the next stage of some or all other machines. If $x_i(k)$ denotes
the starting time of the kth stage on machine i, and a_{ij} denotes the duration of the
operation at which machine M_j prepares the component necessary for machine M_i
in the $(k+1)$st stage then

$$x_i(k+1) = \max(x_1(k) + a_{i1}, \ldots, x_n(k) + a_{in}) \quad (i = 1, \ldots, n; k = 0, 1, \ldots)$$

or, in max-algebraic notation,

$$x(k+1) = A \otimes x(k) \quad (k = 0, 1, \ldots)$$

where $A = (a_{ij})$. We say that the system reaches a *steady regime* if it eventually
moves forward in regular steps, that is, if for some λ and k_0 we have $x(k + 1) =
\lambda \otimes x(k)$ for all $k \geq k_0$. Obviously, a steady regime is reached immediately if $x(0)$
is an eigenvector of A corresponding to an eigenvalue λ. However, if the choice of
a start-time vector is restricted we may need to find out for which vectors a steady
regime will eventually be reached. Since $x(k) = A^k \otimes x(0)$ for every natural k, we
get the following generic question:

Q: *Given $A \in \overline{\mathbb{R}}^{n \times n}$ and $x \in \overline{\mathbb{R}}^n$ is there an integer $k \geq 0$ such that $A^k \otimes x$ is an
eigenvector of A? That is, does*

$$\left. \begin{array}{c} A^{k+1} \otimes x = \lambda \otimes A^k \otimes x, \\[2mm] A^k \otimes x \neq \varepsilon, \end{array} \right\} \tag{8.1}$$

hold for some $\lambda \in \overline{\mathbb{R}}$?

Clearly, λ in (8.1) is one of the eigenvalues of A and therefore $\lambda = \lambda(A)$ if A is
irreducible.

In general, if $\lambda > \varepsilon$ and $A^k \otimes x$ is an eigenvector of A associated with λ then

$$A^{k+1} \otimes x = A \otimes (A^k \otimes x) = \lambda \otimes (A^k \otimes x) \neq \varepsilon$$

P. Butkovič, *Max-linear Systems: Theory and Algorithms*,
Springer Monographs in Mathematics 151,
DOI 10.1007/978-1-84996-299-5_8, © Springer-Verlag London Limited 2010

and hence $A^{k+1} \otimes x$ is also an eigenvector of A. However, if $\lambda = \varepsilon$ then $A^{k+1} \otimes x$ may not be an eigenvector even if $A^k \otimes x$ is, for instance when

$$A = \begin{pmatrix} \varepsilon & 0 \\ \varepsilon & \varepsilon \end{pmatrix}, \qquad x = \begin{pmatrix} 0 \\ 0 \end{pmatrix}, \qquad k = 1,$$

in which case $A \otimes x = (0, \varepsilon)^T$, $A^2 \otimes x = (\varepsilon, \varepsilon)^T$. We will therefore require $\lambda > \varepsilon$ in (8.1).

Recall that $A \in \overline{\mathbb{R}}^{n \times n}$ may have up to n eigenspaces, corresponding to a different eigenvalue each (Chap. 4, Theorem 4.5.4, Corollary 4.5.7). Being motivated by the task Q, we define for $A = (a_{ij}) \in \overline{\mathbb{R}}^{n \times n}$ and $x \in \overline{\mathbb{R}}^n$ the *orbit* of A with *starting vector* x as the sequence

$$O(A, x) = \{A^k \otimes x\}_{k=0,1,\dots}.$$

If $O(A, x)$ contains an eigenvector of a matrix B then we say that an *eigenspace of B is reachable by orbit* $O(A, x)$. If $A = B$ then we say *an eigenspace of A is reachable with starting vector x*.

Although the answer to Q may be negative, some periodic behavior can always be guaranteed provided that the production matrix is irreducible. This is due to one of the fundamental results of max-algebra, the Cyclicity Theorem (Theorem 8.3.5):

For any irreducible matrix $A \in \overline{\mathbb{R}}^{n \times n}$ there exist positive integers p and T such that

$$A^{k+p} = (\lambda(A))^p \otimes A^k \tag{8.2}$$

holds for every integer $k \geq T$.

The smallest value of p satisfying (8.2) is called the *period* of A. If p is the period of A then the least value of T satisfying (8.2) is called the *transient* of the sequence $\{A^k\}_{k=0}^{\infty}$.

It is easily seen that $A^k \otimes x \neq \varepsilon$ for all k if A is irreducible and $x \neq \varepsilon$ (Lemma 1.5.2). It follows that for any irreducible matrix A a generalized periodic regime will be reached with any starting vector $x \neq \varepsilon$:

$$\left. \begin{aligned} A^{k+p} \otimes x &= (\lambda(A))^p \otimes A^k \otimes x, \\ A^k \otimes x &\neq \varepsilon. \end{aligned} \right\} \tag{8.3}$$

We will use the notions of a *period* and *transient* for matrix orbits in a way similar to matrix sequences. The arising operational task then is to find the period of an orbit:

Q1: *Given $A \in \overline{\mathbb{R}}^{n \times n}$ and $x \in \overline{\mathbb{R}}^n$, find the period of $O(A, x)$, that is, the least integer p such that for some T (8.3) is satisfied for all $k \geq T$.*

Given $A \in \overline{\mathbb{R}}^{n \times n}$ and a positive integer p, the *p-attraction space*, $\mathrm{Attr}(A, p)$, is the set of all vectors $x \in \overline{\mathbb{R}}^n$, for which there exists an integer T such that (8.3) holds for every $k \geq T$ [16]. Using this concept we may formulate another related question:

Q2: *Given* $A \in \overline{\mathbb{R}}^{n \times n}$, $x \in \overline{\mathbb{R}}^n$ *and a positive integer* p *decide whether* $x \in \mathrm{Attr}(A, p)$.

Note that Q2 for $p = 1$ is identical with Q. Also, observe that due to Theorem 4.5.10, conditions (8.3) may be written as:

$$A^p \otimes (A^k \otimes x) = (\lambda(A)^p) \otimes (A^k \otimes x),$$
$$A^k \otimes x \neq \varepsilon,$$

that is, (8.3) for a given matrix A and vector x means to find the smallest value of p for which an eigenspace of A^p is reachable by $O(A, x)$. Note also that by Theorem 7.6.1 every attraction space is a finitely generated subspace.

It may be of practical interest to characterize matrices, called *robust*, for which a steady regime is reached with any start-time vector, that is, matrices $A \in \overline{\mathbb{R}}^{n \times n}$ such that an eigenspace of A is reachable with any vector $x \in \overline{\mathbb{R}}^n$, $x \neq \varepsilon$. Hence we will also be interested in the following:

Q3: *Given* $A \in \overline{\mathbb{R}}^{n \times n}$, *is it robust?*

In this chapter we will address Q1 and Q2 for irreducible matrices and Q3 for both irreducible and reducible matrices. We will also analyze a number of related questions, such as estimates of the transient and computation of periodic powers of a matrix.

If $\lambda(A) > \varepsilon$ and (8.3) is multiplied by $(\lambda(A))^{-k-p}$ then the obtained identity reads

$$B^{k+p} \otimes x = B^k \otimes x \neq \varepsilon,$$

where B is the definite matrix $(\lambda(A))^{-1} \otimes A$. Therefore in Q1–Q3 we may assume without loss of generality that A is definite.

A first step towards our goal is to present in Sect. 8.1 the diagonal scaling of matrices as a tool for the visualization of spectral properties of matrices. Then, in Sect. 8.2, we study how eigenspaces change with matrix powers.

Sections 8.3, 8.4 and 8.5 present periodic properties of matrices and methods for solving questions Q1 and Q2. They have been prepared in cooperation with Sergeĭ Sergeev.

Finally, robustness (question Q3) is studied in Sect. 8.6.

8.1 Visualization of Spectral Properties by Matrix Scaling

Recall that for $x = (x_1, \ldots, x_n)^T \in \overline{\mathbb{R}}^n$ we denote $\mathrm{diag}(x) = \mathrm{diag}(x_1, \ldots, x_n)$; if $x \in \mathbb{R}^n$ and $X = \mathrm{diag}(x)$ then

$$X^{-1} = \mathrm{diag}(x_1^{-1}, \ldots, x_n^{-1}).$$

A useful tool in our discussion will be that of a *matrix scaling* [41, 81, 127, 129] introduced in Sect. 1.5, that is, an operator that assigns to a square matrix A a matrix $X^{-1} \otimes A \otimes X$, where X is a diagonal matrix. Using matrix scaling it is possible to simplify the structure of a matrix, yet preserving many of its properties. In particular, it enables us to "visualize" some features, such as entries corresponding to the arcs on critical cycles.

First we show that matrix scaling does not change essential spectral properties of matrices [60, 80] and then we show the visualization effect. Recall that $\mathrm{pd}(A)$ stands for the principal dimension of A, that is, the dimension of the principal eigenspace of A.

Lemma 8.1.1 *Let* $A, B \in \overline{\mathbb{R}}^{n \times n}$ *and* $B = X^{-1} \otimes A \otimes X$, *where* $X = \mathrm{diag}(x_1, \ldots, x_n)$, $x_1, \ldots, x_n \in \mathbb{R}$.

(a) *A is irreducible if and only if B is irreducible.*
(b) *$\lambda(A) = \lambda(B)$.*
(c) *$N_c(A)$ and $N_c(B)$ are equal and have the same equivalence classes.*
(d) *$\mathrm{pd}(A) = \mathrm{pd}(B)$.*
(e) *For all integers $k \geq 1$ and $x \in \overline{\mathbb{R}}^n$ we have:*

$$B^k = X^{-1} \otimes A^k \otimes X.$$

(f) *$\Gamma(B) = X^{-1} \otimes \Gamma(A) \otimes X$ and $\Delta(B) = X^{-1} \otimes \Delta(A) \otimes X$.*
(g) *For all integers $p \geq 1$ we have: $z \in \overline{\mathbb{R}}^n$ satisfies*

$$A^{k+p} \otimes z = A^k \otimes z$$

if and only if $y = X^{-1} \otimes z$ satisfies

$$B^{k+p} \otimes y = B^k \otimes y.$$

Proof D_B has the same node set and arc set as D_A and so the first statement follows. By Lemma 1.5.5 $w(\sigma, A) = w(\sigma, B)$ for every cycle σ and so (b) and consequently also (c) and (d) follow (recall that by Corollary 4.4.5 $\mathrm{pd}(A)$ is equal to the number of critical components of A). Clearly,

$$(X^{-1} \otimes A \otimes X)^k = X^{-1} \otimes A^k \otimes X,$$

which proves (e).

For (f) we have using (e):

$$\Gamma(B) = \sum_{j \in N}^{\oplus} (X^{-1} \otimes A \otimes X)^j = \sum_{j \in N}^{\oplus} (X^{-1} \otimes A^j \otimes X)$$

$$= X^{-1} \otimes \left(\sum_{j \in N}^{\oplus} A^j \right) \otimes X = X^{-1} \otimes \Gamma(A) \otimes X.$$

Similarly for $\Delta(B)$.

Statement (g) is proved readily using (e).   □

Lemma 8.1.1 implies that the tasks Q1–Q3 are invariant with respect to matrix scaling. In what follows we will use a special type of scaling, namely a scaling that visualizes spectral properties of matrices.

We say that $A = (a_{ij}) \in \overline{\mathbb{R}}^{n \times n}$ is *visualized* if

$$a_{ij} \leq \lambda(A) \quad \text{for all } i, j \in N$$

and

$$a_{ij} = \lambda(A) \quad \text{for all } (i, j) \in E_c(A).$$

A visualized matrix is called *strictly visualized* if

$$a_{ij} = \lambda(A) \quad \text{if and only if} \quad (i, j) \in E_c(A).$$

A matrix A with $\lambda(A) = \varepsilon$ cannot be scaled to a visualized one unless $A = \varepsilon$. We will show in Theorem 8.1.4 below that every matrix with $\lambda(A) > \varepsilon$ can be transformed to a strictly visualized one using matrix scaling. However, we will also present a weaker scaling result in Theorem 8.1.3, as it is much simpler and in many cases sufficient.

Observe that $X^{-1} \otimes A \otimes X$ is visualized [strictly visualized] if and only if $X^{-1} \otimes A_\lambda \otimes X$ is visualized [strictly visualized]. Therefore we may assume without loss of generality that the matrix we need to scale to a visualized or strictly visualized one, is definite (but we will not always need to do so).

We start with a technical lemma. Let us denote $\Delta(A) = (\Delta_{ij})$.

Lemma 8.1.2 *If $A \in \overline{\mathbb{R}}^{n \times n}$ is definite then*

$$a_{ij} \otimes \Delta_{ji} \leq 0$$

for all $i, j \in N$ and

$$a_{ij} \otimes \Delta_{ji} = 0 \quad \Longleftrightarrow \quad (i, j) \in E_c(A).$$

Proof Since A is definite, $a_{ii} \leq 0 = \Delta_{ii}$ for any $i \in N$ and $a_{ii} = 0$ if and only if $(i, i) \in E_c(A)$.

Suppose now $i \neq j$. Then $\Delta_{ij} = \gamma_{ij}$. Recall that $\Gamma(A) = (\gamma_{ij})$ is the matrix of the greatest weights of paths in D_A (Sect. 1.6.2). Therefore $a_{ij} \otimes \Delta_{ji}$ is the weight of a heaviest cycle in D_A containing arc (i, j). Since A is definite, this value is nonpositive for any $(i, j) \in E$ and it is zero exactly when $(i, j) \in E_c(A)$. $\qquad \square$

Theorem 8.1.3 [41, 43, 136, 139] *Let $A \in \overline{\mathbb{R}}^{n \times n}$, $\lambda(A) > \varepsilon$ and*

$$(\Delta_{ij}) = \Delta\big((\lambda(A))^{-1} \otimes A\big).$$

(a) *If $x \in V^*(A)$ and $X = \mathrm{diag}(x)$ then $X^{-1} \otimes A \otimes X$ is visualized; this is true in particular for $x = \sum_{k \in N}^{\oplus} \Delta_{.k}$.*

(b) *If A is irreducible, $\alpha_k > 0$ for $k \in N$, $\sum_{k \in N} \alpha_k = 1$ and $x = \sum_{k \in N} \alpha_k \Delta_{.k}$ (conventional convex combination with positive coefficients), then $X^{-1} \otimes A \otimes X$ is strictly visualized.*

Proof (a) By Theorem 1.6.18 $x_i^{-1} \otimes a_{ij} \otimes x_j \leq \lambda(A)$ for all $i, j \in N$; equality for $(i, j) \in E_c(A)$ follows from Lemma 1.6.19.

(b) We assume without loss of generality that A is definite. Recall that $\Delta(A)$ is finite for A irreducible. By Lemma 1.6.20 x is a finite solution to $A \otimes x \leq x$. Hence $x \in V^*(A)$ by Theorem 1.6.18 and by part (a) $X^{-1} \otimes A \otimes X$ is visualized.

For strong visualization we need to show that $(i, j) \notin E_c(A)$ implies $a_{ij} \otimes x_j < x_i$. This inequality is equivalent to

$$a_{ij} + \sum_{k \in N} \alpha_k \Delta_{jk} < \sum_{k \in N} \alpha_k \Delta_{ik}$$

or

$$\sum_{k \in N} \alpha_k (a_{ij} + \Delta_{jk}) < \sum_{k \in N} \alpha_k \Delta_{ik}. \tag{8.4}$$

Since every $\Delta_{.k}$ is a solution to $A \otimes x \leq x$ we have

$$a_{ij} + \Delta_{jk} \leq \Delta_{ik} \tag{8.5}$$

for all $k \in N$ and for $k = i$ this inequality is strict because $\Delta_{ii} = 0$ and $a_{ij} \otimes \Delta_{ji} < 0$ by Lemma 8.1.2. If we now multiply each inequality (8.5) by α_k and add them all up, we get (8.4). $\square$

We will now prove that actually every matrix A with finite $\lambda(A)$ can be scaled to a strictly visualized one. This result will not be used in this book and the rest of this section may be skipped without loss of continuity. To prove the strict visualization result, we transform this problem from the principal interpretation to "max-times algebra" (that is, from $\mathcal{G}_0$ to $\mathcal{G}_2$, see Sect. 1.4). It is essential that all statements referred to in the proof of part (b) of Theorem 8.1.3, that is, part (a) of Theorem 8.1.3, Theorem 1.6.18 and Lemmas 1.6.19, 8.1.2, 1.6.20 have immediate analogues in $\mathcal{G}_2$. The proofs of the first four follow the lines of the proofs of these statements in $\mathcal{G}_0$, except that $\otimes$ stands for multiplication rather than for addition. In the case of Lemma 1.6.20 the reasoning is slightly different as the system of inequalities now reads

$$a_{ij} x_j \leq \lambda x_i.$$

But as it is again a system of linear inequalities (although a different one), the set of nonnegative solutions is convex.

We are ready to prove the main result of this section, that is, part (b) of Theorem 8.1.3, modified by the removal of the irreducibility assumption:

Theorem 8.1.4 *If $A \in \overline{\mathbb{R}}^{n \times n}$ and $\lambda(A) > \varepsilon$ then there exists $x \in \mathbb{R}^n$ such that $X^{-1} \otimes A \otimes X$ is strictly visualized, where $X = \mathrm{diag}(x)$.*

Proof We assume without loss of generality that A is definite. Take (in conventional notation) $B = (b_{ij}) = (2^{a_{ij}})$. Then the inequalities

$$a_{ij} + x_j \leq x_i, \quad i, j \in N, \ x_1, \ldots, x_n \text{ finite} \tag{8.6}$$

are equivalent (in conventional notation) to

$$b_{ij} y_j \le y_i, \quad i, j \in N, \ y_1, \ldots, y_n \text{ positive}. \tag{8.7}$$

A solution to (8.7) can be converted to a solution of (8.6) by setting $x_j = \log_2 y_j$, $j \in N$. The same applies when the inequalities are strict.

Let $\Delta(B) = (\Delta_{ij})$ in $\mathcal{G}_2$, $\alpha_k > 0$ for $k \in N$, $\sum_{k \in N} \alpha_k = 1$ and $y = \sum_{k \in N} \alpha_k \Delta_{.k}$ (conventional convex combination with positive coefficients). The vector y is positive as every row of the nonnegative matrix $\Delta(B)$ has at least one positive entry (namely 1 on the diagonal). It is now proved exactly as in part (b) of Theorem 8.1.3 that $Y^{-1} \otimes B \otimes Y$ is strictly visualized (in $\mathcal{G}_2$), where $Y = \mathrm{diag}(y)$. Hence $X^{-1} \otimes A \otimes X$ is strictly visualized in the principal interpretation, where $X = \mathrm{diag}(x)$ and $x_j = \log y_j$ for all $j \in N$. $\qquad\square$

Remark 8.1.5 Note that unlike (8.6), the system (8.7) is also homogeneous (in the conventional sense) and it is therefore feasible to take for y in Theorem 8.1.4 any linear combination (in particular the sum) of the columns of $\Delta(B)$ with positive coefficients instead of a convex combination.

Example 8.1.6 Consider

$$A = \begin{pmatrix} 0 & \varepsilon \\ 1 & 0 \end{pmatrix}$$

in $\mathcal{G}_0$, thus $\lambda(A) = 0$. Following the notation in the proof of Theorem 8.1.4 we have

$$B = \begin{pmatrix} 1 & 0 \\ 2 & 1 \end{pmatrix}$$

in $\mathcal{G}_2$. Hence $\Delta(B) = B$ and using Remark 8.1.5 we take for y the (conventional) sum of the columns of B, that is, $y = (1, 3)^T$. Hence $x = (0, \log_2 3)^T$ and

$$X^{-1} \otimes A \otimes X = \begin{pmatrix} 0 & \varepsilon \\ 1 - \log_2 3 & 0 \end{pmatrix}.$$

More information on matrix scaling including a complete description of all matrix scalings producing visualized or strictly visualized matrices can be found in [139].

Recall that a cycle in a weighted digraph is called a *zero cycle* if all arcs of this cycle have zero weight.

Corollary 8.1.7 *If $A \in \overline{\mathbb{R}}^{n \times n}$ is definite, $B = X^{-1} \otimes A \otimes X$ and B is visualized, then a cycle σ is critical in D_A if and only if σ is a zero cycle in D_B. Consequently, $C(A) = C(B)$ and thus every cycle in the critical digraph of A is critical.*

8.2 Principal Eigenspaces of Matrix Powers

In the analysis of principal eigenspaces of matrix powers a crucial role is played by
Theorem 8.2.1 below. This theorem applies to definite, nonpositive matrices. Note
that the statements proved in the previous section show how a general matrix A
can be transformed to a definite, nonpositive matrix B with the same set of critical
cycles (which, in the case of B, is the set of zero cycles).

We start with key definitions. If D' is a strongly connected component of a digraph D then the greatest common divisor of all directed cycles in D' is called the
cyclicity of D'. The *cyclicity* of D, notation $\sigma(D)$, is the least common multiple of
the cyclicities of all strongly connected components of D. The cyclicity of a digraph
consisting of a single node and no arc is 1 by definition. The cyclicity of a digraph
can be found in linear time [74]. The digraph D is called *primitive* if $\sigma(D) = 1$ and
imprimitive otherwise. The *cyclicity* of $A \in \overline{\mathbb{R}}^{n \times n}$, notation $\sigma(A)$, is the cyclicity of
its critical digraph $C(A)$. We will use the adjectives "primitive" and "imprimitive"
for matrices in the same way as for their critical digraphs.

A matrix $A = (a_{ij}) \in \overline{\mathbb{R}}^{n \times n}$ is called 0-*irreducible* if the zero digraph Z_A is
strongly connected. Since a strongly connected digraph with two or more nodes
contains at least one cycle, every 0-irreducible nonpositive matrix of order two or
more is definite.

Theorem 8.2.1 below is an application of ([18], Theorem 3.4.5).

Theorem 8.2.1 (Brualdi–Ryser) *Let $A \in \overline{\mathbb{R}}^{n \times n}, n > 1$ be a 0-irreducible and nonpositive matrix and let σ be the cyclicity of A. Let k be a positive integer. Then there
is a permutation matrix P such that $P^{-1} \otimes A^k \otimes P$ has r 0-irreducible diagonal
blocks where $r = \gcd(k, \sigma)$ and all elements outside these blocks are negative. The
cyclicity of each of these blocks is σ/r.*

Corollary 8.2.2 *Let $A \in \overline{\mathbb{R}}^{n \times n}$ be a matrix with $\lambda(A) > \varepsilon$. Suppose that $C(A)$ has
only one critical component and let σ be the cyclicity of A.*

(a) *If k is a positive integer then $C(A^k)$ has r critical components, where
$r = \gcd(k, \sigma)$. The cyclicity of each of these components is σ/r.*
(b) *$\mathrm{pd}(A^k) = 1$ for every $k \geq 1$ if and only if $\sigma = 1$.*

Proof Follows from Theorems 8.1.7, 8.2.1 and Lemma 4.1.3. □

As another application we immediately have the following classical result when
([18], Theorem 3.4.5) is applied to D_A:

Corollary 8.2.3 *Let $A \in \overline{\mathbb{R}}^{n \times n}$ be irreducible, $n > 1$ and $\sigma = \sigma(D_A)$.*

(a) *$A^k, k \geq 1$, is equivalent to a blockdiagonal matrix with $\gcd(k, \sigma)$ irreducible
diagonal blocks.*
(b) *A^k is irreducible for every positive integer k if and only if D_A is primitive.*

Since $C(A)$ is a subgraph of D_A, we have that if $C(A)$ is primitive then also D_A is primitive, yielding the following (recall that the primitivity of a matrix is determined by the primitivity of $C(A)$ rather than D_A):

Corollary 8.2.4 *If $A \in \overline{\mathbb{R}}^{n \times n}$ is primitive and irreducible then A^k is irreducible for all $k \geq 1$.*

It will also be useful to know that the definiteness of an irreducible matrix A is preserved blockwise in the powers of A.

Lemma 8.2.5 *If $A \in \overline{\mathbb{R}}^{n \times n}$ is definite and irreducible then every diagonal block of A^k is definite for all $k \geq 1$.*

Proof Let $x \in V(A)$ and $A^k[J]$ be a diagonal block of A^k for some $J \subseteq N$. Then x is finite and $A^k[J] \otimes x[J] = x[J]$. Since $A^k[J]$ is irreducible, it has only one eigenvalue and hence is definite.   □

Spectral properties of matrix powers play an important role in solving reachability problems. Next we summarize some of these properties.

Theorem 8.2.6 [34] *Let k, n be positive integers and $A = (a_{ij}) \in \overline{\mathbb{R}}^{n \times n}$.*

(a) $\Gamma((\lambda(A^k))^{-1} \otimes A^k) \leq \Gamma((\lambda(A))^{-1} \otimes A)$.
(b) $N_c(A) = N_c(A^k)$ *and the equivalence classes of $N_c(A^k)$ are either equal to the equivalence classes of $N_c(A)$ or are their refinements.*
(c) *If g_j, g'_j ($j \in N_c(A)$) are the fundamental eigenvectors of A and A^k respectively, then $g_j \geq g'_j$ for all $j \in N_c(A)$.*
(d) *If σ_i is the cyclicity of the ith connected component of $C(A)$ then this component splits into $\gcd(\sigma_i, k)$ connected components of $C(A^k)$. The cyclicity of each of these components is $\sigma_i / \gcd(\sigma_i, k)$.*
(e) $pd(A^k) = \sum_i \gcd(\sigma_i, k)$.

Proof (a) Denote $(\lambda(A))^{-1} \otimes A$ as B. Then the LHS is

$$\Gamma(B^k) = B^k \oplus B^{2k} \oplus \cdots \oplus B^{nk} \leq \Gamma(B)$$

because by (1.20) $B^r \leq \Gamma(B)$ for every natural $r > 0$.

(b) $N_c(A) \supseteq N_c(A^k)$ follows from part (a) immediately since in a metric matrix all diagonal elements are nonpositive and the jth diagonal entry is zero if and only if j is a critical node.

Now let $j \in N_c(A)$ and $\sigma = (j = j_0, j_1, \ldots, j_r = j)$ be any critical cycle in A (and B) containing j, thus $w(\sigma, B) = 0$. Let us denote

$$\pi = (j = j_0, j_{k(\mathrm{mod}\, r)}, j_{2k(\mathrm{mod}\, r)}, \ldots, j_{rk(\mathrm{mod}\, r)} = j)$$

and B^k by $C = (c_{ij})$. Then for all $i = 0, 1, \ldots, r - 1$ we have (all indices are mod r and, for convenience, we write here $c(i, j)$ rather than c_{ij}, similarly $b(i, j)$):

$$c(j_{ik}, j_{ik+k}) \geq b(j_{ik}, j_{ik+1}) + b(j_{ik+1}, j_{ik+2}) + \cdots + b(j_{ik+k-1}, j_{ik+k})$$

since $c(j_{ik}, j_{ik+k})$ is the weight of a heaviest path of length k from j_{ik} to j_{ik+k} with respect to B and the RHS is the weight of one such path. Therefore

$$w\big(\pi, (\lambda(A^k))^{-1} \otimes A^k\big) = w(\pi, B^k) \geq (w(\sigma, B))^k = 0.$$

Hence, equality holds, as there are no positive cycles in $(\lambda(A^k))^{-1} \otimes A^k$. This implies that π is a critical cycle with respect to A^k and so $j \in N_c(A^k)$.

If w is the weight of an arc (u, v) on a critical cycle for A^k then there is a path from u to v having the total weight w with respect to A. Therefore all nodes on a critical cycle for A^k belong to one critical cycle for A. Hence the refinement statement.

(c) Follows from part (a) immediately.

(d) It now follows from Theorems 8.1.7 and Theorem 8.2.1.

(e) Follows from part (d) immediately.   □

8.3 Periodic Behavior of Matrices

8.3.1 Spectral Projector and the Cyclicity Theorem

For $A \in \overline{\mathbb{R}}^{n \times n}$ it will be practical in Sects. 8.3–8.5 to denote the Kleene star $\Delta(A)$ (see Sect. 1.6) by $A^* = (a_{ij}^*)$ (so A^* does not denote here the conjugate matrix). The rows of A^* will be denoted by $\rho_1(A^*), \ldots, \rho_n(A^*)$, the columns by $\tau_1(A^*), \ldots, \tau_n(A^*)$, or just $\rho_1, \ldots, \rho_n, \tau_1, \ldots, \tau_n$.

Recall that if $A \in \overline{\mathbb{R}}^{n \times n}$ and $\lambda(A) \leq 0$ (and in particular when A is definite) then

$$\Gamma(A) = A \oplus A^2 \oplus \cdots \oplus A^n$$

and

$$A^* = I \oplus A \oplus A^2 \oplus \cdots \oplus A^{n-1} = I \oplus \Gamma(A).$$

The columns of $\Gamma(A)$ are denoted $g_1, \ldots, g_n$. Hence $\tau_j = g_j$ for all $j \in N_c(A)$ and τ_j differs from g_j for $j \notin N_c(A)$ only on the diagonal position where τ_j has a zero whereas g_j has a negative value.

Let $Q(A) = (q_{ij}) \in \overline{\mathbb{R}}^{n \times n}$ be the matrix with entries

$$q_{ij} = \sum_{k \in N_c(A)}^{\oplus} a_{ik}^* \otimes a_{kj}^*, \quad i, j \in N. \tag{8.8}$$

Hence $Q(A) = \sum_{k \in N_c(A)}^{\oplus} Q^{[k]}(A)$, where $Q^{[k]}(A)$, or just $Q^{[k]}$ is the outer product $\tau_k(A^*) \otimes \rho_k(A^*)$.

Proposition 8.3.1 [8] *Let* $A \in \overline{\mathbb{R}}^{n \times n}$ *be definite and* $Q = Q(A)$. *Then*

$$A \otimes Q = Q \otimes A = Q = Q^2.$$

Proof Since the columns of $Q^{[k]}$ are multiples of τ_k, for every $k \in N_c(A)$ all columns of $Q^{[k]}$ are principal eigenvectors of A and so $A \otimes Q = Q$. By symmetry the same is true about the rows of $Q^{[k]}$ and so also $Q \otimes A = Q$ (see Remark 4.3.6).

We also have $(Q^{[k]})^2 = \tau_k \otimes (\rho_k \otimes \tau_k) \otimes \rho_k = \tau_k \otimes \rho_k = Q^{[k]}$ since $\rho_k \otimes \tau_k$ is the kth diagonal entry of $(A^*)^2 = A^*$ which is 0. To prove $Q^2 = Q$ it is sufficient to show that $Q^{[k]} \otimes Q^{[l]} \leq Q^{[k]} \oplus Q^{[l]}$. The proof of this inequality is left to Exercise 8.7.4. $\square$

By Proposition 8.3.1, if A is definite and $x \in \overline{\mathbb{R}}^n$ then $Q \otimes x \in V(A, 0)$ and $Q^2 \otimes x = Q \otimes x$. Therefore $Q(A)$ is called the *spectral projector* of A.

The following observation will be useful:

Proposition 8.3.2 *Let* $A \in \overline{\mathbb{R}}^{n \times n}$ *be definite and* $Q = Q(A)$. *Then for any* $i \in N_c(A)$ *the* ith *row (column) of* Q *is equal to* ρ_i (τ_i).

Proof Since $(A^*)^2 = A^*$ (see Proposition 1.6.15), we have for all $i, j = 1, \ldots, n$:

$$q_{ij} = \sum_{k \in N_c(A)}^{\oplus} a_{ik}^* \otimes a_{kj}^* \leq \sum_{k \in N}^{\oplus} a_{ik}^* \otimes a_{kj}^* = a_{ij}^*.$$

If $i \in N_c(A)$ and $j \in N$ then

$$q_{ij} = \sum_{k \in N_c(A)}^{\oplus} a_{ik}^* \otimes a_{kj}^* \geq a_{ii}^* \otimes a_{ij}^* = a_{ij}^*$$

and the statement for rows follows. The statement for the columns is proved similarly. $\square$

Spectral projectors are closely related to the periodicity questions, as the following fundamental result suggests, proved both in [8] and [102]. For finite, strongly definite matrices it also appears in [60], Sect. 27.3, where $Q(A)$ is called the *orbital matrix*.

Theorem 8.3.3 *Let* $A \in \overline{\mathbb{R}}^{n \times n}$ *be irreducible, primitive and definite. Then there is an integer* R *such that* $A^r = Q(A)$ *for all* $r \geq R$.

The statement of Theorem 8.3.3 is also true for a blockdiagonal matrix whose every block is primitive and definite. This will be important for the subsequent theory and we therefore present it in detail. For a set $S \subseteq \mathbb{Z}$ and $k \in \mathbb{Z}$ we denote

$$k + S = S + k = \{k + s; s \in S\}.$$

Theorem 8.3.4 *Let $A \in \overline{\mathbb{R}}^{n \times n}$ be a blockdiagonal matrix whose every block is primitive and definite. Then there is an integer R such that $A^r = Q(A)$ for all $r \geq R$.*

Proof Let $A_1, \ldots, A_s$ be the diagonal blocks of A and $N_1, \ldots, N_s$ be the corresponding partition of N, that is, $A_i = A(N_i)$, $i = 1, \ldots, s$. Let us denote $|N_i| = n_i$ for all i. Hence $N_c(A_i) \subseteq \{1, \ldots, n_i\}$. By Theorem 8.3.3 for each $i \in \{1, \ldots, s\}$ there is an R_i such that for all $r \geq R_i$ we have:

$$A_i^r = Q(A_i) = \sum_{k \in N_c(A_i)}^{\oplus} Q^{[k]}(A_i),$$

where $Q^{[k]}(A_i)$ is the outer product $\tau_k(A_i^*) \otimes \rho_k(A_i^*)$.

Denote $n^{(i)} = \sum_{1 \leq j < i} n_j$, $i = 1, \ldots, s$ (thus $n^{(1)} = 0$). Then

$$N_c(A) = \bigcup_{i=1}^{s} N_c'(A_i),$$

where

$$N_c'(A_i) = N_c(A_i) + n^{(i)}.$$

Thus we have

$$Q(A) = \sum_{k \in N_c(A)}^{\oplus} Q^{[k]}(A) = \sum_{i=1,\ldots,s}^{\oplus} \sum_{k \in N_c'(A_i)}^{\oplus} Q^{[k]}(A)$$

where $Q^{[k]}(A)$ is the outer product $\tau_k(A^*) \otimes \rho_k(A^*)$.

Let $r \geq R = \max(R_1, \ldots, R_s)$. The matrix A^r is blockdiagonal and its diagonal blocks are $A_1^r, \ldots, A_s^r$. Consequently, A^* is also blockdiagonal and its diagonal blocks are $(A_1)^*, \ldots, (A_s)^*$. Hence for $i \in \{1, \ldots, s\}$ and $k \in N_c'(A_i)$, we have $k = k' + n^{(i)}$, $k' \in N_c(A_i)$ and

$$\tau_k(A^*) = \begin{pmatrix} \varepsilon \\ \tau_{k'}(A_i^*) \\ \varepsilon \end{pmatrix}, \rho_k(A^*) = (\varepsilon | \rho_{k'}(A_i^*) | \varepsilon),$$

where in both expressions the first ε is of dimension $n^{(i)}$ and the second of dimension $n - n^{(i)} - n_i$. It follows that the $n \times n$ matrix

$$B_{k;i} = \tau_k(A^*) \otimes \rho_k(A^*)$$

has all entries ε except for $B_{k;i}[N_i]$, which is

$$\tau_{k'}(A_i^*) \otimes \rho_{k'}(A_i^*) = Q^{[k']}(A_i).$$

Therefore

$$\sum_{k \in N_c'(A_i)}^{\oplus} B_{k;i}[N_i] = \sum_{k' \in N_c(A_i)}^{\oplus} Q^{[k']}(A_i) = Q(A_i) = A_i^r.$$

Finally, we deduce:

$$Q(A) = \sum_{i=1,\dots,s}^{\oplus} \sum_{k \in N'_c(A_i)}^{\oplus} Q^{[k]}(A) = \sum_{i=1,\dots,s}^{\oplus} \sum_{k \in N'_c(A_i)}^{\oplus} B_{k;i}$$

$$= \sum_{i=1,\dots,s}^{\oplus} \begin{pmatrix} \varepsilon & \cdots & \varepsilon & \cdots & \varepsilon \\ & \ddots & & & \\ \varepsilon & \cdots & A_i^r & \cdots & \varepsilon \\ & & & \ddots & \\ \varepsilon & \cdots & \varepsilon & \cdots & \varepsilon \end{pmatrix}$$

$$= \begin{pmatrix} A_1^r & \varepsilon & \varepsilon & \varepsilon \\ \varepsilon & A_2^r & \varepsilon & \varepsilon \\ \varepsilon & \varepsilon & \ddots & \varepsilon \\ \varepsilon & \varepsilon & \varepsilon & A_s^r \end{pmatrix} = A^r. \qquad \square$$

If σ is the cyclicity of A, it follows from Proposition 8.2.6, part (d), that all components of $C(A^\sigma)$ are primitive and thus A^σ is primitive. The matrix A^σ may not be irreducible, but is blockdiagonal with $\gcd(\sigma, \sigma(D_A)) = \sigma(D_A)$ blocks (Corollary 8.2.3). Each block is irreducible (Corollary 8.2.3), definite (Lemma 8.2.5) and primitive. By Theorem 8.3.4

$$(A^\sigma)^r = (A^\sigma)^{r+1} = Q(A^\sigma)$$

for all r sufficiently large (observe that $\Lambda(A^\sigma) = \{0\}$ and so $V(A^\sigma)$ indeed is an eigenspace). This also implies that for any k large enough, $k \equiv s \bmod \sigma$, we have for some r:

$$A^k = A^{s+r\sigma} = A^{s+r\sigma+\sigma} = A^{k+\sigma}.$$

Recall that if $\lambda(A) > \varepsilon$ then $(\lambda(A))^{-1} \otimes A$ is definite and clearly

$$((\lambda(A))^{-1} \otimes A)^k = (\lambda(A))^{-k} \otimes A^k.$$

We can now deduce one of the fundamental results of max-algebra (note that we did not prove the minimality of $\sigma(A)$):

Theorem 8.3.5 (Cyclicity Theorem) *For every irreducible matrix $A \in \overline{\mathbb{R}}^{n \times n}$ the cyclicity of A is the period of A, that is, the smallest natural number p for which there is an integer T such that*

$$A^{k+p} = (\lambda(A))^p \otimes A^k \tag{8.9}$$

for every $k \geq T$.

Recall that the smallest value of T for which (8.9) holds is called the transient of $\{A^k\}$ and will be denoted by $T(A)$. A matrix A for which there is a p and T

such that (8.9) holds for $k \geq T$ is called *ultimately periodic*. Thus every irreducible matrix is ultimately periodic.

Theorem 8.3.5 has been proved for finite matrices in [60]. A proof for general matrices was presented in [50], see also [51] for an overview without proofs. A proof in a different setting covering the case of finite matrices is given in [118]. The general irreducible case is also proved in [6, 8, 92, 102]. A generalization to the reducible case is studied in [93, 114] (see Theorem 8.6.9). Periodic behavior of matrix powers is also studied in [75].

Recall that the entries of A^r are denoted by $a_{ij}^{(r)}$, in contrast to a_{ij}^r, which denote the rth powers of a_{ij}.

It will be important that the entries $a_{ij}^{(r)}$, where either i or j is critical, may become periodic much faster than the noncritical part of A:

Theorem 8.3.6 [117, 133] *Let $A \in \overline{\mathbb{R}}^{n \times n}$ be irreducible. Critical rows and columns of A^r are periodic for $r \geq n^2$, that is, there exists a positive integer q such that for all $i \in N_c(A)$ and $j \in N$, or for all $j \in N_c(A)$ and $i \in N$ we have:*

$$a_{ij}^{(r+q)} = (\lambda(A))^q \otimes a_{ij}^{(r)}.$$

Proof Without loss of generality we prove this statement for the rows of definite matrices only. Let $i \in N_c(A)$. Then there is a critical cycle of length l_i to which i belongs. Hence $a_{ii}^{(kl_i)} = 0$ for all $k \geq 1$. Since for all $m < k$ and $s = 1, \ldots, n$ we have

$$a_{is}^{(ml_i)} = a_{ii}^{((k-m)l_i)} \otimes a_{is}^{(ml_i)} \leq a_{is}^{(kl_i)},$$

it follows that

$$a_{is}^{(kl_i)} = \sum_{m=1,\ldots,k}^{\oplus} a_{is}^{(ml_i)}. \tag{8.10}$$

Entries $a_{is}^{(kl_i)}$ are maximal weights of paths of length k with respect to the matrix A^{l_i}. Since the weights of all cycles are less than or equal to 0 and paths of length n or more are not elementary, the maximum is achieved at $k \leq n$ (see Lemma 1.5.4). Using (8.10) we obtain that $a_{is}^{((t+1)l_i)} = a_{is}^{(tl_i)}$ for all $t \geq n$. Further for any d, $0 \leq d \leq l_i - 1$,

$$a_{is}^{(tl_i+d)} = \sum_{j \in N}^{\oplus} a_{ij}^{(tl_i)} \otimes a_{js}^{(d)},$$

and it follows that $a_{is}^{((t+1)l_i+d)} = a_{is}^{(tl_i+d)}$ for all $t \geq n$. Hence $a_{is}^{(k)}$ is periodic for $k \geq nl_i$, and all these sequences, for any $i \in N_c(A)$ and any s, become periodic for $k \geq n^2$. $\square$

We will denote by $T_c(A)$ the least integer such that the critical rows and columns of A^r are periodic for $r \geq T_c(A)$. It follows from Theorem 8.3.6 that $T_c(A) \leq n^2$.

Remark 8.3.7 The statement of Theorem 8.3.3 has found a remarkable generalization in [138] where it has been proved that for any (reducible) matrix A all powers $A^r, r \geq 3n^2$, can be expressed as a max-algebraic sum of terms of the form $C \otimes S^r \otimes R$, called CSR products. All these terms can be found in $O(n^4 \log n)$ time. Here C and R are extracted from the columns and rows of a certain Kleene star (the same for both) and $C \otimes R$ is the spectral projector $Q(A)$ if A is irreducible. The matrix S is diagonally similar to the Boolean incidence matrix of a certain critical digraph. It is shown that the powers have a well-defined ultimate behavior, where certain terms are totally or partially suppressed, thus leading to ultimate $C \otimes S^r \otimes R$ terms and the corresponding ultimate expansion. This generalizes the Cyclicity Theorem to reducible matrices. The expansion is then used to derive an $O(n^4 \log n)$ method for solving the question whether the orbit of a reducible matrix is ultimately periodic with any starting vector.

8.3.2 Cyclic Classes and Ultimate Behavior of Matrix Powers

Imprimitive digraphs have interesting combinatorial structure which plays a key role in solving the reachability problem (Sect. 8.6 below). We briefly introduce this structure similarly as in [18], Sect. 3.4.

Note first that the length of every cycle is the sum of the lengths of elementary cycles and therefore the greatest common divisor of all cycles is equal to the greatest common divisor of elementary cycles.

Let D be a strongly connected digraph with cyclicity σ. Let i and j be any two nodes of D and σ_i and σ_j be the greatest common divisor of cycles containing i and j, respectively. Let α be a cycle of some length r containing i. By strong connectivity, there is a path from i to j, say β, of length s and a path from j to i, say γ, of length t. Clearly, combinations of β and γ and that of α, β and γ yield cycles containing j, of length $s + t$ and $r + s + t$. Since σ_j is a divisor of both, it is also a divisor of r. Since α was arbitrary, σ_j divides the length of every cycle containing i and thus σ_j divides σ_i. By symmetry also σ_i divides σ_j and so $\sigma_i = \sigma_j$. Since i and j were arbitrary, we have $\sigma_i = \sigma_j = \sigma$. If β' is another path from i to j, say of length s', then σ divides both $s + t$ and $s' + t$ and thus $s \equiv s' \bmod \sigma$.

We therefore deduce that by fixing a node, say i, we can partition the set of nodes N into σ mutually disjoint nonempty subsets $C_1, \ldots, C_\sigma$ as follows:

$$C_k = \{ j \in N; \text{the length of each } i - j \text{ path is } k \bmod \sigma \},$$

for $k = 1, \ldots, \sigma$. Clearly, the length of any (and therefore all) paths with the starting node in C_k and endnode in C_l is $l - k \bmod \sigma$. We also have $i \in C_\sigma$. Every arc in D leaves a node in C_k and enters a node in C_{k+1} for some k, $1 \leq k \leq \sigma$, where $C_{\sigma+1} = C_1$. We will use notation $[i]$ for the class containing node i. Clearly, for any i and j there is an integer t, $0 \leq t \leq \sigma - 1$ such that for the length l of any path starting in $[i]$ and terminating in $[j]$ we have $l \equiv t \bmod \sigma$. We will write $[i] \rightarrow_t [j]$. Clearly, if $[i] \rightarrow_t [j]$ then $[j] \rightarrow_{\sigma-t} [i]$.

The sets $[1], \ldots, [\sigma]$ will be called *cyclic classes* of D.

We will now apply cyclic classes to critical digraphs of matrices. They may consist of several connected components, in which case we will treat each component separately.

Lemma 8.3.8 *Let $A \in \overline{\mathbb{R}}^{n \times n}$ be definite and irreducible, $\sigma = \sigma(A)$, and let $t \geq 0$ be such that $t\sigma \geq T(A)$. Then the following hold for every integer $l \geq 0$ and $k = 1, \ldots, n$:*

$$A_{k \cdot}^{t\sigma+l} = \sum_{i \in N_c(A)}^{\oplus} a_{ki}^{(t\sigma)} \otimes A_{i \cdot}^{t\sigma+l},$$

$$A_{\cdot k}^{t\sigma+l} = \sum_{i \in N_c(A)}^{\oplus} a_{ik}^{(t\sigma)} \otimes A_{\cdot i}^{t\sigma+l}. \tag{8.11}$$

Proof The matrix $B = A^\sigma$ is primitive, definite and blockdiagonal. Due to Theorem 8.3.4, for any $r \geq T(B)$ we have

$$b_{kj}^{(r)} = \sum_{i \in N_c(A)}^{\oplus} b_{ki}^* \otimes b_{ij}^* \tag{8.12}$$

for $k, j = 1, \ldots, n$. By Proposition 8.3.2, if $u \in N_c(A)$ or $v \in N_c(A)$, then $b_{uv}^* = b_{uv}^{(r)}$ for all $r \geq T(B)$. Hence for any $t\sigma \geq T(A)$ (8.12) implies

$$a_{kj}^{(t\sigma)} = \sum_{i \in N_c(A)}^{\oplus} a_{ki}^{(t\sigma)} \otimes a_{ij}^{(t\sigma)}. \tag{8.13}$$

In the matrix notation, this is equivalent to:

$$A_{k \cdot}^{t\sigma} = \sum_{i \in N_c(A)}^{\oplus} a_{ki}^{(t\sigma)} \otimes A_{i \cdot}^{t\sigma}$$

and similarly for the columns:

$$A_{\cdot k}^{t\sigma} = \sum_{i \in N_c(A)}^{\oplus} a_{ik}^{(t\sigma)} \otimes A_{\cdot i}^{t\sigma}.$$

Multiplying the last two identities by any power A^l, we obtain (8.11). $\qquad\square$

In the proof of the next theorem we will use the following "Bellman-type" principle

$$a_{ij}^{(r)} \otimes a_{jk}^{(s)} \leq a_{ik}^{(r+s)}, \quad \forall i, j, k, r, s, \tag{8.14}$$

which immediately follows from the fact that $A^r \otimes A^s = A^{r+s}$.

Theorem 8.3.9 *Let $A \in \overline{\mathbb{R}}^{n \times n}$ be a definite and irreducible matrix, $\sigma = \sigma(A)$ and let $i, j \in N_c(A)$ be such that $[i] \to_l [j]$, for some $l \geq 0$.*

(a) *For any $r \geq T_c(A)$ there exists an integer $t \geq 0$ such that*

$$a_{ij}^{(t\sigma+l)} A_{\cdot i}^r = A_{\cdot j}^{r+l}, \quad a_{ij}^{(t\sigma+l)} A_{j\cdot}^r = A_{i\cdot}^{r+l}. \tag{8.15}$$

(b) *If A is visualized, then for all $r \geq T_c(A)$*

$$A_{\cdot i}^r = A_{\cdot j}^{r+l}, \quad A_{j\cdot}^r = A_{i\cdot}^{r+l}. \tag{8.16}$$

Proof Let $i, j \in N_c(A)$. If $[i] \to_l [j]$ then $[j] \to_s [i]$, where $l + s = \sigma$. Hence there exists a critical path of length $t\sigma + l$, for some integer $t \geq 0$, connecting i to j, and a critical path of length $u\sigma + s$, for some integer $u \geq 0$, connecting j to i. Thus

$$a_{ij}^{(t\sigma+l)} \otimes a_{ji}^{(u\sigma+s)} = 0, \tag{8.17}$$

and in the visualized case

$$a_{ij}^{(t\sigma+l)} = a_{ji}^{(u\sigma+s)} = 0. \tag{8.18}$$

Combining this with (8.14) we obtain:

$$A_{\cdot i}^r = A_{\cdot i}^r \otimes a_{ij}^{(t\sigma+l)} \otimes a_{ji}^{(u\sigma+s)} \leq A_{\cdot i}^{r+t\sigma+l} \otimes a_{ji}^{(u\sigma+s)} \leq A_{\cdot i}^{r+(t+u+1)\sigma}.$$

Since $r \geq T_c(A)$, by Theorem 8.3.6 we have $A_{\cdot i}^r = A_{\cdot i}^{r+(t+u+1)\sigma}$, hence all inequalities hold with equality. Now multiply the equality

$$A_{\cdot i}^r = A_{\cdot i}^{r+t\sigma+l} \otimes a_{ji}^{(u\sigma+s)}$$

by $a_{ij}^{(t\sigma+l)}$:

$$a_{ij}^{(t\sigma+l)} \otimes A_{\cdot i}^r = A_{\cdot i}^{r+t\sigma+l} \otimes a_{ji}^{(u\sigma+s)} \otimes a_{ij}^{(t\sigma+l)}$$

and the statement for columns now follows by (8.17) and Theorem 8.3.6. The proof for the rows is similar and part (b) follows from (8.18). □

Letting $l = 0$ in Theorem 8.3.9 we obtain the following.

Corollary 8.3.10 *Let $A \in \overline{\mathbb{R}}^{n \times n}$ be definite, irreducible and $r \geq T_c(A)$. All rows of A^r with indices in the same cyclic class are equal, and the statement holds similarly for the columns.*

Theorem 8.3.9 says that for any power A^r for $r \geq T_c(A)$ (and in particular for $r \geq n^2$), the critical columns (or rows) can be obtained from the critical columns (or rows) of the spectral projector $Q(A^\sigma)$ by permuting the sets of columns (or rows) which correspond to the cyclic classes of the critical digraph. Lemma 8.3.8 adds to this that all noncritical columns (or rows) of any periodic power are in the subspace spanned by the critical columns (or rows). Since all columns of $Q(A^\sigma)$ are eigenvectors of A^σ, we conclude the following.

Theorem 8.3.11 *If $A \in \overline{\mathbb{R}}^{n \times n}$ is definite and irreducible then all powers A^r for $r \geq T(A)$ have the same column span, which is the eigenspace $V(A^\sigma)$.*

Theorem 8.3.11 enables us to say that $V(A^\sigma)$ is the *ultimate column span* of A. Similarly, we have the *ultimate row span* which is $V((A^T)^\sigma)$. These subspaces are generated by critical columns (or rows) of the Kleene star $(A^\sigma)^*$. For a basis of this subspace, we can take any set of columns $(A^\sigma)^*$ (equivalently $Q(A^\sigma)$ or $A^{t\sigma}$ for $t\sigma \geq T(A)$), whose indices form a minimal set of representatives of all cyclic classes of $C(A)$ or, equivalently, any maximal set of nonequivalent eigennodes of $N_c(A)$ (see Lemma 4.3.2).

8.4 Solving Reachability

Let $A \in \overline{\mathbb{R}}^{n \times n}$ be definite and p a positive integer. Recall that the *p-attraction space* $\text{Attr}(A, p)$ is the set of all vectors for which there exists an integer r such that $A^r \otimes x = A^{r+p} \otimes x \neq \varepsilon$ (and hence this is also true for all integers greater than or equal to r). Actually we may speak of any $r \geq T(A)$, due to the following observation.

Proposition 8.4.1 *Let A be irreducible and definite, p positive integer and $x \in \overline{\mathbb{R}}^n$. Then*

$$A^s \otimes x = A^{s+p} \otimes x$$

for some $s \geq T(A)$ if and only if

$$A^r \otimes x = A^{r+p} \otimes x$$

for all $r \geq T(A)$.

Proof Let x satisfy $A^s \otimes x = A^{s+p} \otimes x$ for some $s \geq T(A)$, then it also satisfies

$$A^l \otimes x = A^{l+p} \otimes x$$

for all $l > s$ (to see this, multiply the first equation by A^{l-s}).

Due to periodicity, for all k, $T(A) \leq k \leq s$, there exists $l > s$ such that $A^k = A^l$. Hence $A^k \otimes x = A^{k+p} \otimes x$ also holds if $T(A) \leq k \leq s$.   □

Corollary 8.4.2 $\text{Attr}(A, p) = \text{Attr}(A^p, 1)$.

Proof By Proposition 8.4.1, $\text{Attr}(A, p)$ is the solution set to the system $A^r \otimes x = A^{r+p} \otimes x$ for any $r \geq T(A)$, in particular for multiples of p, which proves the statement.   □

An equation of $A^r \otimes x = A^{r+p} \otimes x$ whose index is in $N_c(A)$ will be called *critical*, and the subsystem consisting of all critical equations will be called the *critical subsystem*.

Lemma 8.4.3 *Let A be irreducible and definite and let $r \geq T(A)$. Then $A^r \otimes x = A^{r+p} \otimes x$ is equivalent to its critical subsystem.*

Proof Consider a noncritical equation $A^r_{k.} \otimes x = A^{r+p}_{k.} \otimes x$. Using Lemma 8.3.8 it can be written as

$$\sum_{i \in N_c(A)}^{\oplus} a^{(r)}_{ki} \otimes A^r_{i.} \otimes x = \sum_{i \in N_c(A)}^{\oplus} a^{(r)}_{ki} \otimes A^{r+p}_{i.} \otimes x,$$

hence it is a max-combination of equations in the critical subsystem. $\square$

We are ready to present a method for deciding whether $x \in \mathrm{Attr}(A, p)$, as well as other related problems which we formulate below. We assume in all that $A \in \overline{\mathbb{R}}^{n \times n}$ is a given irreducible and definite matrix and $\sigma = \sigma(A)$.

For ease of reference we denote:

P1. For a given $x \in \overline{\mathbb{R}}^n$ and positive integer p, decide whether $x \in \mathrm{Attr}(A, p)$.
P2. For a given k, $0 \leq k < \sigma$, compute the periodic power A^s where $s \equiv k \bmod \sigma$.
P3. For a given $x \in \overline{\mathbb{R}}^n$ compute the period of $O(A, x)$.

Observe that P1 is identical with Q2 and P3 with Q1 formulated at the beginning of this chapter. The proof of the next statement is constructive and provides algorithms for solving P1–P3. Note that a similar argument was used in the max-min setting [130].

Theorem 8.4.4 [133] *For any irreducible matrix $A \in \overline{\mathbb{R}}^{n \times n}$, the problems P1–P3 can be solved in $O(n^3 \log n)$ time.*

Proof Suppose that k and p are given. First note that using the Karp and Floyd–Warshall algorithms (see Chap. 1) we can compute both $\lambda(A)$ and a finite eigenvector of A, and find all critical nodes in $O(n^3)$ time (see Theorem 8.1.3 and Corollary 8.1.7). Further we can identify all cyclic classes of $C(A)$ by Balcer–Veinott condensation in $O(n^2)$ operations [9]. We can now assume that A is definite and visualized.

By Theorem 8.3.6 the critical rows and columns become periodic for $r \geq n^2$. To find the critical rows and columns of the required power $s \geq T(A)$, we first compute A^r for one (arbitrary) exponent $r \geq n^2$ which can be done in $O(\log n)$ matrix squaring $(A, A^2, A^4, \ldots)$ and takes $O(n^3 \log n)$ time. Then following Theorem 8.3.9, we shift the rows and columns of A^r to obtain the critical rows and columns of A^s (to do this as described in (8.16) we assume that $r \in [i], s \in [j]$ and $[i] \longrightarrow_l [j]$ for some i, j, l). This requires $O(n^2)$ operations. In a similar way we find the critical rows of A^{s+p}.

By Lemma 8.4.3 we can solve P1 by the verification of the critical subsystem of $A^s \otimes x = A^{s+p} \otimes x$ which takes $O(n^2)$ operations. Using linear dependence of Lemma 8.3.8 the remaining noncritical submatrix of A^s and A^{s+p} for any $s \geq T(A)$ such that $s \equiv k \bmod \sigma$, can be computed in $O(n^3)$ time. This solves P2.

As the noncritical rows of A are generated by the critical ones, the period of $O(A, x)$ is determined by the critical components. For a visualized matrix we know that $A_{i.}^{r+t} = A_{j.}^{r}$ for all $i, j \in N_c(A)$ such that $[i] \to_t [j]$. This implies $(A^{r+t} \otimes x)_i = (A^r \otimes x)_j$ for $[i] \to_t [j]$, that is, to determine the period we only need the critical subvector of $A^r \otimes x$ for any fixed $r \geq n^2$. Indeed, for any $i \in N_c(A)$ and $r \geq n^2$ the sequence $\{(A^{r+t} \otimes x)_i\}_{t \geq 0}$ can be represented as a sequence of critical indices of $A^r \otimes x$ determined by a permutation on cyclic classes of the strongly connected component $\overline{C}$ to which i belongs. That is if in $\overline{C}$ we have

$$[i_1] \longrightarrow [i_2] \longrightarrow \cdots \longrightarrow [i_{\overline{\sigma}}] \longrightarrow [i_1],$$

where $\overline{\sigma} = \sigma(\overline{C})$, then we take a sequence $\{j_r\}_{r=1}^{\overline{\sigma}}$ such that $j_r \in [i_r]$. This sequence can be taken randomly since by Corollary 8.3.10 all rows and columns of A^r with indices from the same cyclic class are equal. Now we consider the sequence $x_{j_1}, \ldots, x_{j_{\overline{\sigma}}}$ and find its period. Even by checking all possible periods it takes no more than $\overline{\sigma}^2 \leq n^2$ operations. The period of $A^r \otimes x$ is then the least common multiple of periods found for each strongly connected component. It remains to note that all operations above do not require more than $O(n^3)$ time. This solves P3. $\qquad \square$

Example 8.4.5 We will examine problems P2 and P3 on the following strictly visualized 9×9 matrix:

$$A = \begin{pmatrix} -1 & 0 & -1 & -1 & -9 & -7 & -10 & -4 & -8 \\ 0 & -1 & 0 & -1 & -10 & -1 & -10 & -9 & -4 \\ -1 & -1 & -1 & 0 & -2 & -3 & -2 & -6 & -6 \\ 0 & -1 & -1 & -1 & -10 & -6 & -10 & -6 & -1 \\ -10 & -2 & -8 & -1 & -1 & 0 & -1 & -10 & -1 \\ -5 & -5 & -10 & -9 & -1 & -1 & 0 & -3 & -6 \\ -9 & -10 & -7 & -10 & 0 & -1 & -1 & -8 & -8 \\ -75 & -80 & -77 & -83 & -80 & -77 & -82 & -2 & -0.5 \\ -84 & -81 & -77 & -80 & -78 & -77 & -78 & -0.5 & -2 \end{pmatrix}.$$

The critical components of A, see Fig. 8.1, have node sets $\{1, 2, 3, 4\}$ and $\{5, 6, 7\}$. The cyclicities are $\sigma_1 = 2$, $\sigma_2 = 3$, so $\sigma(A) = \mathrm{lcm}(2, 3) = 6$. Let us denote $M = \{8, 9\}$.

The matrix can be decomposed into blocks

$$A = \begin{pmatrix} A_{11} & A_{12} & A_{1M} \\ A_{21} & A_{22} & A_{2M} \\ A_{M1} & A_{M2} & A_{MM} \end{pmatrix},$$

where the submatrices A_{11} and A_{22} correspond to two critical components of $C(A)$, see Fig. 8.1. They are

$$A_{11} = \begin{pmatrix} -1 & 0 & -1 & -1 \\ 0 & -1 & 0 & -1 \\ -1 & -1 & -1 & 0 \\ 0 & -1 & -1 & -1 \end{pmatrix}$$

Fig. 8.1 Critical digraph in
Example 8.4.5

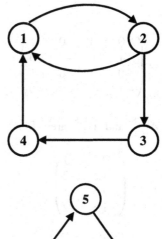

and

$$A_{22} = \begin{pmatrix} -1 & 0 & -1 \\ -1 & -1 & 0 \\ 0 & -1 & -1 \end{pmatrix}.$$

The noncritical principal submatrix

$$A_{MM} = \begin{pmatrix} -2 & -0.5 \\ -0.5 & -2 \end{pmatrix}.$$

It can be checked that the powers of A become periodic after $T(A) = 154$.
We will consider the following instances of problems P2 and P3:

P2. Compute A^r for $r \geq T(A)$ and $r \equiv 2 \bmod 6$.
P3. For a given $x \in \overline{\mathbb{R}}^9$, find the period of $\{A^k \otimes x\}$.

Solving P2. We perform 7 squarings $A, A^2, A^4, \ldots$ to raise A to the power $128 > 9 \times 9$. This brings us to the matrix

$$A^{128} = \begin{pmatrix} A_{11}^{(128)} & A_{12}^{(128)} & A_{1M}^{(128)} \\ A_{21}^{(128)} & A_{22}^{(128)} & A_{2M}^{(128)} \\ A_{M1}^{(128)} & A_{M2}^{(128)} & A_{MM}^{(128)} \end{pmatrix},$$

where

$$A_{11}^{(128)} = \begin{pmatrix} 0 & -1 & 0 & -1 \\ -1 & 0 & -1 & 0 \\ 0 & -1 & 0 & -1 \\ -1 & 0 & -1 & 0 \end{pmatrix}, \qquad A_{22}^{(128)} = \begin{pmatrix} -1 & -1 & 0 \\ 0 & -1 & -1 \\ -1 & 0 & -1 \end{pmatrix},$$

all entries of $A_{12}^{(128)}$ and $A_{21}^{(128)}$ are -1 and

$$A_{1M}^{(128)} = \begin{pmatrix} -2.5 & -1 \\ -1.5 & -2 \\ -2.5 & -1 \\ -1.5 & -2 \end{pmatrix}, \qquad A_{2M}^{(128)} = \begin{pmatrix} -1.5 & -2 \\ -2.5 & -2 \\ -2.5 & -1 \end{pmatrix},$$

$$A_{M1}^{(128)} = \begin{pmatrix} -76 & -75.5 \\ -75 & -76.5 \\ -76 & -75.5 \\ -75 & -76.5 \end{pmatrix}^T, \qquad A_{M2}^{(128)} = \begin{pmatrix} -76 & -76.5 \\ -76 & -76.5 \\ -76 & -76.5 \end{pmatrix}^T.$$

We are lucky since $128 \equiv 2 \mod 6$, thus we already have true critical columns and rows of A^r. However, the noncritical principal submatrix of A^{128} is

$$A_{MM}^{(128)} = \begin{pmatrix} -64 & -65.5 \\ -65.5 & -64 \end{pmatrix}.$$

It can be checked that this is *not* the noncritical submatrix of A^r that we seek (recall that $T(A) = 154$). Hence, it remains to compute the principal noncritical submatrix $A_{MM}^{(r)}$.

We note that A^{132} has critical rows and columns of the spectral projector $Q(A)$, since 132 is a multiple of $\sigma = 6$. In A^{132}, the critical rows and columns $1-4$ are the same as those of A^{128}, since $\sigma_1 = 2$ and both 128 and 132 are even. The critical rows $5-7$ can be computed from those of A^{128} by cyclic permutation $(5, 6, 7)$. Since $A_{M1}^{(128)}$ and $A_{M2}^{(128)}$ happen to have equal columns, all blocks in A^{132} are the same as in A^{128} above (after a similar block decomposition of A^{132}), except for

$$A_{22}^{(132)} = \begin{pmatrix} 0 & -1 & -1 \\ -1 & 0 & -1 \\ -1 & -1 & 0 \end{pmatrix}, \qquad A_{2M}^{(132)} = \begin{pmatrix} -2.5 & -2 \\ -2.5 & -1 \\ -1.5 & -2 \end{pmatrix}.$$

Now the remaining noncritical submatrix of A^r can be computed using linear dependence of Lemma 8.3.8, which now reads

$$A_{\cdot k}^{(r)} = \sum_{i=1,\dots,7}^{\oplus} a_{ik}^{(132)} \otimes A_{\cdot i}^{(128)}, \qquad k = 8, 9.$$

This yields

$$A_{MM}^{(r)} = \begin{pmatrix} -76.5 & -77 \\ -78 & -76.5 \end{pmatrix}.$$

Solving P3. We examine the orbit period of $A^k \otimes x$ for $x = x^1, x^2, x^3, x^4$, where

$$x^1 = (1, \quad 2, \quad 3, \quad 4, \quad 5, \quad 6, \quad 7, \quad 8, \quad 9)^T,$$
$$x^2 = (1, \quad 2, \quad 3, \quad 4, \quad 0, \quad 0, \quad 0, \quad 0, \quad 0)^T,$$
$$x^3 = (0, \quad 0, \quad 1, \quad 1, \quad 0, \quad 0, \quad 1, \quad 1, \quad 1)^T,$$
$$x^4 = (0, \quad 0, \quad 1, \quad 1, \quad 0, \quad 0, \quad 0, \quad 0, \quad 0)^T.$$

Let us compute $y = A^{128} \otimes x$ for $x = x^1, x^2, x^3, x^4$:

$$y^1 = A^{128} \otimes x^1 = (8, \quad 7, \quad 8, \quad 7, \quad 7, \quad 7, \quad 8, \quad \times, \quad \times)^T,$$
$$y^2 = A^{128} \otimes x^2 = (3, \quad 4, \quad 3, \quad 4, \quad 3, \quad 3, \quad 3, \quad \times, \quad \times)^T,$$
$$y^3 = A^{128} \otimes x^3 = (1, \quad 1, \quad 1, \quad 1, \quad 1, \quad 0, \quad 0, \quad \times, \quad \times)^T,$$
$$y^4 = A^{128} \otimes x^4 = (1, \quad 1, \quad 1, \quad 1, \quad 0, \quad 0, \quad 0, \quad \times, \quad \times)^T.$$

Here $\times$ correspond to noncritical entries that are not needed. The cyclic classes in the first critical component have node sets $C_1 = \{1,3\}$, $C_2 = \{2,4\}$, and the cyclic classes in the second have node sets $C_3 = \{5\}$, $C_4 = \{6\}$ and $C_5 = \{7\}$, see Fig. 8.2.

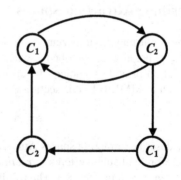

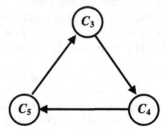

Fig. 8.2 Cyclic classes in Example 8.4.5

From Theorem 8.4.4 it follows that the coordinate sequences

$$\{(A^r \otimes x)_i, \; r \geq T(A)\}$$

are

$$y_1, y_2, y_1, y_2, \ldots, \quad \text{for } i = 1, 2, 3, 4,$$

$$y_5, y_6, y_7, y_5, y_6, y_7, \ldots, \quad \text{for } i = 5, 6, 7.$$

Note that the first sequence has been taken randomly from four possibilities:

$$y_1, y_2, y_1, y_2, \ldots,$$

$$y_3, y_4, y_3, y_4, \ldots,$$

$$y_1, y_4, y_1, y_4, \ldots,$$

$$y_3, y_2, y_3, y_2, \ldots.$$

The second sequence is uniquely determined since all cyclic classes in C_2 are one-element.

From $y^1, \ldots, y^4$ above we deduce that the orbit of x^1 is of the largest possible period 6, the orbit of x^2 is of period 2 (that is, $x^2 \in \mathrm{Attr}(A, 2)$), the orbit of x^3 is of period 3 (that is, $x^3 \in \mathrm{Attr}(A, 3)$), and the orbit of x^4 is of period 1 (that is, $x^4 \in \mathrm{Attr}(A, 1)$).

8.5 Describing Attraction Spaces

For applications it may be important to decide not only whether a vector is in an attraction space but also to describe the whole attraction space as efficiently as possible and thus to provide a choice of starting time vectors leading to stability of processes such as MMIPP. In this section we discuss the systems

$$A^r \otimes x = A^{r+1} \otimes x, \tag{8.19}$$

which fully describe attraction spaces $\mathrm{Attr}(A, 1)$ provided that r is sufficiently big. We will therefore call such systems *attraction systems*. The task of finding A^r for such r has been solved in Sect. 8.4. The results of this section enable us to simplify these systems for irreducible matrices A.

Note that if the critical digraph is strongly connected then for an irreducible matrix A there is a $v \in \overline{\mathbb{R}}^n - \{\varepsilon\}$ such that $V(A) = \{\alpha \otimes v; \alpha \in \mathbb{R}\}$. The attraction space is then described by the essentially one-sided system

$$A^r \otimes x = v \otimes y,$$

where y is a single variable. Therefore the unique scaled basis of the attraction space can be found using the results of Sect. 7.2.4. If, moreover, all nodes of A are critical, then Corollary 7.2.6 offers an even simpler way of finding the basis, see Remark 8.5.6. The case when there is only one critical cycle has been analysed in more detail in [16].

8.5.1 The Core Matrix

Let $A \in \overline{\mathbb{R}}^{n \times n}$ be irreducible. We will assume that the critical nodes of D_A are the first (say) c nodes. Suppose also that $C(A)$ consists of n_c strongly connected components C_μ with cyclicities σ_μ, for $\mu = 1, \ldots, n_c$. Let $\bar{c}$ be the number of noncritical nodes. Further it will be convenient to consider, together with these components, also noncritical, that is, trivial, components C_μ for $\mu = n_c + 1, \ldots, n_c + \bar{c}$, whose node sets N_μ consist of just one noncritical node, and the sets of arcs are empty.

Consider the block decomposition of A^r for $r \geq 1$, induced by the subsets N_μ for $\mu = 1, \ldots, n_c + \bar{c}$. The submatrix of A^r extracted from the rows in N_μ and columns in N_ν will be denoted by $A_{\mu\nu}^{(r)}$. If A is visualized and definite, we define the corresponding *core matrix* $A^{\text{Core}} = (\alpha_{\mu\nu})$, $\mu, \nu = 1, \ldots, n_c + \bar{c}$ by

$$\alpha_{\mu\nu} = \max\{a_{ij}; i \in N_\mu, j \in N_\nu\}. \tag{8.20}$$

The entries of $(A^{\text{Core}})^*$ will be denoted by $\alpha_{\mu\nu}^*$. Their role is shown in the next theorem.

Theorem 8.5.1 [137] *Let $A \in \overline{\mathbb{R}}^{n \times n}$ be an irreducible, definite, visualized matrix and $r \geq T_c(A)$. Let $\mu, \nu \in \{1, \ldots, n_c + \bar{c}\}$ be such that at least one of these indices is critical. Then the maximal entry of the block $A_{\mu\nu}^{(r)}$ is equal to $\alpha_{\mu\nu}^*$ and therefore this entry appears in every row and column of $A_{\mu\nu}^{(r)}$.*

Proof The entry $\alpha_{\mu\nu}^*$ is the maximal weight over paths from μ to ν in $D_{A^{\text{Core}}}$. Take one such path, say $(\mu_1, \ldots, \mu_l)$ of maximal weight, where $\mu_1 := \mu$ and $\mu_l = \nu$. With this path we can associate a path π in D_A defined by $\pi = \tau_1 \circ \sigma_1 \circ \tau_2 \circ \cdots \circ \sigma_{l-1} \circ \tau_l$, where τ_i are paths containing only critical arcs, which entirely belong to the components C_{μ_i}, and σ_i are arcs of maximal weight from C_{μ_i} to $C_{\mu_{i+1}}$. Such a path π exists since any two nodes in the same component C_μ can be connected to each other by critical paths if μ is critical, and C_μ consists just of one node if μ is noncritical. The weights of τ_i are 0, hence the weight of π is equal to $\alpha_{\mu\nu}^*$. It follows from the definition of $\alpha_{\mu\nu}$ and $\alpha_{\mu\nu}^*$ that $\alpha_{\mu\nu}^*$ is the greatest weight of a path from C_μ to C_ν. As at least one of the indices μ, ν is critical, there is freedom in the choice of the paths τ_1 or τ_l which can be of arbitrary length. Assume without loss of generality that μ is critical. Then for any r exceeding the length of $\sigma_1 \circ \tau_2 \circ \cdots \circ \sigma_{l-1} \circ \tau_l$ which we denote by $l_{\mu\nu}$, the block $A_{\mu\nu}^{(r)}$ contains an entry equal to $\alpha_{\mu\nu}^*$, which is the greatest entry of the block. Taking the maximum $T'(A)$ of $l_{\mu\nu}$ over all ordered pairs (μ, ν) with μ or ν critical, we obtain the claim for $r \geq T'(A)$. Evidently, $T'(A)$ can be replaced by $T_c(A)$. $\square$

Further we observe that the dimensions of periodic powers can be reduced. The rows and columns with indices in the same cyclic class coincide in any power A^r, where $r \geq T_c(A)$ and A is definite and visualized (Theorem 8.3.9). Hence after an appropriate permutation of the rows and columns, the blocks of A^r, for $\mu, \nu = 1, \ldots, n_c + \overline{c}$ and $r \geq T_c(A)$, are of the form

$$A_{\mu\nu}^{(r)} = \begin{pmatrix} \tilde{a}_{s_1 t_1}^{(r)} \otimes O_{s_1 t_1} & \cdots & \tilde{a}_{s_1 t_m}^{(r)} \otimes O_{s_1 t_m} \\ \vdots & \ddots & \vdots \\ \tilde{a}_{s_k t_1}^{(r)} \otimes O_{s_k t_1} & \cdots & \tilde{a}_{s_k t_m}^{(r)} \otimes O_{s_k t_m} \end{pmatrix}, \tag{8.21}$$

where k (resp. m) are cyclicities of C_μ (resp. C_ν), indices $s_1, \ldots, s_k$ and $t_1, \ldots, t_m$ correspond to cyclic classes of C_μ and C_ν, respectively, and $O_{s_i t_j}$ are zero matrices of appropriate dimensions. We assume that C_μ has just one cyclic class if μ is noncritical.

Formula (8.21) defines the matrix $\tilde{A}^{(r)} \in \overline{\mathbb{R}}^{(\tilde{c}+\overline{c}) \times (\tilde{c}+\overline{c})}$, where $\tilde{c}$ is the total number of cyclic classes, as the matrix with entries $\tilde{a}_{s_i t_j}^{(r)}$. By (8.21), this matrix has blocks

$$\tilde{A}_{\mu\nu}^{(r)} = \begin{pmatrix} \tilde{a}_{s_1 t_1}^{(r)} & \cdots & \tilde{a}_{s_1 t_m}^{(r)} \\ \vdots & \ddots & \vdots \\ \tilde{a}_{s_k t_1}^{(r)} & \cdots & \tilde{a}_{s_k t_m}^{(r)} \end{pmatrix}. \tag{8.22}$$

It follows that $\tilde{A}^{(r_1+r_2)} = \tilde{A}^{(r_1)} \otimes \tilde{A}^{(r_2)}$ for all $r_1, r_2 \geq T_c(A)$. In other words, the multiplication of any two powers $A^{(r_1)}$ and $A^{(r_2)}$ for $r_1, r_2 \geq T_c(A)$ reduces to the multiplication of $\tilde{A}^{(r_1)}$ and $\tilde{A}^{(r_2)}$.

Let $\sigma = \sigma(A)$. If we take $r = \sigma t + l \geq T(A)$ (instead of $T_c(A)$ above) and denote $\tilde{A} := \tilde{A}^{(\sigma t+1)}$, then due to the periodicity we obtain

$$\tilde{A}^{(\sigma t+l)} = \tilde{A}^{((\sigma t+1)l)} = \tilde{A}^l = \tilde{A}^{\sigma t+l}, \tag{8.23}$$

so that $\tilde{A}^{(r)}$ can be regarded as the rth power of $\tilde{A}$, for all $r \geq T(A)$.

8.5.2 Circulant Properties

A matrix $A = (a_{ij}) \in \overline{\mathbb{R}}^{m \times n}$ will be called *circulant*, if $a_{ij} = a_{ps}$ whenever $p = i + t$ (mod m) and $s = j + t$ (mod n) for all $i \in M, j \in N, t \geq 1$. For instance

$$A = \begin{pmatrix} 0 & 1 & 2 & 0 & 1 & 2 & 0 & 1 & 2 \\ 2 & 0 & 1 & 2 & 0 & 1 & 2 & 0 & 1 \\ 1 & 2 & 0 & 1 & 2 & 0 & 1 & 2 & 0 \\ 0 & 1 & 2 & 0 & 1 & 2 & 0 & 1 & 2 \\ 2 & 0 & 1 & 2 & 0 & 1 & 2 & 0 & 1 \\ 1 & 2 & 0 & 1 & 2 & 0 & 1 & 2 & 0 \end{pmatrix} \tag{8.24}$$

is circulant. Note that if $m = n$ then there exist scalars $\alpha_1, \ldots, \alpha_n$ such that $a_{ij} = \alpha_d$ whenever $j - i = d \pmod{n}$ and a circulant matrix then has the form:

$$A = \begin{pmatrix} \alpha_1 & \alpha_2 & \alpha_3 & \cdots & & \alpha_n \\ \alpha_n & \alpha_1 & \alpha_2 & \ddots & & \alpha_{n-1} \\ \alpha_{n-1} & \alpha_n & \alpha_1 & \ddots & & \alpha_{n-2} \\ \vdots & \ddots & \ddots & \ddots & & \vdots \\ \alpha_2 & \alpha_3 & \cdots & & \cdots & \alpha_1 \end{pmatrix}. \qquad (8.25)$$

A matrix $A \in \overline{\mathbb{R}}^{m \times n}$ will be called *block $k \times k$ circulant* if there exist scalars $\alpha_1, \ldots, \alpha_k$ and a block decomposition $A = (A_{ij})$, $i, j = 1, \ldots, k$ such that $A_{ij} = \alpha_d \otimes O_{ij}$ if $j - i = d \pmod{k}$, where O_{ij} are zero matrices.

A matrix $A = (a_{ij}) \in \overline{\mathbb{R}}^{m \times n}$ will be called *d-periodic* when $a_{ij} = a_{is}$ if $(s - j)$ $\bmod\, n$ is a multiple of d, and $a_{ji} = a_{si}$ if $(s - j)\bmod m$ is a multiple of d.

The matrix (8.24) indicates that a rectangular $m \times n$ circulant matrix consists of ordinary $d \times d$ circulant blocks, where $d = \gcd(m, n)$. In particular, it is d-periodic. Also, there exist conventional permutation matrices P and Q such that $B = PAQ$ is block $d \times d$ circulant:

$$B = \begin{pmatrix} 0 & 0 & 0 & 1 & 1 & 1 & 2 & 2 & 2 \\ 0 & 0 & 0 & 1 & 1 & 1 & 2 & 2 & 2 \\ 2 & 2 & 2 & 0 & 0 & 0 & 1 & 1 & 1 \\ 2 & 2 & 2 & 0 & 0 & 0 & 1 & 1 & 1 \\ 1 & 1 & 1 & 2 & 2 & 2 & 0 & 0 & 0 \\ 1 & 1 & 1 & 2 & 2 & 2 & 0 & 0 & 0 \end{pmatrix}.$$

Observe that if $A \in \overline{\mathbb{R}}^{m \times n}$ is circulant and m and n are coprime then A is constant. We formalize these observations in the following.

Proposition 8.5.2 *Let $A \in \overline{\mathbb{R}}^{m \times n}$ be circulant and $d = \gcd(m, n)$.*

(a) *A is d-periodic.*
(b) *There exist conventional permutation matrices P and Q such that PAQ is a block $d \times d$ circulant.*

Proof (a) There are integers t_1 and t_2 such that $d = t_1 m + t_2 n$. Using the definition of a circulant matrix we obtain $a_{ij} = a_{is}$, if $s = j + t_1 m \pmod{n}$, and hence if $s = j + d \pmod{n}$. Similarly for the rows, we obtain that $a_{ji} = a_{si}$, if $s = j + t_2 n \pmod{m}$, and hence if $s = j + d \pmod{m}$.

(b) As A is d-periodic, all rows such that $i + d = j \pmod{m}$ are equal, so that $\{1, \ldots, m\}$ can be divided into d groups with m/d indices each, in such a way that $A_{i.} = A_{j.}$, if i and j belong to the same group. We can find a permutation matrix P such that $A' = PA$ will have rows $A'_{1.} = \cdots = A'_{d.} = A_{1.}$, $A'_{d+1.} = \cdots = A'_{2d.} = A_{2.}$, and so on. Similarly, we can find a permutation matrix Q such that $A'' = PAQ$ will have columns $A''_{\cdot 1} = \cdots = A''_{\cdot d} = A'_{\cdot 1}$, $A''_{\cdot d+1} = \cdots = A''_{\cdot 2d} = A'_{\cdot 2}$, and so on. Then A'' has blocks A''_{ij} for $i, j = 1, \ldots, d$ of dimension $n/d \times m/d$, where $A''_{ij} = a_{ij} \otimes O_{ij}$, and O_{ij} is a zero matrix. As A is d-periodic, the submatrix extracted from the first d rows and columns is circulant. Hence A'' is block $d \times d$ circulant.   □

Proposition 8.5.3 *Let $A \in \overline{\mathbb{R}}^{n \times n}$ be an irreducible, definite and visualized matrix which admits block decomposition (8.21), $\sigma = \sigma(A)$ and $r \geq T(A)$. Let C_μ, C_ν be two (possibly equal) components of $C(A)$, and $d = \gcd(\sigma_\mu, \sigma_\nu)$.*

(a) *$\tilde{A}_{\mu\nu}^{(r)}$ is circulant.*

(b) *For any critical μ and ν, there is a permutation P such that $(P^T \tilde{A} P)_{\mu\nu}^{(r)}$ is a block $d \times d$ circulant matrix.*

(c) *If r is a multiple of σ, then $\tilde{A}_{\mu\mu}^{(r)}$ are circulant Kleene stars, where all off-diagonal entries are negative.*

Proof (a) Using (8.16) and notation (8.22) we see that for all (i, j) and (k, l) such that $k = i + t \pmod{\sigma_\mu}$ and $l = j + t \pmod{\sigma_\nu}$,

$$\tilde{a}_{s_k t_l}^{(r)} = \tilde{a}_{s_i t_l}^{(r+t)} = \tilde{a}_{s_i t_j}^{(r)}.$$

(b) If $\mu = \nu$ then $P = I$, and if $\mu \neq \nu$ then P is any permutation matrix such that its subpermutations for N_μ and N_ν are given by P and Q of Proposition 8.5.2.

(c) Part (a) shows that $\tilde{A}_{\mu\mu}^{(r)}$ are circulants for any $r \geq T(A)$ and critical μ. If r is a multiple of σ, then $\tilde{A}_{\mu\mu}^{(r)}$ are submatrices of $\tilde{A}^\sigma = Q(\tilde{A}^\sigma)$ and hence of $(\tilde{A}^\sigma)^*$. This implies, using Corollary 1.6.16, that they are Kleene stars. As the μth component of $C(\tilde{A})$ is just a cycle of length σ_μ, the corresponding component of $C(\tilde{A}^\sigma)$ consists of σ_μ loops, showing that the off-diagonal entries of $\tilde{A}_{\mu\mu}^{(r)}$ are negative.   □

8.5.3 Max-linear Systems Describing Attraction Spaces

Let $A \in \overline{\mathbb{R}}^{n \times n}$ be definite and irreducible. It follows from Sect. 8.4, in particular Theorem 8.4.4, that the coefficients of the system $A^r \otimes x = A^{r+p} \otimes x$ for integers $p \geq 1$ and $r \geq T(A)$ can be found using $O(n^3 \log n)$ operations, by means of matrix squaring and permutation of cyclic classes. Due to Corollary 8.4.2 we may assume without loss of generality that $p = 1$.

Next we show how the specific circulant structure of A^r at $r \geq T(A)$ can be exploited, to derive a more efficient system of equations for the attraction space $\mathrm{Attr}(A, 1)$. Due to Theorem 8.5.1 the core matrix

$$A^{\mathrm{Core}} = \{\alpha_{\mu\nu}; \ \mu, \nu = 1, \ldots, n_c + \bar{c}\},$$

and its Kleene star

$$(A^{\mathrm{Core}})^* = \{\alpha_{\mu\nu}^*; \ \mu, \nu = 1, \ldots, n_c + \bar{c}\}$$

will be of special importance. We will use the notation

$$M_v^{(r)}(i) = \{j \in N_v;\ a_{ij}^{(r)} = \alpha_{\mu v}^*\},\quad i \in N_\mu,\ \forall v : C_v \neq C_\mu,$$
$$K^{(r)}(i) = \{t > c;\ a_{it}^{(r)} = \alpha_{\mu v(t)}^*\},\quad i \in N_\mu,$$

(8.26)

where C_μ and C_v are strongly connected components of $C(A)$, N_μ and N_v are their node sets and $v(t)$ in the second definition denotes the index of the noncritical component which consists of the node t. The sets $M_v^{(r)}(i)$ defined in (8.26) are nonempty for any $r \geq T_c(A)$, due to Theorems 8.3.9 and 8.5.1.

The results of Sect. 8.5.2 yield the following properties of $M_v^{(r)}(i)$ and $K^{(r)}(i)$.

Proposition 8.5.4 *Let $A \in \overline{\mathbb{R}}^{n \times n}$ be an irreducible, definite and visualized matrix, $r \geq T_c(A)$ and $\mu, v \in \{1, \ldots, n_c\}$.*

1. *If $[i] \to_t [j]$ and $i, j \in N_\mu$ then $M_v^{(r+t)}(i) = M_v^{(r)}(j)$ and $K^{(r+t)}(i) = K^{(r)}(j)$.*
2. *Each $M_v^{(r)}(i)$ is the union of some cyclic classes of C_v.*
3. *Let $i \in N_\mu$ and $d = \gcd(\sigma_\mu, \sigma_v)$. Then, if $[p] \subseteq M_v^{(r)}(i)$ and $[p] \to_d [s]$ then $[s] \subseteq M_v^{(r)}(i)$.*
4. *Let $i, j \in N_\mu$ and $p, s \in N_v$. Let $[i] \to_t [j]$ and $[p] \to_t [s]$. Then $[p] \subseteq M_v^{(r)}(i)$ if and only if $[s] \subseteq M_v^{(r)}(j)$.*

Next we establish the cancellation rules which will enable us to simplify systems of equations for the attraction space Attr$(A, 1)$.

Recall first that by Lemma 7.4.1, if $a < c$, then

$$\{x;\ a \otimes x \oplus b = c \otimes x \oplus d\} = \{x;\ b = c \otimes x \oplus d\}. \tag{8.27}$$

Consider now a chain of equations

$$\sum_{i \in N}^{\oplus} a_{1i} x_i \oplus c_1 = \sum_{i \in N}^{\oplus} a_{2i} x_i \oplus c_2 = \cdots = \sum_{i \in N}^{\oplus} a_{ni} x_i \oplus c_n. \tag{8.28}$$

Suppose that $\alpha_1, \ldots, \alpha_n \in \mathbb{R}$ are such that $a_{li} \leq \alpha_i$ for all l and i, and $S_l = \{i;\ a_{li} = \alpha_i\}$ for $l = 1, \ldots, n$. Let S_l be such that $\bigcup_{l=1}^{n} S_l = \{1, \ldots, n\}$. By repeatedly applying the elementary cancellation law (8.27), we obtain that (8.28) is equivalent to

$$\sum_{i \in S_1}^{\oplus} \alpha_i x_i \oplus c_1 = \sum_{i \in S_2}^{\oplus} \alpha_i x_i \oplus c_2 = \cdots = \sum_{i \in S_n}^{\oplus} \alpha_i x_i \oplus c_n. \tag{8.29}$$

We will refer to the equivalence between (8.28) and (8.29), as to the *chain cancellation*.

We may now formulate the following key result.

Theorem 8.5.5 [137] *Let $A \in \overline{\mathbb{R}}^{n \times n}$ be an irreducible, definite and visualized matrix and $r \geq T(A)$ be a multiple of $\sigma = \sigma(A)$. Then the system $A^r \otimes x = A^{r+1} \otimes x$*

is equivalent to

$$\sum_{k\in[i]}^{\oplus} x_k \oplus \sum_{v\neq\mu}^{\oplus}\left(\alpha_{\mu v}^* \otimes \sum_{k\in M_v^{(r)}(i)}^{\oplus} x_k\right) \oplus \sum_{t\in K^{(r)}(i)}^{\oplus}\alpha_{\mu v(t)}^* \otimes x_t$$

$$= \sum_{k\in[j]}^{\oplus} x_k \oplus \sum_{v\neq\mu}^{\oplus}\left(\alpha_{\mu v}^* \otimes \sum_{k\in M_v^{(r)}(j)}^{\oplus} x_k\right) \oplus \sum_{t\in K^{(r)}(j)}^{\oplus}\alpha_{\mu v(t)}^* \otimes x_t, \quad (8.30)$$

where $\mu = 1, \ldots, n_c$ and $[i]$ and $[j]$ range over all pairs of cyclic classes in C_μ such that $[i] \rightarrow_1 [j]$.

Proof By Lemma 8.4.3 $A^r \otimes x = A^{r+1} \otimes x$ is equivalent to its critical subsystem. Consider critical equations of $A^r \otimes x = A^{r+1} \otimes x$:

$$\sum_k^{\oplus} a_{ik}^{(r)} \otimes x_k = \sum_k^{\oplus} a_{ik}^{(r+1)} \otimes x_k, \quad i = 1, \ldots, c. \quad (8.31)$$

Take $i, j \in \{1, \ldots, c\}$ such that $[i] \rightarrow_1 [j]$. Then by Theorem 8.3.9,

$$a_{ik}^{(r+1)} = a_{jk}^{(r)},$$

hence the critical subsystem of $A^r \otimes x = A^{r+1} \otimes x$ is as follows:

$$\sum_k^{\oplus} a_{ik}^{(r)} \otimes x_k = \sum_k^{\oplus} a_{jk}^{(r)} \otimes x_k, \quad \forall i, j : [i] \rightarrow_1 [j]. \quad (8.32)$$

Proposition 8.5.3, part (c), implies that all principal submatrices of A^r extracted from critical components have circulant block structure. In this structure, all entries of the diagonal blocks are equal to 0, and the entries of all off-diagonal blocks are strictly less than 0. Hence we can apply the chain cancellation (equivalence between (8.28) and (8.29)) and obtain the first terms on both sides of (8.30). By Theorem 8.5.1 each block $A_{\mu v}$ contains an entry equal to $\alpha_{\mu v}^*$. For a noncritical $v(t)$, this readily implies that the corresponding subcolumn $A_{\mu v(t)}$ contains an entry $\alpha_{\mu v(t)}^*$. Applying the chain cancellation we obtain the last terms on both sides of (8.30). From the block circulant structure of $A_{\mu v}$ with both μ and v critical, see Propositions 8.5.3 or 8.5.4, we deduce that each column of such a block also contains an entry equal to $\alpha_{\mu v}^*$. Applying the chain cancellation we obtain the remaining terms in (8.30). $\qquad\square$

It follows that (8.19) is equivalent to (8.30). As $\text{Attr}(A, t) = \text{Attr}(A^t, 1)$, this system can also be used to describe more general attraction spaces. It is only necessary to substitute $C(A^t)$ for $C(A)$ and the entries of $((A^t)^{\text{Core}})^*$ for $\alpha_{\mu v}^*$ (the dimension of this matrix will be different in general, see Theorem 8.2.6 part (d)).

We note that (8.30) naturally breaks into several chains of equations corresponding to individual strongly connected components of $C(A)$. Let $\mu \in \{1, \ldots, n_c\}$ and consider the subsystem of (8.32) corresponding to C_μ. It is a single chain of equations. Denote the common value of all sides in this chain by z_μ. Then, the subsystem can be written in the form $P \otimes x = (z_\mu, \ldots, z_\mu)^T$, where each row of P corresponds to one side of the chain. Therefore the whole system (8.32) can equivalently be written as $R \otimes x = H \otimes z$, where $H = (h_{i\mu}) \in \overline{\mathbb{R}}^{\tilde{c} \times n_c}$ ($\tilde{c}$ is the total number of cyclic classes) has entries

$$
h_{i\mu} = \begin{cases} 0, & \text{if } i \in N_\mu, \\ \varepsilon, & \text{otherwise.} \end{cases} \tag{8.33}
$$

Remark 8.5.6 If all nodes of A are critical and the critical digraph is strongly connected then the sets of variables on individual sides in (8.30) are pairwise disjoint, corresponding to individual cyclic classes. In this case the unique scaled basis of the attraction space of A is described by Corollary 7.2.6.

Theorem 8.5.5 can be used for finding the attraction system in a way different from matrix scaling and permutation of cyclic classes [134]. This method is more efficient if the number of strongly connected components of $C(A)$ and the number of noncritical nodes are small relative to n.

Example 8.5.7 Consider the following 9×9 definite, strictly visualized matrix:

$$
A = \begin{pmatrix}
-8 & 0 & -1 & -8 & -8 & -9 & -4 & -5 & -1 \\
-4 & -5 & 0 & -2 & -6 & 0 & -7 & -3 & -9 \\
-7 & -9 & -8 & 0 & -8 & -4 & -6 & -9 & -10 \\
-8 & -8 & -10 & -7 & 0 & -4 & -6 & -10 & -1 \\
-2 & -8 & -7 & -4 & -8 & 0 & -3 & -1 & -10 \\
0 & -1 & -2 & -7 & -10 & -6 & -3 & -6 & -1 \\
-10 & -7 & -7 & -7 & -6 & -1 & -5 & 0 & -9 \\
-8 & -3 & -6 & -8 & -6 & -8 & -5 & -10 & 0 \\
-4 & -3 & -5 & -6 & -6 & -10 & 0 & -6 & -9
\end{pmatrix}.
$$

The critical digraph of this matrix consists of two strongly connected components, comprising 6 and 3 nodes respectively. They are shown in Figs. 8.3 and 8.4, together with their cyclic classes. Note that $\sigma(A) = \mathrm{lcm}(\gcd(6, 3), 3) = 3$. The components of $C(A)$ induce block decomposition

$$
A = \begin{pmatrix} A_{11} & A_{12} \\ A_{21} & A_{22} \end{pmatrix}, \tag{8.34}
$$

Fig. 8.3 Critical digraph in
Example 8.5.7

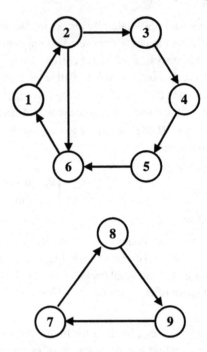

Fig. 8.4 Cyclic classes in
Example 8.5.7

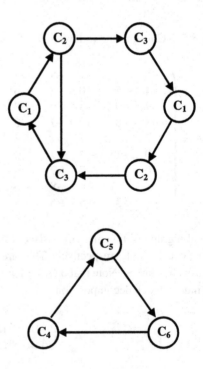

where

$$
A_{11} = \begin{pmatrix}
-8 & 0 & -1 & -8 & -8 & -9 \\
-4 & -5 & 0 & -2 & -6 & 0 \\
-7 & -9 & -8 & 0 & -8 & -4 \\
-8 & -8 & -10 & -7 & 0 & -4 \\
-2 & -8 & -7 & -4 & -8 & 0 \\
0 & -1 & -2 & -7 & -10 & -6
\end{pmatrix},
$$

$$
A_{22} = \begin{pmatrix}
-5 & 0 & -9 \\
-5 & -10 & 0 \\
0 & -6 & -9
\end{pmatrix}.
$$

(8.35)

The core matrix and its Kleene star are equal to

$$
A^{\text{Core}} = (A^{\text{Core}})^* = \begin{pmatrix} 0 & -1 \\ -1 & 0 \end{pmatrix}.
$$

(8.36)

By calculating $A, A^2, \dots$ we obtain that the powers of A become periodic after $T(A) = 6$. In the block decomposition of A^6 induced by (8.34), we have the following circulants:

$$
A_{11}^{(6)} = \begin{pmatrix}
0 & -1 & -2 & 0 & -1 & -2 \\
-2 & 0 & -1 & -2 & 0 & -1 \\
-1 & -2 & 0 & -1 & -2 & 0 \\
0 & -1 & -2 & 0 & -1 & -2 \\
-2 & 0 & -1 & -2 & 0 & -1 \\
-1 & -2 & 0 & -1 & -2 & 0
\end{pmatrix}, \quad
A_{12}^{(6)} = \begin{pmatrix}
-2 & -1 & -1 \\
-1 & -2 & -1 \\
-1 & -1 & -2 \\
-2 & -1 & -1 \\
-1 & -2 & -1 \\
-1 & -1 & -2
\end{pmatrix},
$$

(8.37)

$$
A_{21}^{(6)} = \begin{pmatrix}
-3 & -1 & -2 & -3 & -1 & -2 \\
-2 & -3 & -1 & -2 & -3 & -1 \\
-1 & -2 & -3 & -1 & -2 & -3
\end{pmatrix}, \quad
A_{22}^{(6)} = \begin{pmatrix}
0 & -3 & -2 \\
-2 & 0 & -3 \\
-3 & -2 & 0
\end{pmatrix}.
$$

(8.38)

The corresponding blocks of reduced power $\tilde{A}^{(6)}$ are

$$
\tilde{A}_{11}^{(6)} = \begin{pmatrix}
0 & -1 & -2 \\
-2 & 0 & -1 \\
-1 & -2 & 0
\end{pmatrix}, \quad
\tilde{A}_{12}^{(6)} = \begin{pmatrix}
-2 & -1 & -1 \\
-1 & -2 & -1 \\
-1 & -1 & -2
\end{pmatrix},
$$

$$
\tilde{A}_{11}^{(6)} = \begin{pmatrix}
-3 & -1 & -2 \\
-2 & -3 & -1 \\
-1 & -2 & -3
\end{pmatrix}, \quad
\tilde{A}_{12}^{(6)} = \begin{pmatrix}
0 & -3 & -2 \\
-2 & 0 & -3 \\
-3 & -2 & 0
\end{pmatrix}.
$$

Note that $\tilde{A}_{11}^{(6)}$ and $\tilde{A}_{22}^{(6)}$ are Kleene stars, with all off-diagonal entries negative.

Using (8.37) and (8.38), we see that the attraction system consists of two chains of equations, namely

$$x_1 \oplus x_4 \oplus (x_8 - 1) \oplus (x_9 - 1) = x_2 \oplus x_5 \oplus (x_7 - 1) \oplus (x_9 - 1)$$
$$= x_3 \oplus x_6 \oplus (x_7 - 1) \oplus (x_8 - 1)$$

and

$$(x_2 - 1) \oplus (x_5 - 1) \oplus x_7 = (x_3 - 1) \oplus (x_6 - 1) \oplus x_8$$
$$= (x_1 - 1) \oplus (x_4 - 1) \oplus x_9.$$

Note that only 0 and -1, the coefficients of $(A^{\text{Core}})^*$ (which is equal to A^{Core} in this example), appear in this system.

8.6 Robustness of Matrices

8.6.1 Introduction

In this section we deal with Q3, that is, with the task of recognizing robust matrices. We start with a few basic observations and then analyze the problem first for irreducible and then for reducible matrices.

Let $A = (a_{ij}) \in \overline{\mathbb{R}}^{n \times n}$ and recall that $\text{Attr}(A, 1)$ is the set of all starting vectors from which the orbit reaches the eigenspace, that is

$$\text{Attr}(A, 1) = \{x \in \overline{\mathbb{R}}^n; O(A, x) \cap V(A) \neq \{\varepsilon\}\}.$$

Clearly,

$$V(A) - \{\varepsilon\} \subseteq \text{Attr}(A, 1) \subseteq \overline{\mathbb{R}}^n - \{\varepsilon\}$$

and so robust matrices are exactly those for which $\text{Attr}(A, 1) = \overline{\mathbb{R}}^n - \{\varepsilon\}$.

It may happen that $\text{Attr}(A, 1) = V(A) - \{\varepsilon\}$, for instance when A is the irreducible matrix

$$\begin{pmatrix} -1 & 0 \\ 0 & -1 \end{pmatrix}.$$

Here $\lambda(A) = 0$ and by Theorem 4.4.4

$$V(A) - \{\varepsilon\} = \{\alpha \otimes (0, 0)^T; \alpha \in \mathbb{R}\}.$$

Since

$$A \otimes \begin{pmatrix} a \\ b \end{pmatrix} = (\max(a - 1, b), \max(a, b - 1))^T,$$

we have that $A \otimes \binom{a}{b}$ is an eigenvector of A if and only if $a = b$, that is, $A \otimes x$ is an eigenvector of A if and only if x is an eigenvector of A. Hence $\mathrm{Attr}(A, 1) = V(A) - \{\varepsilon\}$.

$\mathrm{Attr}(A, 1)$ may also be different from both $V(A) - \{\varepsilon\}$ and $\overline{\mathbb{R}}^n - \{\varepsilon\}$: Consider the irreducible matrix

$$A = \begin{pmatrix} -1 & 0 & -1 \\ 0 & -1 & -1 \\ -1 & -1 & 0 \end{pmatrix}.$$

Here $\lambda(A) = 0$ and $x = (-2, -2, 0)^T$ is not an eigenvector of A but $A \otimes x = (-1, -1, 0)^T$ is, showing that $\mathrm{Attr}(A, 1) \neq V(A) - \{\varepsilon\}$. At the same time if $y = (0, -1, 0)^T$ then $A^k \otimes y$ is y for k even and $(-1, 0, 0)^T$ for k odd, showing that $y \notin \mathrm{Attr}(A, 1)$.

Lemma 8.6.1 *If $A, B \in \overline{\mathbb{R}}^{n \times n}$ and $A \equiv B$ then A is robust if and only if B is robust.*

Proof $B = P^{-1} \otimes A \otimes P$ for some permutation matrix P. Hence

$$B^{k+1} \otimes x = P^{-1} \otimes A^{k+1} \otimes P \otimes x = P^{-1} \otimes \lambda \otimes A^k \otimes P \otimes x$$

$$= \lambda \otimes B^k \otimes x. \qquad \square$$

Due to Lemma 8.6.1 we may without loss of generality investigate robustness of matrices arising from a given matrix by a simultaneous permutation of the rows and columns.

We finish this introduction by excluding a pathological case:

Lemma 8.6.2 *A matrix with an ε column is not robust. This is true in particular if one of its eigenvalues is ε.*

Proof If (say) the kth column of A is ε then $A \otimes x = \varepsilon$ for any $x \in \overline{\mathbb{R}}^n$ such that $x_i = \varepsilon$ for $i \neq k$. Hence A is not robust.

The second statement follows from Lemma 4.5.11. $\qquad \square$

8.6.2 Robust Irreducible Matrices

Characterization of robustness for irreducible matrices using the results of the previous sections is relatively easy. We will also deduce a few corollaries of the following main result.

Theorem 8.6.3 *Let $A \in \overline{\mathbb{R}}^{n \times n}$ be column $\mathbb{R}$-astic and $|\Lambda(A)| = 1$ (that is, $\Lambda(A) = \{\lambda(A)\}$). Then A is robust if and only if the period of A is 1.*

Proof Suppose that the period of A is 1. Let $x \in \overline{\mathbb{R}}^n - \{\varepsilon\}$ and $k \geq T(A)$. Then $A^k \otimes x \in \overline{\mathbb{R}}^n - \{\varepsilon\}$ by Lemma 1.5.2, $A^{k+1} \otimes x = \lambda \otimes A^k \otimes x$ and so A is robust (and all columns of A^k are eigenvectors of A).

Now let A be robust and x be the jth column of A. Then $x \in \overline{\mathbb{R}}^n - \{\varepsilon\}$ and thus there is an integer k_j such that $A^{k+1} \otimes x = \lambda(A) \otimes A^k \otimes x$ for all $k \geq k_j$. So, if $k_0 = \max(k_1, \ldots, k_n)$ then $A^{k+2} = \lambda(A) \otimes A^{k+1}$ for all $k \geq k_0$, and thus the period of A is 1. $\qquad\square$

Recall that every irreducible $n \times n$ matrix $(n > 1)$ is column $\mathbb{R}$-astic (Lemma 1.5.1), but not conversely.

Note that if A is the 1×1 matrix (ε) then A is irreducible, $\sigma(A) = 1$ but A is not robust. This is an exceptional case that has to be excluded in the statements that follow.

Corollary 8.6.4 [34, 102] *Let $A \in \overline{\mathbb{R}}^{n \times n}$, $A \neq \varepsilon$ be irreducible. Then A is robust if and only if A is primitive.*

Proof Every irreducible matrix has a unique eigenvalue and if $A \neq \varepsilon$ then it is also $\mathbb{R}$-astic. The period of A is $\sigma(A)$ by Theorem 8.3.5 and the statement now follows from Theorem 8.6.3. $\qquad\square$

Corollary 8.6.5 *Let $A \in \overline{\mathbb{R}}^{n \times n}$, $A \neq \varepsilon$ be irreducible. If A is primitive, $x \neq \varepsilon$ then $A^k \otimes x$ is finite for all sufficiently large k.*

Proof If A is primitive then A is robust, thus for $x \in \overline{\mathbb{R}}^n$, $x \neq \varepsilon$, and all sufficiently large k we have $A^k \otimes x \in V(A) - \{\varepsilon\} = V^+(A)$ since A is irreducible. $\qquad\square$

Example 8.6.6 For the irreducible matrix A of Example 4.3.7 we have that the cyclicity of the critical component with the node set $\{1, 2\}$ is 2, and that of the component on $\{4, 5, 6\}$ is $\gcd\{1, 3\} = 1$. Hence $\sigma(A) = \sigma(C(A)) = \operatorname{lcm}\{1, 2\} = 2$ and so A is not robust.

The following classical sufficient condition for robustness now easily follows:

Corollary 8.6.7 [65] *Let $A = (a_{ij}) \in \overline{\mathbb{R}}^{n \times n}$, $A \neq \varepsilon$ be irreducible. Then A is robust if $a_{ii} = \lambda(A)$ for every $i \in N_c(A)$.*

Proof If $a_{ii} = \lambda(A)$ for every $i \in N_c(A)$ then a cycle of length one exists in every component of the critical digraph, hence A is primitive and so A is robust. $\qquad\square$

We also deduce that the powers of a robust irreducible matrix remain irreducible:

Corollary 8.6.8 *Let $A \in \overline{\mathbb{R}}^{n \times n}$ be irreducible and robust. Then A^k is irreducible for every positive integer k.*

Proof The statement follows from Corollary 8.2.4. $\qquad\square$

8.6.3 Robust Reducible Matrices

Robustness of reducible matrices is not very strongly related to ultimate periodicity (unlike for irreducible matrices). However and although it will not be directly used in this book, for the sake of completeness we present a (slightly reformulated) generalization of the Cyclicity Theorem to reducible matrices:

Theorem 8.6.9 [114] (General Cyclicity Theorem) *A matrix* $A \in \overline{\mathbb{R}}^{n \times n}$ *is ultimately periodic if and only if each irreducible diagonal block of the FNF of A has the same eigenvalue.*

The rest of this subsection is based on [44].

Recall that if $A = (a_{ij}) \in \overline{\mathbb{R}}^{n \times n}$ is in the FNF (4.7) and $N_1, \ldots, N_r$ are the classes of A then we have denoted $R = \{1, \ldots, r\}$. If $i \in R$ then we now also denote $T_i = \{j \in R; N_j \longrightarrow N_i\}$ and $M_i = \bigcup_{j \in T_i} N_j$. A class N_i of A is called *trivial* if N_i contains only one index, say k, and $a_{kk} = \varepsilon$.

We start with a lemma. Without loss of generality we assume in the rest of this subsection that A is in the FNF (4.7).

Lemma 8.6.10 [85] *If every nontrivial class of $A \in \overline{\mathbb{R}}^{n \times n}$ has eigenvalue 0 and period 1 then $A^{k+1} = A^k$ for some k.*

Proof We prove the statement by induction on the number of classes.

If A has only one class then either this class is trivial or A is irreducible. In both cases the statement follows immediately.

If A has at least two classes then by Lemma 4.1.3 we can assume without loss of generality:

$$A = \begin{pmatrix} A_{11} & \varepsilon \\ A_{21} & A_{22} \end{pmatrix}$$

and thus

$$A^k = \begin{pmatrix} A_{11}^k & \varepsilon \\ B_k & A_{22}^k \end{pmatrix},$$

where

$$B_k = \sum_{i+j=k-1}^{\oplus} A_{22}^i \otimes A_{21} \otimes A_{11}^j.$$

By the induction hypothesis there are k_1 and k_2 such that

$$A_{11}^{k_1+1} = A_{11}^{k_1} \text{ and } A_{22}^{k_2+1} = A_{22}^{k_2}.$$

It is sufficient now to prove that

$$B_k = \sum^{\oplus} \{A_{22}^i \otimes A_{21} \otimes A_{11}^j; i \le k_2, j \le k_1, i = k_2 \text{ or } j = k_1\} \qquad (8.39)$$

holds for all $k \ge k_1 + k_2 + 1$.

For all i, j we have

$$A_{22}^i \otimes A_{21} \otimes A_{11}^j = A_{22}^{i'} \otimes A_{21} \otimes A_{11}^{j'},$$

where $i' = \min(i, k_2)$, $j' = \min(j, k_1)$. If $i + j + 1 = k \geq k_1 + k_2 + 1$ then either $i \geq k_2$ or $j \geq k_1$. Hence either $i' = k_2$ or $j' = k_1$ and therefore $\leq$ in (8.39) follows. For $\geq$ let $i = k_2$ (say) and $j \leq k_1$. Since $k \geq k_1 + k_2 + 1 \geq j + i + 1$, we have $k - j - 1 \geq i = k_2$ and thus

$$A_{22}^i \otimes A_{21} \otimes A_{11}^j = A_{22}^{k-j-1} \otimes A_{21} \otimes A_{11}^j \leq B_k. \qquad \square$$

We are ready to prove one of the key results of this book.

Theorem 8.6.11 [44] *Let* $A \in \overline{\mathbb{R}}^{n \times n}$ *be column* $\mathbb{R}$*-astic and in the FNF* (4.7), $N_1, \ldots, N_r$ *be the classes of* A *and* $R = \{1, \ldots, r\}$. *Then* A *is robust if and only if the following hold*:

1. *All nontrivial classes* $N_1, \ldots, N_r$ *are spectral.*
2. *If* $i, j \in R$, N_i, N_j *are nontrivial and* $i \notin T_j$ *and* $j \notin T_i$ *then* $\lambda(N_i) = \lambda(N_j)$.
3. $\sigma(A_{jj}) = 1$ *for all* $j \in R$.

Proof If $r = 1$ then A is irreducible and the statement follows by Theorem 8.6.4. We will therefore assume $r \geq 2$ in this proof.

Let A be robust, we prove that 1.–3. hold.

1. Let $i \in R$, $A_{ii} \neq \varepsilon$ and $x \in \overline{\mathbb{R}}^n$ be defined by taking any $x_s \in \mathbb{R}$ for $s \in M_i$ and $x_s = \varepsilon$ for $s \notin M_i$. Then $A^{k+1} \otimes x = \lambda \otimes A^k \otimes x$ for some k and $\lambda \in \Lambda(A)$. Let $z = A^k \otimes x$. Then $z[M_i]$ is finite since $A[M_i]$ has no ε row and

$$A[M_i] \otimes z[M_i] = (A \otimes z)[M_i] = \lambda \otimes z[M_i],$$

hence $z[M_i] \in V^+(A[M_i])$. By Lemma 8.6.2 $\lambda > \varepsilon$ and so by Theorem 4.4.4 then $\lambda(N_t) \leq \lambda(N_i)$ for all $t \in T_i$. Hence N_i is spectral.
2. Suppose $i, j \in R$, N_i, N_j are nontrivial and $i \notin T_j$, $j \notin T_i$. Let $x \in \overline{\mathbb{R}}^n$ be defined by taking any

$$x[N_i] \in V^+(A[N_i]),$$

$$x[N_j] \in V^+(A[N_j])$$

and $x_s = \varepsilon$ for $s \in N - N_i \cup N_j$. Then $A^{k+1} \otimes x = \lambda \otimes A^k \otimes x$ for some k and $\lambda \in \Lambda(A)$. Denote $z = A^k \otimes x$. Then $z[N_j]$ is finite. Since $i \notin T_j$ we have $a_{uv} = \varepsilon$ for all $u \in N_i$ and $v \in N_j$. Hence

$$\lambda \otimes z[N_j] = (A \otimes z)[N_j] = A[N_j] \otimes z[N_j]$$

and so by Theorem 4.4.4 $\lambda(N_j) = \lambda$. Similarly it is proved that $\lambda(N_i) = \lambda$.

3. Let $j \in R$ and $A[N_j] \neq \varepsilon$ (otherwise the statement follows trivially). Let $x \in \overline{\mathbb{R}}^n$ be any vector such that $x \neq \varepsilon$ and $x_s = \varepsilon$ for $s \notin N_j$. Then $A^{k+1} \otimes x = \lambda \otimes A^k \otimes x$ for some k and $\lambda \in \Lambda(A)$. Let $z = A^k \otimes x$. Since

$$z[N_j] = (A[N_j])^k \otimes x[N_j],$$

we may assume without loss of generality that $z[N_j] \neq \varepsilon$. At the same time

$$A[N_j] \otimes z[N_j] = (A \otimes z)[N_j] = \lambda \otimes z[N_j]$$

and thus $z[N_j] \in V(A[N_j])$. Hence $A[N_j]$ is irreducible and robust. Thus by Theorem 8.6.4 we have $\sigma(A[N_j]) = \sigma(A_{jj}) = 1$.

Suppose now that conditions 1.–3. are satisfied. We prove then that A is robust by induction on the number of classes of A. As already observed at the beginning of this proof, the case $r = 1$ follows from Theorem 8.6.4. Suppose now that $r \geq 2$ and let $x \in \overline{\mathbb{R}}^n, x \neq \varepsilon$. Let

$$U = \{i \in N; (\exists j)i \longrightarrow j, x_j \neq \varepsilon\}.$$

We have

$$(A^k \otimes x)[U] = (A[U])^k \otimes x[U]$$

and

$$(A^k \otimes x)_i = \varepsilon$$

for $i \notin U$. Therefore we may assume without loss of generality that $U = N$. Let M be a final class in C_A, clearly $x[M] \neq \varepsilon$ by the definition of U. Let us denote

$$S = \{i \in N; (\exists j \in M)(i \longrightarrow j)\}$$

and

$$S' = N - S.$$

By Lemma 4.1.3 we may assume without loss of generality that

$$A = \begin{pmatrix} A_{11} & \varepsilon & \varepsilon \\ A_{21} & A_{22} & A_{23} \\ \varepsilon & \varepsilon & A_{33} \end{pmatrix},$$

where the individual blocks correspond (in this order) to the sets $M, S - M$ and S' respectively. Let us define $x^k = A^k \otimes x$ for all integers $k \geq 0$. We also set

$$x_1^k = x^k[M],$$
$$x_2^k = x^k[S - M],$$
$$x_3^k = x^k[S'].$$

Obviously,

$$x_1^{k+1} = A_{11} \otimes x_1^k,$$

$$x_2^{k+1} = A_{21} \otimes x_1^k \oplus A_{22} \otimes x_2^k \oplus A_{23} \otimes x_3^k,$$

$$x_3^{k+1} = A_{33} \otimes x_3^k.$$

Assume first that M is nontrivial. Then $\lambda(A_{11}) \neq \varepsilon$ and by taking (if necessary) $(\lambda(A_{11}))^{-1} \otimes A$ instead of A, we may assume without loss of generality that $\lambda(A_{11}) = 0$. By assumption 3 and Theorem 8.3.5 we have $A_{11}^{k_1+1} = A_{11}^{k_1}$ for some k_1. By assumption 2 every class of A_{33} has eigenvalue 0. Since each of these classes has also period 1 by assumption 3, it follows from Lemma 8.6.10 that $A_{33}^{k_3+1} = A_{33}^{k_3}$ for some k_3. We may also assume without loss of generality that

$$x_1^0 = x_1^1 = x_1^2 = \cdots$$

and

$$x_3^0 = x_3^1 = x_3^2 = \cdots .$$

Therefore

$$x_2^{k+1} = A_{21} \otimes x_1^0 \oplus A_{22} \otimes x_2^k \oplus A_{23} \otimes x_3^0.$$

Let $v = A_{21} \otimes x_1^0 \oplus A_{23} \otimes x_3^0$. We deduce that

$$x_2^k = A_{22}^k \otimes x_2^0 \oplus (A_{22}^{k-1} \oplus \cdots \oplus A_{22}^0) \otimes v \tag{8.40}$$

for all k. Moreover, $\lambda(A_{22}) \leq \lambda(A_{11}) = 0$ since M is spectral by assumption 1. Hence

$$A_{22}^{k-1} \oplus \cdots \oplus A_{22}^0 = \Gamma(A_{22})$$

for all $k \geq n$. Note that x_1^0 is finite as an eigenvector of the irreducible matrix A_{11}. Also, since every node in S has access to M, the vector

$$\Gamma(A_{22}) \otimes A_{21} \otimes x_1^0$$

is finite and hence also $\Gamma(A_{22}) \otimes v$ is finite. If $\lambda(A_{22}) < 0$ then $A_{22}^k \otimes x_2^0 \longrightarrow -\infty$ as $k \longrightarrow \infty$ and we deduce that $x_2^k = \Gamma(A_{22}) \otimes v$ for all k large enough. If $\lambda(A_{22}) = 0$ then

$$A_{22}^{k_2+1} = A_{22}^{k_2}$$

by the induction hypothesis and thus

$$x_2^k = A_{22}^{k_2} \otimes x_2^0 \oplus \Gamma(A_{22}) \otimes v$$

for all $k \geq \max(k_1, k_2, k_3)$.

Fig. 8.5 Condensation
digraph of a robust matrix

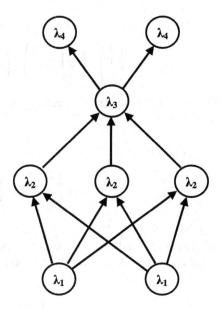

It remains to consider the case when A_{11} is trivial. Then $x_1^k = \varepsilon$ for all $k \geq 1$ and we have

$$\begin{pmatrix} x_2^{k+1} \\ x_3^{k+1} \end{pmatrix} = \begin{pmatrix} A_{22} & A_{23} \\ \varepsilon & A_{33} \end{pmatrix} \otimes \begin{pmatrix} x_2^k \\ x_3^k \end{pmatrix}$$

for all $k \geq 1$. We apply the induction hypothesis to the matrix

$$\begin{pmatrix} A_{22} & A_{23} \\ \varepsilon & A_{33} \end{pmatrix}$$

and deduce that $x^{k+1} = x^k$ for k sufficiently large. This completes the proof. $\quad\square$

An example of the condensation digraph of a robust reducible matrix can be seen in Fig. 8.5, where the nodes correspond to primitive classes with unique eigenvalues $\lambda_1, \lambda_2, \lambda_3, \lambda_4$ and $\lambda_1 < \lambda_2 < \lambda_3 < \lambda_4$.

Example 8.6.12 Let

$$A = \begin{pmatrix} 2 & \varepsilon & \varepsilon \\ \varepsilon & 1 & \varepsilon \\ 0 & 0 & 0 \end{pmatrix},$$

thus $r = 3$, $\Lambda(A) = \{0, 1, 2\}$, $N_j = \{j\}$, $j = 1, 2, 3$. If

$$x = \begin{pmatrix} 0 \\ 0 \\ 0 \end{pmatrix},$$

then $O(A, x)$ is

$$\begin{pmatrix} 2 \\ 1 \\ 0 \end{pmatrix}, \begin{pmatrix} 4 \\ 2 \\ 2 \end{pmatrix}, \begin{pmatrix} 6 \\ 3 \\ 4 \end{pmatrix}, \begin{pmatrix} 8 \\ 4 \\ 6 \end{pmatrix}, \dots,$$

which obviously will never reach an eigenvector. The reason is that $1 \notin T_2$, $2 \notin T_1$ but $\lambda(N_1) \neq \lambda(N_2)$.

Example 8.6.13 Let

$$A = \begin{pmatrix} 2 & \varepsilon & \varepsilon \\ \varepsilon & \varepsilon & \varepsilon \\ 0 & 0 & 0 \end{pmatrix},$$

thus $r = 3$, $\Lambda(A) = \{0, 2\}$, $N_j = \{j\}$, $j = 1, 2, 3$. This matrix is robust since both nontrivial classes (N_1 and N_3) are spectral, $\sigma(A_{ii}) = 1$ ($i = 1, 2, 3$) and there are no nontrivial classes N_i, N_j such that $i \notin T_j$ and $j \notin T_i$. Indeed, if

$$x = \begin{pmatrix} 0 \\ 0 \\ 0 \end{pmatrix},$$

then $O(A, x)$ is

$$\begin{pmatrix} 2 \\ \varepsilon \\ 0 \end{pmatrix}, \begin{pmatrix} 4 \\ \varepsilon \\ 2 \end{pmatrix}, \begin{pmatrix} 6 \\ \varepsilon \\ 4 \end{pmatrix}, \begin{pmatrix} 8 \\ \varepsilon \\ 6 \end{pmatrix}, \dots,$$

hence an eigenvector is reached in the first step.

8.6.4 M-robustness

Note that in this subsection the symbol M has a reserved meaning. Requirements 1.–3. of Theorem 8.6.11 imply that every robust matrix A either has only one superblock or $|\Lambda(A)| = 1$. Obviously this restricts the concept of robustness for reducible matrices quite significantly. Therefore we present an alternative concept of robustness and provide a criterion which will enable us to characterize a wider class of matrices displaying robustness properties reflecting the rich spectral structure of reducible matrices.

We start with a simple observation.

Lemma 8.6.14 *Let* $A = \begin{pmatrix} A' & \varepsilon \\ \cdots & A[M] \end{pmatrix}$ *be column* $\mathbb{R}$-*astic,* $x \in \overline{\mathbb{R}}^n$, $M \subseteq N$ *and* $y = A^k \otimes x$. *If* $x[N - M] = \varepsilon$ *then* $y[N - M] = \varepsilon$.

Proof Straightforward.   □

Let $x \in \overline{\mathbb{R}}^n$. Recall that the set $\{j \in N; x_j > \varepsilon\}$ is called the *support* of x, notation $\text{Supp}(x)$. Lemma 8.6.14 implies that if M is the support of an eigenvector and $\text{Supp}(x) \subseteq M$ for some $x \in \overline{\mathbb{R}}^n$ then $\text{Supp}(A^k \otimes x) \subseteq M$ for all positive integers k. This motivates the following definitions:

Let $A = (a_{ij}) \in \overline{\mathbb{R}}^{n \times n}$ be in an FNF. Then $M \subseteq N$ is called *regular* if for some λ there is an $x \in V(A, \lambda)$ with $x[M]$ finite and $x[N - M] = \varepsilon$. We also denote $\lambda = \lambda(M)$.

Remark 8.6.15 Even if M is regular there still may exist an $x \in V(A, \lambda(M))$ with $x_j = \varepsilon$ for some $j \in M$.

Since for a given matrix the finiteness structure of all eigenvectors is well described (Theorem 4.6.4), we aim to characterize matrices for which an eigenvector in $V(A, \lambda(M))$ for a given regular set M is reached with any starting vector whose support is a subset of M.

It follows from the description of $V(A)$ (Sect. 4.6) that M is regular if and only if there exist spectral indices $i_1, \ldots, i_s$ for some s such that

$$M = \{i \in N; i \to N_{i_1} \cup \cdots \cup N_{i_s}\}.$$

Let $M \subseteq N$. We denote

$$\overline{\mathbb{R}}^n(M) = \{x \in \overline{\mathbb{R}}^n - \{\varepsilon\}; (\forall j \in N - M)(x_j = \varepsilon)\}.$$

Let $A = (a_{ij}) \in \overline{\mathbb{R}}^{n \times n}$ be a column $\mathbb{R}$-astic matrix in an FNF and $M \subseteq N$ be regular. Then A will be called *M-robust* if

$$(\forall x \in \overline{\mathbb{R}}^n(M)) (\exists k) A^k \otimes x \in V(A, \lambda(M)).$$

Theorem 8.6.16 *Let $A = (a_{ij}) \in \overline{\mathbb{R}}^{n \times n}$ be a column $\mathbb{R}$-astic matrix in an FNF, $M \subseteq N$ be regular and $B = A[M]$. Then A is M-robust if and only if $\sigma(B) = 1$.*

Proof Without loss of generality let $A = \begin{pmatrix} A[N-M] & \varepsilon \\ \cdots & B \end{pmatrix}$.

Suppose that A is M-robust. Take $x = A_j$, $j \in M$. Then $x \in \overline{\mathbb{R}}^n(M)$ because A (and therefore also B) is column $\mathbb{R}$-astic and there is a k_j such that $A^k \otimes A_j \in V(A, \lambda(M))$ for all $k \geq k_j$. Since $A_j = \begin{pmatrix} \varepsilon \\ A_j[M] \end{pmatrix}$, we have

$$A \otimes (A^k \otimes A_j) = \begin{pmatrix} \varepsilon \\ B \otimes (B^k \otimes A_j[M]) \end{pmatrix} = \lambda(M) \otimes \begin{pmatrix} \varepsilon \\ B^k \otimes A_j[M] \end{pmatrix}.$$

Hence, for $k \geq \max_{j \in M} k_j$ there is

$$B^{k+2} = \lambda(M) \otimes B^{k+1},$$

that is, $\sigma(B) = 1$ with $\lambda = \lambda(M)$.

Suppose now $B^{k+1} = \lambda \otimes B^k$ for some λ and for all $k \geq k_0$. If the FNF of B is

$$B = \begin{pmatrix} B_1 & & \varepsilon \\ \vdots & \ddots & \\ . & \cdots & B_r \end{pmatrix}$$

then

$$B^k = \begin{pmatrix} B_1^k & & \varepsilon \\ \vdots & \ddots & \\ . & \cdots & B_r^k \end{pmatrix}$$

and so $B_i^{k+1} = \lambda \otimes B_i^k$ $(i = 1, \ldots, r)$. But since every B_i is irreducible, $\lambda = \lambda(B_i) = \lambda(M)$ $(i = 1, \ldots, r)$. Let $M = M_1 \cup \cdots \cup M_r$ be the partition of M determined by the FNF of B. Let $x \in \overline{\mathbb{R}}^n(M)$,

$$x = \begin{pmatrix} x[N - M] = \varepsilon \\ x[M_1] \\ \vdots \\ x[M_r] \end{pmatrix}$$

and let

$$s = \min\{i; x[M_i] \neq \varepsilon\}.$$

Denote $y = A^k \otimes x$,

$$y = \begin{pmatrix} y[N - M] \\ y[M_1] \\ \vdots \\ y[M_r] \end{pmatrix}.$$

Clearly, $y[N - M] = \varepsilon$ and

$$y[M_s] = B^k \otimes x[M_s] \neq \varepsilon$$

since B_s is irreducible (note that using Corollary 8.6.5 it would be possible to prove here that $y[M_i]$ is finite for all $i \geq s$). Hence $y \in \overline{\mathbb{R}}^n(M)$. At the same time

$$B^{k+1} \otimes x[M] = \lambda \otimes B^k \otimes x[M]$$

and

$$y = \begin{pmatrix} \varepsilon \\ B^k \otimes x[M] \end{pmatrix}.$$

Therefore

$$A \otimes y = \begin{pmatrix} \varepsilon \\ B \otimes B^k \otimes x[M] \end{pmatrix} = \lambda(M) \otimes \begin{pmatrix} \varepsilon \\ B^k \otimes x[M] \end{pmatrix} = \lambda(M) \otimes y.$$

We conclude that $y \in V(A, \lambda(M))$. $\qquad\qquad\qquad\qquad\qquad\qquad\qquad\qquad\square$

8.7 Exercises

Exercise 8.7.1 Are any of the matrices in Exercises 1.7.11 and 1.7.12 robust? [Both are robust]

Exercise 8.7.2 Use matrix scaling to obtain a visualized matrix from the matrix

$$A = \begin{pmatrix} 1 & -4 & 6 & 0 \\ 1 & 2 & 4 & 2 \\ 1 & -1 & 2 & 3 \\ -2 & 5 & 4 & 0 \end{pmatrix}$$

and then deduce the cyclicity of A.

$$\left[\begin{pmatrix} 1 & -6 & 4 & -1 \\ 3 & 2 & 4 & 3 \\ 3 & -1 & 2 & 4 \\ -1 & 4 & 3 & 0 \end{pmatrix}, \sigma(A) = 3 \right]$$

Exercise 8.7.3 For the matrix

$$A = \begin{pmatrix} 4 & 4 & 3 & 8 & 1 \\ 3 & 3 & 4 & 5 & 4 \\ 5 & 3 & 4 & 7 & 3 \\ 2 & 1 & 2 & 3 & 0 \\ 6 & 6 & 4 & 8 & 1 \end{pmatrix}$$

of Exercise 4.8.1 find the critical digraph $C(A)$, all strongly connected components of $C(A)$ and their cyclicities and the cyclicity of A. Is A robust? [$N_1 = \{1, 3, 4\}$, $N_2 = \{2, 5\}$; $\sigma(N_1) = 1$, $\sigma(N_2) = 2$, $\sigma(A) = 2$, A is not robust]

Exercise 8.7.4 Let $A \in \overline{\mathbb{R}}^{n \times n}$ be definite and denote by $\rho_1, \ldots, \rho_n$ ($\tau_1, \ldots, \tau_n$) the rows (columns) of $\Delta(A)$. Prove then that $Q^{[k]} \otimes Q^{[l]} \le Q^{[k]} \oplus Q^{[l]}$, where $Q^{[r]}$ is the outer product $\tau_r \otimes \rho_r^T$ (see Proposition 8.3.1).

Exercise 8.7.5 The digraphs of Figs. 8.6, 8.7 and 8.8 are condensation digraphs of reducible matrices A, B, C in the FNF, whose all diagonal blocks are primitive. The integers in the digraphs stand for the unique eigenvalues of the corresponding diagonal blocks. Decide about each matrix whether it is robust. [A is not, B and C are]

Exercise 8.7.6 Prove that if A is an irreducible matrix then every subeigenvector of A is in the attraction space of A. [Hint: Use the Cyclicity Theorem]

Fig. 8.6 Condensation
digraph for the matrix A of
Exercise 8.7.5

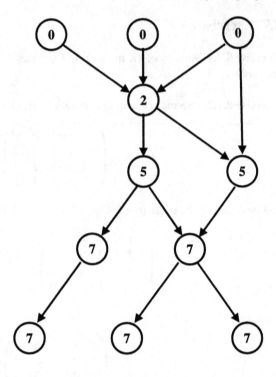

Fig. 8.7 Condensation
digraph for the matrix B of
Exercise 8.7.5

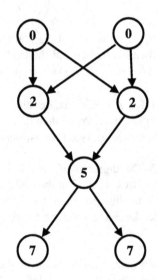

Fig. 8.8 Condensation digraph for the matrix C of Exercise 8.7.5

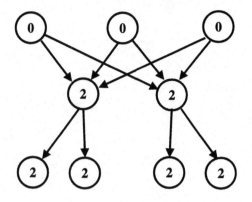

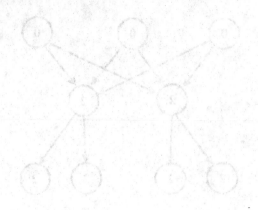

Chapter 9
Generalized Eigenproblem

This chapter deals with the *generalized eigenproblem* (GEP) in max-algebra defined as follows:

Given $A, B \in \overline{\mathbb{R}}^{m \times n}$, *find all* $\lambda \in \overline{\mathbb{R}}$ *(generalized eigenvalues) and* $x \in \overline{\mathbb{R}}^{n}, x \neq \varepsilon$ *(generalized eigenvectors) such that*

$$A \otimes x = \lambda \otimes B \otimes x. \qquad (9.1)$$

When $\lambda \in \overline{\mathbb{R}}$ and $x \in \overline{\mathbb{R}}^{n}, x \neq \varepsilon$ satisfying (9.1) exist then we say that *GEP is solvable* or also that (A, B) *is solvable*. Obviously, the eigenproblem is obtained from the GEP when $B = I$ or $\lambda = \varepsilon$ and we will therefore assume in this chapter that $\lambda > \varepsilon$.

It is likely that GEP is much more difficult than the eigenproblem. This is indicated by the fact that the GEP for a pair of real matrices may have no generalized eigenvalue, a finite number or a continuum of generalized eigenvalues [70]. It is known [135] that the union of any system of closed (possibly one-element) intervals is the set of generalized eigenvalues for suitably taken A and B.

GEP has been studied in [15] and [70]. The first of these papers solves the problem completely when $m = 2$ and special cases for general m and n, the second solves some other special cases. No solution method seems to be known either for finding a λ or an $x \neq \varepsilon$ satisfying (9.1) for general real matrices. Obviously, once λ is fixed, the GEP reduces to a two-sided max-linear system (Chap. 7). We therefore concentrate on the question of finding the generalized eigenvalues. First we will study basic properties and solvable special cases of GEP. In Sect. 9.3 we then present a method for narrowing the search for generalized eigenvalues for a pair of real square matrices. It is based on the solvability conditions for two-sided systems formulated using symmetrized semirings (Sect. 7.5).

A motivation for the GEP is given in Sect. 1.3.2.

Given $A, B \in \overline{\mathbb{R}}^{m \times n}$ we denote the set of generalized eigenvalues by $\Lambda(A, B)$, the set containing ε and all generalized eigenvectors corresponding to $\lambda \in \overline{\mathbb{R}}$ by $V(A, B, \lambda)$ and the set of all generalized eigenvectors by $V(A, B)$, that is:

P. Butkovič, *Max-linear Systems: Theory and Algorithms*,
Springer Monographs in Mathematics 151,
DOI 10.1007/978-1-84996-299-5_9, © Springer-Verlag London Limited 2010

$$V(A, B, \lambda) = \left\{ x \in \overline{\mathbb{R}}^n; A \otimes x = \lambda \otimes B \otimes x \right\}, \lambda \in \overline{\mathbb{R}},$$

$$V(A, B) = \left\{ x \in \overline{\mathbb{R}}^n; A \otimes x = \lambda \otimes B \otimes x, \lambda \in \overline{\mathbb{R}} \right\}$$

and

$$\Lambda(A, B) = \left\{ \lambda \in \overline{\mathbb{R}}; V(A, B, \lambda) \neq \{\varepsilon\} \right\}.$$

9.1 Basic Properties of the Generalized Eigenproblem

In this section we present some properties of the GEP provided that A and B are finite matrices [70]. We therefore assume that $A = (a_{ij})$, $B = (b_{ij}) \in \mathbb{R}^{m \times n}$ are given matrices and, as before, we denote $M = \{1, \ldots, m\}$ and $N = \{1, \ldots, n\}$. We will also denote:

$$C = (c_{ij}) = (a_{ij} \otimes b_{ij}^{-1})$$

and

$$D = (d_{ij}) = (b_{ij} \otimes a_{ij}^{-1}).$$

Theorem 9.1.1 *If (A, B) is solvable and $\lambda \in \Lambda(A, B)$ then C satisfies*

$$\max_{i \in M} \min_{j \in N} c_{ij} \leq \lambda \leq \min_{i \in M} \max_{j \in N} c_{ij}. \tag{9.2}$$

Proof No row of $\lambda \otimes B$ strictly dominates the corresponding row of A, so for every i there is a j such that $a_{ij} \geq \lambda \otimes b_{ij}$, i.e. $\lambda \leq c_{ij}$. Hence for all i we have $\lambda \leq \max_j c_{ij}$, thus $\lambda \leq \min_i \max_j c_{ij}$. Similarly, no row of A strictly dominates the corresponding row of $\lambda \otimes B$, yielding for all i: $\lambda \geq \min_j c_{ij}$, thus $\lambda \geq \max_i \min_j c_{ij}$. ☐

The interval $[\max_{i \in M} \min_{j \in N} c_{ij}, \min_{i \in M} \max_{j \in N} c_{ij}]$ is called the *feasible interval* for the generalized eigenproblem (9.1).

Example 9.1.2 If $A = \left(\begin{smallmatrix} 1 & 2 \\ -1 & 0 \end{smallmatrix} \right)$ and $B = \left(\begin{smallmatrix} 0 & 1 \\ 0 & 1 \end{smallmatrix} \right)$ then (A, B) is not solvable because $C = \left(\begin{smallmatrix} 1 & 1 \\ -1 & -1 \end{smallmatrix} \right)$ does not satisfy (9.2).

Recall that for a square matrix A the symbol $\lambda(A)$ stands for the maximum cycle mean of A. We now also denote by $\lambda'(A)$ the *minimum cycle mean*.

Corollary 9.1.3 *If $m = n$, (A, B) is solvable and $\lambda \in \Lambda(A, B)$ then C satisfies*

$$\lambda'(C) \leq \lambda \leq \lambda(C).$$

Proof A cycle in D_C whose every arc has the weight equal to a row maximum in C exists. The arc weights on this cycle are all at least the smallest row maximum, thus $\lambda(C) \geq \min_{i \in M} \max_{j \in N} c_{ij}$. The second inequality now follows from Theorem 9.1.1 and the other inequality by swapping max and min. ∎

Recall that the conjugate of B is $B^* = (b^*_{ij}) = (b^{-1}_{ji})$. Then the ith element of the diagonal of $A \otimes B^*$ equals

$$\max_j(a_{ij} + b^*_{ji}) = \max_j(a_{ij} \otimes b^{-1}_{ij}) = \max_j c_{ij}.$$

Similarly, the ith element of the diagonal of $A \otimes' B^*$ equals $\min_j c_{ij}$. Hence by Theorem 9.1.1 we have:

Corollary 9.1.4 *If (A, B) is solvable then the greatest element of the diagonal of $A \otimes' B^*$ does not exceed the least element of the diagonal of $A \otimes B^*$.*

By Corollary 9.1.3 we also have:

Corollary 9.1.5 *If (A, B) is solvable and $\lambda \in \Lambda(A, B)$ then*

$$\lambda'(A \otimes' B^*) \leq \lambda \leq \lambda(A \otimes B^*).$$

The next statement is a remarkable observation on generalized eigenvalues, yet there is no description of the unique possible value for the eigenvalue.

Theorem 9.1.6 [15] *If both (A, B) and (A^T, B^T) are solvable then both these problems have a unique and identical eigenvalue, that is, there is a real number λ such that*

$$\Lambda(A, B) = \{\lambda\} = \Lambda(A^T, B^T)$$

provided that $\Lambda(A, B) \neq \emptyset$ and $\Lambda(A^T, B^T) \neq \emptyset$.

Proof Suppose that

$$A \otimes x = \lambda \otimes B \otimes x$$

and

$$A^T \otimes y = \mu \otimes B^T \otimes y$$

for some λ, μ, x, y. Then

$$\lambda \otimes y^T \otimes B \otimes x = y^T \otimes A \otimes x = x^T \otimes A^T \otimes y$$
$$= \mu \otimes x^T \otimes B^T \otimes y = \mu \otimes y^T \otimes B \otimes x.$$

Since $y^T \otimes B \otimes x$ are finite it follows that $\lambda = \mu$. ∎

Corollary 9.1.7 *If $A, B \in \mathbb{R}^{n \times n}$ are symmetric then $|\Lambda(A, B)| \leq 1$.*

The following simple corollary provides in some cases a powerful tool of proving that the generalized eigenproblem is not solvable:

Corollary 9.1.8 *If $A, B \in \mathbb{R}^{n \times n}$ and (A^T, B^T) has more than one generalized eigenvalue then (A, B) is not solvable.*

9.2 Easily Solvable Special Cases

9.2.1 Essentially the Eigenproblem

If either A or B is a generalized permutation matrix then (9.1) is easily solvable. If (say) B is a generalized permutation matrix then B has the inverse B^{-1} and after multiplying (9.1) by B^{-1} the GEP is transformed to the eigenproblem. Unfortunately, since in max-algebra matrices other than generalized permutation matrices do not have an inverse (see Theorem 1.1.3), this case is fairly limited.

9.2.2 When A and B Have a Common Eigenvector

Proposition 9.2.1 [70] *A common eigenvector of A and B is a generalized eigenvector for A and B; more precisely, if $A, B \in \overline{\mathbb{R}}^{n \times n}$, $\lambda \otimes \mu^{-1} \in \mathbb{R}$, then*

$$V(A, \lambda) \cap V(B, \mu) \subseteq V(A, B, \lambda \otimes \mu^{-1}).$$

Proof If $x \in V(A, \lambda) \cap V(B, \mu)$ and $\lambda > \varepsilon$ then $\mu \in \mathbb{R}$ and

$$A \otimes x = \lambda \otimes x = \lambda \otimes \mu^{-1} \otimes B \otimes x.$$

If $\lambda = \varepsilon$ then $\lambda \otimes \mu^{-1} = \varepsilon$ and the statement trivially follows.   □

An example of pairs of matrices having a common eigenvector are commuting matrices (Theorem 4.7.2). Hence we have:

Theorem 9.2.2 *If $A, B \in \mathbb{R}^{n \times n}$ and $A \otimes B = B \otimes A$ then both (A, B) and (A^T, B^T) are solvable, with identical, unique generalized eigenvalue.*

Proof A and B have a common eigenvector corresponding to finite eigenvalues by Theorem 4.7.2 and so by Proposition 9.2.1 (A, B) is solvable. At the same time A^T and B^T are also commuting and by a repeated argument we have that (A^T, B^T) is solvable. The equality of all generalized eigenvalues now follows by Theorem 9.1.6.   □

9.2.3 When One of A, B Is a Right-multiple of the Other

Theorem 9.2.3 [70] *If one of $A, B \in \overline{\mathbb{R}}^{m \times n}$ is a right-multiple of the other then (A, B) is solvable.*

Proof Suppose e.g. $A = B \otimes P$, where $P \in \overline{\mathbb{R}}^{n \times n}$. Let $\lambda \in \Lambda(P)$ and $x \in V(P, \lambda)$, $x \neq \varepsilon$. Then

$$A \otimes x = B \otimes P \otimes x = B \otimes (\lambda \otimes x) = \lambda \otimes B \otimes x. \qquad \square$$

Example 9.2.4 Suppose

$$A = \begin{pmatrix} 4 & 6 \\ 7 & 9 \end{pmatrix}, \qquad B = \begin{pmatrix} 0 & 1 \\ 3 & 1 \end{pmatrix}, \qquad P = \begin{pmatrix} 4 & 6 \\ -2 & 0 \end{pmatrix}.$$

Then $\lambda(P) = 4$,

$$\Gamma(\lambda^{-1} \otimes P) = \begin{pmatrix} 0 & 2 \\ -6 & -4 \end{pmatrix}$$

and

$$x = \begin{pmatrix} 0 \\ -6 \end{pmatrix}, \qquad A \otimes x = \begin{pmatrix} 4 \\ 7 \end{pmatrix}, \qquad B \otimes x = \begin{pmatrix} 0 \\ 3 \end{pmatrix}.$$

We can also prove a sufficient condition for λ to attain the upper bound in (9.2) when (say) A is a right-multiple of B and $A, B \in \mathbb{R}^{m \times n}$. Recall that $C = (c_{ij})$ is the matrix $(a_{ij} \otimes b_{ij}^{-1})$, $D = (d_{ij}) = (b_{ij} \otimes a_{ij}^{-1})$ and let us denote

$$L = \max_i \min_j c_{ij}$$

and

$$U = \min_i \max_j c_{ij}.$$

It follows from the proof of Theorem 9.2.3 and from Theorem 9.1.1 that $\lambda(P) \in [L, U]$ for every P satisfying $A = B \otimes P$. If $A = B \otimes P$ then we have:

$$A = B \otimes (B^* \otimes' A).$$

Let us denote $B^* \otimes' A$ by $\overline{P} = (\overline{p}_{ij})$ and $\overline{\lambda} = \lambda(\overline{P})$; thus $L \leq \overline{\lambda} \leq U$. The following technical lemma will help us to characterize in Theorem 9.2.6 when the upper bound U is attained.

Lemma 9.2.5 *If $A, B \in \mathbb{R}^{m \times n}$ and $L' = \max_j \min_i c_{ij}$ then $L' \leq \overline{\lambda}$.*

Proof

$$\bar{\lambda} = \lambda(\overline{P}) \geq \max_i \overline{p}_{ii} = \max_i \min_j (b_{ij}^* \otimes a_{ji})$$

$$= \max_i \min_j (a_{ji} \otimes b_{ji}^{-1}) = \max_i \min_j c_{ji} = \max_j \min_i c_{ij} = L'.$$ □

Theorem 9.2.6 [70] *If $A, B \in \mathbb{R}^{m \times n}$, D has a saddle point and there is a matrix P such that $A = B \otimes P$ then $\bar{\lambda} = U$ where $\bar{\lambda} = \lambda(\overline{P}) = \lambda(B^* \otimes' A)$.*

Proof $D = (d_{ij})$ has a saddle point means

$$\max_i \min_j d_{ij} = \min_j \max_i d_{ij}.$$

Therefore the inverses of both sides are equal:

$$U = \min_i \max_j c_{ij} = \max_j \min_i c_{ij} = L'.$$

Hence by Lemma 9.2.5: $L' = \bar{\lambda} = U$.   □

The following dual statement is proved in a dual way:

Theorem 9.2.7 [70] *Let $A, B \in \mathbb{R}^{m \times n}$. If there is a matrix P such that $A = B \otimes' P$ and C has a saddle point then $\bar{\lambda}' = L$ where $\bar{\lambda}' = \lambda'(\overline{P}) = \lambda'(B^* \otimes' A)$.*

Even if one of A, B is a right-multiple of the other, the eigenvalue may not be unique as the following example shows.

Example 9.2.8 With A, B as in Example 9.2.4, we find for the principal solution matrix $\overline{P}$:

$$\overline{P} = \begin{pmatrix} 4 & 6 \\ 3 & 5 \end{pmatrix}, \qquad \lambda(\overline{P}) = 5,$$

$$\Gamma(\lambda^{-1} \otimes \overline{P}) = \begin{pmatrix} -1 & 1 \\ -2 & 0 \end{pmatrix},$$

$$A \otimes \begin{pmatrix} 1 \\ 0 \end{pmatrix} = \begin{pmatrix} 6 \\ 9 \end{pmatrix}$$

and

$$B \otimes \begin{pmatrix} 1 \\ 0 \end{pmatrix} = \begin{pmatrix} 1 \\ 4 \end{pmatrix}.$$

Hence for the same A, B we find two solutions to (9.1), with different values of λ.

9.3 Narrowing the Search for Generalized Eigenvalues

9.3.1 Regularization

In the absence of any method, exact or approximate, for finding generalized eigenvalues for a general pair of matrices, we concentrate now on narrowing the set containing all generalized eigenvalues (if there are any) for finite A and B.

Let $C = (c_{ij})$, $D = (d_{ij}) \in \mathbb{R}^{m \times n}$. The system

$$C \otimes x = D \otimes x \qquad (9.3)$$

is called *regular* if

$$c_{ij} \neq d_{ij}$$

for all i, j. The aim of the method we will present in this section is to identify as
closely as possible the set of generalized eigenvalues for which (9.1) is regular.

Let us first briefly discuss the values of λ for which this requirement is not
satisfied. There are at most mn such values of λ. We will call these values *extreme* and the set of extreme values will be denoted by L. More precisely, for
$A = (a_{ij})$, $B = (b_{ij}) \in \mathbb{R}^{m \times n}$ we set

$$L = \{\lambda \in \mathbb{R}; a_{ij} = \lambda \otimes b_{ij} \text{ for some } i, j\}.$$

Note that the elements of L are entries of the matrix $A - B$. Obviously,

$$|L| \leq mn \qquad (9.4)$$

and (9.1) is regular for all $\lambda \in \mathbb{R} - L$. Recall that solvability of (9.1) can be checked
for each fixed and in particular extreme value of λ using, say, the Alternating
Method.

Remark 9.3.1 The upper bound in (9.4) can slightly be improved: If for some i we
have $c_{ij} > d_{ij}$ for all j then (9.3) has no nontrivial solution. Therefore (9.1) has
no nontrivial solution if λ is too big or too small, in particular for $\lambda > \max L$ and
$\lambda < \min L$. These two conditions may be slightly refined as follows: $a_{ij} > \lambda \otimes b_{ij}$
for all j or $a_{ij} < \lambda \otimes b_{ij}$ for all j must not hold for any $i = 1, \ldots, m$. Hence (9.1)
has no nontrivial solution for $\lambda < \lambda'$ and $\lambda > \lambda''$ where λ' is the mth smallest value
in L and λ'' is the mth greatest value in L (both considered with multiplicities). So
actually only at most $mn - 2m$ extreme values of λ need to be checked individually
by the Alternating Method.

Let us denote the extreme values described in Remark 9.3.1 by $\lambda_1, \ldots, \lambda_t$, where
$\lambda_1 < \cdots < \lambda_t$ and $t \leq mn - 2m$. All these values can easily be found among
the entries of $A - B$ and checked individually for being generalized eigenvalues.
Thus we may now concentrate on the real numbers in open intervals $(\lambda_j, \lambda_{j+1})$,
$j = 1, \ldots, t - 1$. We will call these intervals *regular* and we will also call every real

number *regular* if it belongs to a regular interval. It follows that there are at most $mn - 2m - 1$ regular intervals to be considered. In the rest of this section we assume that one such interval, say J, has been fixed, and we consider (9.1) only for $\lambda \in J$.

9.3.2 A Necessary Condition for Generalized Eigenvalues

Symmetrized semirings have been introduced in Sect. 7.5 and they have been used to derive necessary conditions for the existence of a nontrivial solution to two-sided systems. We now reformulate this to obtain a necessary condition for generalized eigenvalues.

Recall first that $\mathbb{S} = \overline{\mathbb{R}} \times \overline{\mathbb{R}}$ and the operations $\oplus$ and $\otimes$ are extended to $\mathbb{S}$ as follows:

$$(a, a') \oplus (b, b') = (a \oplus b, a' \oplus b'),$$

$$(a, a') \otimes (b, b') = (a \otimes b \oplus a' \otimes b', a \otimes b' \oplus a' \otimes b).$$

Also, $\ominus(a, a') = (a', a)$ and (a, a') is called balanced if $a = a'$. The determinant of $A = (a_{ij}) \in \mathbb{S}^{n \times n}$ has been defined as

$$\det(A) = \sum_{\sigma \in P_n}^{\oplus} \left(\mathrm{sgn}(\sigma) \otimes \prod_{i \in N}^{\otimes} a_{i, \sigma(i)} \right),$$

and we know that

$$|\det(A)| = \mathrm{maper}|A|,$$

see Proposition 7.5.6.

The next statement follows from Theorem 7.5.4 and Corollary 7.5.5. We denote here and in the rest of this section

$$C(\lambda) = A \ominus \lambda \otimes B.$$

Corollary 9.3.2 Let $A, B \in \mathbb{R}^{n \times n}$ and $\lambda \in \mathbb{R}$. Then a necessary condition that the system $A \otimes x = \lambda \otimes B \otimes x$ have a nontrivial solution is that $C(\lambda)$ has balanced determinant.

The idea of narrowing the search for the eigenvalues is based on Corollary 9.3.2: We show how to find *all* λ for which $C(\lambda)$ has balanced determinant. It turns out that this can be done using a polynomial number of operations in terms of n. This method may in some cases identify all eigenvalues, see Examples 9.3.7 and 9.3.8. In general however, it finds only a superset of generalized eigenvalues, see Example 9.3.9.

If λ is regular then $C = A \ominus \lambda \otimes B$ has no balanced entry. The following statement is a reformulation of Theorem 7.5.7 (note that the matrix $\widetilde{C}$ has been defined just before that theorem):

Corollary 9.3.3 *Let $A, B \in \mathbb{R}^{n \times n}$, λ be regular. Then $C(\lambda)$ has balanced determinant if and only if $\widetilde{C(\lambda)}$ is not SNS.*

The problem of checking whether a $(0, 1, -1)$ matrix is SNS or not is equivalent to the even cycle problem in digraphs [18] and therefore polynomially solvable (Remark 1.6.45). Therefore the necessary solvability condition in Corollary 9.3.3 can be checked in polynomial time for any fixed regular value of λ. This will be used later in Sect. 9.3.4. However, $C(\lambda)$ may have balanced determinant for a continuum of values of λ (see Example 9.3.8) and therefore we also need a tool which enables us to make the same decision for an interval. This tool will be presented in Sect. 9.3.4. As a preparation we first show in Sect. 9.3.3 how to find $\mathrm{maper}|C(\lambda)|$ as a function of $\lambda \in J$.

9.3.3 Finding $\mathrm{maper}|C(\lambda)|$

In this subsection we show how to efficiently find the function

$$f(\lambda) = \mathrm{maper}|C(\lambda)|.$$

This will be used in the next section to produce a method for finding all regular values of $\lambda \in J$ for which $\widetilde{C(\lambda)}$ is not SNS.

Recall first that $|C(\lambda)| = (a_{ij} \oplus \lambda \otimes b_{ij}) = (c_{ij}(\lambda))$ and for every $\lambda \in J$ we have

$$a_{ij} \neq \lambda \otimes b_{ij}$$

for all $i, j \in N$. Therefore for every $\lambda \in J$ and for all $i, j \in N$ the entry $c_{ij}(\lambda) = a_{ij} \oplus \lambda \otimes b_{ij}$ is equal to exactly one of a_{ij} and $\lambda \otimes b_{ij}$. Observe that $f(\lambda) = \mathrm{maper}|C(\lambda)|$ is the maximum of $n!$ terms. Each term is a $\otimes$ product of n entries $c_{ij}(\lambda)$, hence of the form $b \otimes \lambda^k$, where $b \in \mathbb{R}$ and k is a natural number between 0 and n. Since $b \otimes \lambda^k$ in conventional notation is simply $k\lambda + b$, we deduce that $f(\lambda)$ is the maximum of a finite number of linear functions and therefore a piecewise linear convex function. Note that the slopes of all linear pieces of $f(\lambda)$ are natural numbers between 0 and n. Recall that $f(\lambda)$ for any particular λ can easily be found by solving the assignment problem for $|C(\lambda)|$. It follows that all linear pieces can therefore efficiently be identified. We now describe one possible way of finding these linear functions: Assume for a while that the linear pieces of smallest and greatest slope are known, let us denote them $f_l(\lambda) = a_l \otimes \lambda^l$ and $f_h(\lambda) = a_h \otimes \lambda^h$, respectively. If $l = h$ then there is nothing to do, so assume $l \neq h$. We start by finding the intersection point of f_l and f_h, that is, say, λ_1 satisfying $f_l(\lambda_1) = f_h(\lambda_1)$. Calculate $f(\lambda_1) = \mathrm{maper}|C(\lambda_1)|$. If $f(\lambda_1) = f_l(\lambda_1) = f_h(\lambda_1)$ then there is no linear piece other than f_l and f_h. Otherwise $f(\lambda_1) > f_l(\lambda_1) = f_h(\lambda_1)$. Let r be the number of λ terms appearing in an optimal permutation (if there are several optimal permutations with various numbers of λ appearances then take any). Since r is the slope of the linear piece we have $l < r < h$. Then $a_r = f(\lambda_1) - r\lambda_1$ and $f_r(\lambda) = a_r \otimes \lambda^r$. This term is a new linear piece and we then repeat this procedure with f_l and f_r and f_r and f_h, and so on. At every step a new linear piece is discovered unless all linear pieces have already been found. Hence the number of iterations is at most $n - 1$.

For finding f_l and f_h it will be convenient to use the *independent ones problem* (IOP) for $0 - 1$ square matrices:

Given a $0 - 1$ matrix $M = (m_{ij}) \in \mathbb{R}^{n \times n}$, find the greatest number of ones in M so that no two are from the same row or column or, equivalently, so that there is a $\pi \in P_n$ selecting all these ones.

Clearly, IOP is a special case of the assignment problem, and therefore easily solvable. Note that in combinatorial terminology IOP is known as the maximum cardinality bipartite matching problem solvable in $O(n^{2.5})$ time [22]. In general we say that a set of positions in a matrix are independent if no two of them belong to the same row or column.

Now we discuss how to find f_l and f_h. The values of l and h are obviously the smallest and biggest number of independent entries in $|C(\lambda)|$ containing λ and these can be found by solving the corresponding IOP. For h this problem can be described by the matrix $M = (m_{ij})$ with $m_{ij} = 1$ when $|c_{ij}(\lambda)| = \lambda \otimes b_{ij}$ and 0 otherwise and for l by $E - M$, where E is the all-one matrix.

Now we show how to find a_l and a_h. Let $d_{ij} = b_{ij}$ if $c_{ij}(\lambda) = \lambda \otimes b_{ij}$ and $d_{ij} = a_{ij}$ if $c_{ij}(\lambda) = a_{ij}$ (note that by regularity of λ only one of these two possibilities occurs for $\lambda \in J$). For finding a_l and a_h we need to determine permutations π and σ that maximize $\sum_{i \in N} d_{i,\pi(i)}$ and $\sum_{i \in N} d_{i,\sigma(i)}$ and select l and h entries containing λ, respectively. To achieve this we interpret the two above mentioned IOPs as assignment problems and describe their solution sets using matrices M_h and M_l obtained by the Hungarian method (that is, nonpositive matrices whose max-algebraic permanent is zero). It remains then to replace all entries in $D = (d_{ij})$ corresponding to nonzero entries in M_h and M_l by $-\infty$ and solve the assignment problem for the obtained matrices.

9.3.4 Narrowing the Search

In this subsection we show how to efficiently find the set of all regular values of λ for which $\det(C(\lambda))$ is balanced. This set will be denoted by S. We use essentially the fact that the decision whether $\det(C(\lambda))$ is balanced can be made efficiently for any individual value of λ (Corollary 9.3.3). The following will be useful:

Lemma 9.3.4 *Let $f(x), g(x), h(x)$ be piecewise linear convex functions on $\mathbb{R}$, $f(x) = g(x) \oplus h(x)$ for all $x \in \mathbb{R}$. Suppose $a, b \in \mathbb{R}$ are such that f is linear on $[a, b]$. If $g(x) = h(x)$ for at least one $x \in (a, b)$ then $g(x) = h(x)$ for all $x \in [a, b]$.*

Proof Suppose $g(x_0) = h(x_0)$, $x_0 \in (a, b)$. Hence $g(x_0) = h(x_0) = f(x_0)$. If $g(x) < f(x)$ for an $x \in [a, b]$, without loss of generality for $x \in [a, x_0)$, then by convexity of g and linearity of f we have that $g(x) > f(x)$ for all $x \in (x_0, b)$, a contradiction. Therefore $g(x) = f(x)$ for all $x \in [a, b]$ and similarly $h(x) = f(x)$ for all $x \in [a, b]$. $\qquad\square$

Recall that as before J is a regular interval. Let us denote

$$\det(C(\lambda)) = \left(d^+(C(\lambda)), d^-(C(\lambda))\right),$$

or just $(d^+(\lambda), d^-(\lambda))$. Then $C(\lambda)$ for $\lambda \in J$ has balanced determinant if and only if

$$d^+(\lambda) = d^-(\lambda). \tag{9.5}$$

It follows from the results of the previous section that the piecewise linear convex function

$$|\det(C(\lambda))| = d^+(\lambda) \oplus d^-(\lambda) = \text{maper}|C(\lambda)|$$

can efficiently be found. By the same argument as for $\text{maper}|C(\lambda)|$ we see that both $d^+(\lambda)$ and $d^-(\lambda)$ are max-algebraic polynomials in λ (hence piecewise linear and convex functions) containing at most $n + 1$ powers of λ between 0 and n. No method other than exhaustive search (requiring $n!$ permutation evaluations) seems to be known for finding $d^+(\lambda)$ and $d^-(\lambda)$ separately for any particular λ [29]; however, for a fixed $\lambda \in \mathbb{R}-L$ by Corollary 9.3.3 we can decide in polynomial time whether $d^+(\lambda) = d^-(\lambda)$ or not. Since $d^+(\lambda) \oplus d^-(\lambda) = \text{maper}|C(\lambda)|$ then if $\text{maper}|C(\lambda)|$ is known, using Lemma 9.3.4 we can easily find *all* values of $\lambda \in J$ satisfying $d^+(\lambda) = d^-(\lambda)$ by checking this equality for any point strictly between any two consecutive breakpoints and for the breakpoints of $\text{maper}|C(\lambda)|$. We summarize these observations in the following:

Theorem 9.3.5 *If the set* $S = \{\lambda \in J; d^+(\lambda) = d^-(\lambda)\}$ *is nonempty then it consists of some of the breakpoints of* $\text{maper}|C(\lambda)|$ *and a number (possibly none) of closed intervals whose endpoints are pairs of adjacent breakpoints of* $\text{maper}|C(\lambda)|$. *All these can be identified in* $O(n^3)$ *time.*

Proof The statement is essentially proved by Lemma 9.3.4. We only need to add that each interval whose endpoints are adjacent breakpoints of $\text{maper}|C(\lambda)|$ can be decided by checking $d^+(\lambda) = d^-(\lambda)$ for one (arbitrary) internal point of the interval and that the number of breakpoints is at most n and therefore the number of intervals is at most $n - 1$. The equality $d^+(\lambda) = d^-(\lambda)$ for a fixed λ can be decided in polynomial time by Theorem 9.3.3. $\qquad\square$

We summarize our work in the following procedure for finding all regular values of λ for which $\det(C(\lambda))$ is balanced:

Algorithm 9.3.6 NARROWING THE EIGENVALUE SEARCH
Input: $A, B \in \mathbb{R}^{n \times n}$ and a regular interval J.
Output: The set $S = \{\lambda \in J; d^+(\lambda) = d^-(\lambda)\}$.

1. $S := \emptyset$.
2. $C(\lambda) := A \ominus \lambda \otimes B$.
3. Find $f(\lambda) = \text{maper}|C(\lambda)|$ as a function of λ, that is, find all breakpoints and linear pieces of $f(\lambda)$.
4. For every breakpoint λ_0 of $f(\lambda)$ do: If $\widetilde{C(\lambda_0)}$ is not SNS then $S := S \cup \{\lambda_0\}$.
5. For any two consecutive breakpoints a, b and arbitrarily taken $\lambda_0 \in (a, b)$ do: If $\widetilde{C(\lambda_0)}$ is not SNS then $S := S \cup (a, b)$.

9.3.5 Examples

In the first two examples below we demonstrate that the described method for narrowing the search for eigenvalues may actually find all eigenvalues. Note that in these examples all matrices are of small sizes and therefore the functions $d^+(\lambda)$ and $d^-(\lambda)$ are explicitly evaluated; however, for bigger matrices this would not be practical and the method described in Sect. 9.3.4 would be used as an efficient tool for finding all regular values of λ for which $d^+(\lambda) = d^-(\lambda)$.

The third example illustrates the situation when the algorithm narrows the feasible interval containing the eigenvalues but a significant proportion of the final interval still consists of real numbers that are not eigenvalues.

Example 9.3.7 Let

$$A = \begin{pmatrix} 3 & 8 & 2 \\ 7 & 1 & 4 \\ 0 & 6 & 3 \end{pmatrix}, \qquad B = \begin{pmatrix} 4 & 4 & 3 \\ 2 & 3 & 4 \\ 3 & 2 & 1 \end{pmatrix}.$$

Then

$$A - B = \begin{pmatrix} -1 & 4 & -1 \\ 5 & -2 & 0 \\ -3 & 4 & 2 \end{pmatrix}$$

and $L = \{-3, -2, -1, 0, 2, 4, 5\}$. For $\lambda < -1$ all terms on the RHS of the first equation in $A \otimes x = \lambda \otimes B \otimes x$ are strictly less than the corresponding terms on the left and therefore there is no nontrivial solution to $A \otimes x = \lambda \otimes B \otimes x$. Similarly, for $\lambda > 4$ all these terms are greater than their counterparts on the left. Hence we only need to investigate regular intervals $(-1, 0), (0, 2)$ and $(2, 4)$ and extreme points $-1, 0, 2, 4$.

For $\lambda \in (-1, 0)$ we have

$$|C(\lambda)| = \begin{pmatrix} 4+\lambda & 8 & 3+\lambda \\ 7 & 3+\lambda & 4 \\ 3+\lambda & 6 & 3 \end{pmatrix},$$

$$d^+(\lambda) = \max(10 + 2\lambda, 14 + \lambda, 9 + 3\lambda),$$
$$d^-(\lambda) = \max(16 + \lambda, 15 + \lambda, 18),$$
$$\mathrm{maper}|C(\lambda)| = 18.$$

Since $d^+(\lambda) \neq d^-(\lambda)$ for $\lambda \in (-1, 0)$, there are no eigenvalues in this interval.

For $\lambda \in (0, 2)$ we have

$$|C(\lambda)| = \begin{pmatrix} 4 + \lambda & 8 & 3 + \lambda \\ 7 & 3 + \lambda & 4 + \lambda \\ 3 + \lambda & 6 & 3 \end{pmatrix},$$

$$d^+(\lambda) = \max(10 + 2\lambda, 15 + 2\lambda, 9 + 3\lambda),$$
$$d^-(\lambda) = \max(16 + \lambda, 14 + 2\lambda, 18),$$
$$\mathrm{maper}|C(\lambda)| = \max(18, 16 + \lambda, 15 + 2\lambda, 9 + 3\lambda).$$

For $\lambda \in (0, 2)$ there is only one breakpoint for $\mathrm{maper}|C(\lambda)|$ at $\lambda_0 = 3/2$. Since $d^+(\lambda) = d^-(\lambda)$ for $\lambda = \lambda_0$, this value is the only candidate for an eigenvalue in $(0, 2)$. It is not difficult to verify that $x = (2, 0, 3.5)^T$ is a corresponding eigenvector.

For $\lambda \in (2, 4)$ we have

$$|C(\lambda)| = \begin{pmatrix} 4 + \lambda & 8 & 3 + \lambda \\ 7 & 3 + \lambda & 4 + \lambda \\ 3 + \lambda & 6 & 1 + \lambda \end{pmatrix},$$

$$d^+(\lambda) = \max(15 + 2\lambda, 16 + \lambda, 9 + 3\lambda),$$
$$d^-(\lambda) = \max(16 + \lambda, 14 + 2\lambda, 8 + 3\lambda),$$
$$\mathrm{maper}|C(\lambda)| = 15 + 2\lambda.$$

Since $d^+(\lambda) \neq d^-(\lambda)$ for $\lambda \in (2, 4)$, there are no eigenvalues in this interval.

Let us consider the extreme point $\lambda = 0$: In this small example we solve the system $A \otimes x = B \otimes x$ by direct analysis but note that in general the Alternating Method would be used. By the cancellation law (Lemma 7.4.1) the two-sided system $A \otimes x = B \otimes x$ is equivalent to the one with

$$A = \begin{pmatrix} \varepsilon & 8 & \varepsilon \\ 7 & \varepsilon & 4 \\ \varepsilon & 6 & 3 \end{pmatrix}, \qquad B = \begin{pmatrix} 4 & \varepsilon & 3 \\ \varepsilon & 3 & 4 \\ 3 & \varepsilon & \varepsilon \end{pmatrix}.$$

Here from the first equation either $x_2 = -4 + x_1$ or $x_2 = -5 + x_3$. In the first case the third equation yields $\max(2 + x_1, 3 + x_3) = 3 + x_1$, thus $x_1 = x_3$. By substituting into the second equation then $x_1 = -4 + x_2$, a contradiction. In the second case the third equation yields again $x_1 = x_3$, which implies a contradiction in the same way. Hence $\lambda = 0$ is not an eigenvalue and a similar analysis would show that neither are the remaining three extreme values.

We conclude that $\Lambda(A, B) = \{3/2\}$.

Example 9.3.8 Let $A = \begin{pmatrix} 4 & 6 \\ 7 & 9 \end{pmatrix}$, $B = \begin{pmatrix} 0 & 1 \\ 3 & 1 \end{pmatrix}$. It is easily seen that $J = (4, 5)$ is the unique regular interval. For $\lambda \in (4, 5)$ we have

$$|C(\lambda)| = \begin{pmatrix} \lambda & 6 \\ 3 + \lambda & 9 \end{pmatrix}$$

and

$$\text{maper}|C(\lambda)| = \max(9 + \lambda, 9 + \lambda) = 9 + \lambda = d^-(\lambda) = d^+(\lambda).$$

Hence every $\lambda \in J$ satisfies the necessary condition. In fact all these values are eigenvalues as $x = (6, \lambda)^T$ is a corresponding eigenvector (for every $\lambda \in J$). This vector is also an eigenvector for $\lambda \in \{4, 5\}$ and thus $\Lambda(A, B) = [4, 5]$.

Example 9.3.9 [132] Let

$$A = \begin{pmatrix} 0 & 1/2 & 1 \\ 1 & 0 & 0 \\ 0 & 0 & 1 \end{pmatrix}, \qquad B = \begin{pmatrix} 0 & -2 & -2 \\ -2 & 0 & 0 \\ 0 & -2 & -2 \end{pmatrix}.$$

Consider only the regular interval $J = (0, 2)$. For $\lambda \in J$ we have

$$|C(\lambda)| = \begin{pmatrix} \lambda & 1/2 & 1 \\ 1 & \lambda & \lambda \\ \lambda & 0 & 1 \end{pmatrix},$$

$$d^+(\lambda) = \max(1 + 2\lambda, 2),$$

and

$$d^-(\lambda) = \max(1 + 2\lambda, 5/2).$$

We deduce that $d^-(\lambda) = d^+(\lambda)$ if and only if $\lambda \geq 3/4$. Hence the algorithm returns $S = [3/4, 2]$. However, there are no eigenvalues in $(1, 2)$. To see this, realize that for $\lambda \in J$ the system (9.1) simplifies using the cancellation rules and then by setting $x_1 = 0$ to:

$$(1/2) \otimes x_2 \oplus 1 \otimes x_3 = \lambda,$$

$$1 = \lambda \otimes x_2 \oplus \lambda \otimes x_3,$$

$$x_2 \oplus 1 \otimes x_3 = \lambda.$$

The second equation is equivalent to $x_2 \oplus x_3 = 1 - \lambda$. Hence, if $\lambda > 1$ and $x = (0, x_2, x_3)^T$ is a solution then both x_2 and x_3 are negative, thus $x_2 \oplus 1 \otimes x_3 < 1 < \lambda$, a contradiction. Note that all $\lambda \in [3/4, 1]$ are eigenvalues since for such λ the vector $(0, 1 - \lambda, \lambda - 1)^T$ is a solution to (9.1).

9.4 Exercises

Exercise 9.4.1 Use Theorem 9.3.3 to give an alternative proof that $\lambda(A)$ is the unique eigenvalue for any irreducible matrix A.

Exercise 9.4.2 Show that the generalized eigenproblem has no nontrivial solution for the matrices

$$A = \begin{pmatrix} 3 & 5 & 4 \\ 7 & 9 & 8 \end{pmatrix}, \qquad B = \begin{pmatrix} 7 & 4 & 1 \\ 3 & 5 & 2 \end{pmatrix}.$$

[The feasible interval is empty]

Exercise 9.4.3 Find all extreme values in the feasible interval for the generalized eigenproblem with matrices

$$A = \begin{pmatrix} 3 & 5 & 4 \\ 0 & 3 & 7 \end{pmatrix}, \qquad B = \begin{pmatrix} 7 & 4 & 1 \\ 3 & 5 & 2 \end{pmatrix}.$$

$[(-3, -2, 1, 3)^T]$

Exercise 9.4.4 Prove the following: Let $A, B \in \mathbb{R}^{n \times n}$. Then (A, B) is solvable if and only if there exist P, Q such that $A \otimes P = B \otimes Q$ and (P, Q) is solvable.

Exercise 9.4.5 Prove or disprove: If $A, B \in \mathbb{R}^{n \times n}$ and $A = B \otimes Q$ then $\lambda(B)$ is the greatest corner of the maxpolynomial maper$(A \oplus \lambda \otimes B)$. [false]

Exercise 9.4.6 Find all generalized eigenvalues if

$$A = \begin{pmatrix} 0 & 1 & 2 \\ 0 & 2 & 4 \end{pmatrix}, \qquad B = \begin{pmatrix} 0 & 0 & 0 \\ 0 & 1 & 2 \end{pmatrix}.$$

[0, 1, 2]

Chapter 10
Max-linear Programs

If $f \in \overline{\mathbb{R}}^n$ then the function $f(x) = f^T \otimes x$ defined on $\overline{\mathbb{R}}^n$ is called a *max-linear function*. In this chapter we develop methods for solving *max-linear programming problems* (briefly, *max-linear programs*), that is, methods for minimizing or maximizing a max-linear function subject to constraints expressed by max-linear equations. Since one-sided max-linear systems are substantially easier to solve than the two-sided, we deal with these two problems separately. Note that if $f(x)$ is a max-linear function then $-f(x)$ may not be of the same type. Therefore unlike in conventional linear programming, in max-linear programming it is not possible to convert minimization of a max-linear function to a maximization of the same type of objective function by considering $-f(x)$ instead of $f(x)$.

The following will be useful and is easily derived from basic properties presented in Chap. 1:

Lemma 10.0.7 *Let* $f(x) = f^T \otimes x$ *be a max-linear function on* $\overline{\mathbb{R}}^n$. *Then*

(a) $f(x)$ *is max-additive and max-homogenous, that is,* $f(\alpha \otimes x \oplus \beta \otimes y) = \alpha \otimes f(x) \oplus \beta \otimes f(y)$ *for every* $x, y \in \overline{\mathbb{R}}^n$ *and* $\alpha, \beta \in \overline{\mathbb{R}}$.
(b) $f(x)$ *is isotone, that is,* $f(x) \leq f(y)$ *for every* $x, y \in \overline{\mathbb{R}}^n, x \leq y$.

Note that in the rest of this chapter we will assume that $f \in \mathbb{R}^n$. This chapter is based on the results presented in [32]. Related software can be downloaded from http://web.mat.bham.ac.uk/P.Butkovic/software/index.htm.

10.1 Programs with One-sided Constraints

Max-linear programs with one-sided constraints have been known for some time [149]. They are of the form

$$f(x) = f^T \otimes x \longrightarrow \min \text{ or } \max$$

P. Butkovič, *Max-linear Systems: Theory and Algorithms*, 243
Springer Monographs in Mathematics 151,
DOI 10.1007/978-1-84996-299-5_10, © Springer-Verlag London Limited 2010

subject to

$$A \otimes x = b, \tag{10.1}$$

where $f = (f_1, \ldots, f_n)^T \in \mathbb{R}^n$, $A = (a_{ij}) \in \mathbb{R}^{m \times n}$ and $b = (b_1, \ldots, b_m)^T \in \mathbb{R}^m$ are given. The systems $A \otimes x = b$ were studied in Chap. 3 and we will denote as before:

$$S = \{x \in \mathbb{R}^n; A \otimes x = b\}$$

and $\bar{x} = (\bar{x}_1, \ldots, \bar{x}_n)^T$, where $\bar{x}_j = \min_{i \in M} b_i \otimes a_{ij}^{-1}$ for $j \in N$. Recall that by Theorem 3.1.1 then $x \leq \bar{x}$ for every $x \in S$ and $x \in S$ if and only if $x \leq \bar{x}$ and

$$\bigcup_{j:x_j = \bar{x}_j} M_j = M,$$

where for $j \in N$ we define

$$M_j = \{i \in M; \bar{x}_j = b_i \otimes a_{ij}^{-1}\}.$$

The task of minimizing (maximizing) $f(x) = f^T \otimes x$ subject to (10.1) will be denoted by $\mathrm{MLP}_1^{\min}$ ($\mathrm{MLP}_1^{\max}$). The sets of optimal solutions will be denoted $S_1^{\min}$ and $S_1^{\max}$ respectively. It follows from Theorem 3.1.1 and from isotonicity of $f(x)$ that $\bar{x} \in S_1^{\max}$, whenever $S \neq \emptyset$. We now present a simple algorithm which solves $\mathrm{MLP}_1^{\min}$.

Algorithm 10.1.1 ONEMAXLINMIN (one-sided max-linear minimization)
Input: $A \in \mathbb{R}^{m \times n}, b \in \mathbb{R}^m$ and $c \in \mathbb{R}^n$.
Output: $x \in S_1^{\min}$.

1. Find $\bar{x}$ and $M_j, j \in N$.
2. Sort $(f_j \otimes \bar{x}_j; j \in N)$, without loss of generality let

$$f_1 \otimes \bar{x}_1 \leq f_2 \otimes \bar{x}_2 \leq \cdots \leq f_n \otimes \bar{x}_n.$$

3. $J := \{1\}, r := 1$.
4. If

$$\bigcup_{j \in J} M_j = M$$

then stop ($x_j = \bar{x}_j$ for $j \in J$ and x_j small enough for $j \notin J$).
5. $r := r + 1$, $J := J \cup \{r\}$.
6. Go to 4.

Note that "small enough" in step 4 may be for instance

$$x_j \leq f_j^{-1} \otimes f_r \otimes \bar{x}_r.$$

Theorem 10.1.2 *The algorithm ONEMAXLINMIN is correct and its computational complexity is $O(mn^2)$.*

Proof Correctness is obvious and computational complexity follows from the fact that the loop 4.−6. is repeated at most n times and each run is $O(mn)$. Step 1 is $O(mn)$ and step 2 is $O(n \log n)$.   □

Note that the problem of minimizing certain objective functions subject to one-sided max-linear constraints is *NP*-complete, see Exercise 10.3.1.

10.2 Programs with Two-sided Constraints

10.2.1 Problem Formulation and Basic Properties

Our main goal in this chapter is to present the necessary theory and methods for finding an $x \in \mathbb{R}^n$ (if it exists) that minimizes (maximizes) the function $f(x) = f^T \otimes x$ subject to

$$A \otimes x \oplus c = B \otimes x \oplus d, \tag{10.2}$$

where $f = (f_1, \ldots, f_n)^T \in \mathbb{R}^n$, $c = (c_1, \ldots, c_m)^T$, $d = (d_1, \ldots, d_m)^T \in \mathbb{R}^m$, $A = (a_{ij})$ and $B = (b_{ij}) \in \mathbb{R}^{m \times n}$ are given matrices and vectors. These two problems will be denoted by MLP$^{\min}$ (MLP$^{\max}$). We now denote

$$S = \left\{ x \in \mathbb{R}^n; A \otimes x \oplus c = B \otimes x \oplus d \right\},$$

$$S^{\min} = \{ x \in S; f(x) \leq f(z) \text{ for all } z \in S \}$$

and

$$S^{\max} = \{ x \in S; f(x) \geq f(z) \text{ for all } z \in S \}.$$

Systems of two-sided max-linear equations are investigated in Chap. 7 and we will follow the terminology introduced there. It has been shown in Sect. 7.4 how the general systems of the form (10.2) can be converted to homogenous systems with separated variables (Lemma 7.4.3) and hence be solved using the (pseudopolynomial) Alternating Method. Since now we assume finiteness of A and B, a two-sided system has a nontrivial solution if and only if it has a finite solution, thus this conversion is slightly more straightforward and is expressed as follows:

Proposition 10.2.1 *Let* $A, B \in \mathbb{R}^{m \times n}, c, d \in \mathbb{R}^m$ *and* $E = (A|c)$, $F = (B|d)$ *be matrices arising from A and B respectively by adding the vectors c and d as the last column. Let*

$$S_h = \left\{ z \in \mathbb{R}^{n+1}; E \otimes z = F \otimes z \right\}.$$

If $x \in S$ *then* $(x|0) \in S_h$ *and conversely, if* $z = (z_1, \ldots, z_{n+1})^T \in S_h$ *then* $z_{n+1}^{-1} \otimes (z_1, \ldots, z_n)^T \in S$.

Proof The statement follows straightforwardly from the definitions. □

In what follows we will need a slight reformulation of the computational complexity formula (7.19):

Theorem 10.2.2 *Let $E = (e_{ij})$, $F = (f_{ij}) \in \mathbb{Z}^{m \times n}$ and $K' = K(E|F)$. There is an algorithm of computational complexity $O(mn(m + n)K')$ that finds an x satisfying*

$$E \otimes z = F \otimes z \tag{10.3}$$

or decides that no such x exists.

Proof It follows from (7.19) immediately. □

Proposition 10.2.1 and Theorem 10.2.2 show that the feasibility question for $\text{MLP}^{\max}$ and $\text{MLP}^{\min}$ can be solved in pseudopolynomial time for instances with integer entries. We will use this result to develop bisection methods for solving $\text{MLP}^{\min}$ and $\text{MLP}^{\max}$. We will prove that these methods need a polynomial number of feasibility checks if all entries are integer and hence overall are also of pseudopolynomial complexity.

The Alternating Method of Sect. 7.3 is an iterative procedure that starts with an arbitrary vector and then only uses the operations of $+$, $-$, max and min applied to the starting vector and the entries of E, F. Hence using Proposition 10.2.1 we deduce:

Theorem 10.2.3 *If all entries in a homogenous max-linear system are integer and the system has a nontrivial solution then this system has an integer solution. The same is true for nonhomogenous max-linear systems.*

Using the cancellation law (7.4.1) we have:

Lemma 10.2.4 *Let $\alpha, \alpha' \in \mathbb{R}$, $\alpha' < \alpha$ and $f(x) = f^T \otimes x$, $f'(x) = f'^T \otimes x$ where $f'_j < f_j$ for every $j \in N$. Then the following holds for every $x \in \mathbb{R}$: $f(x) = \alpha$ if and only if $f(x) \oplus \alpha' = f'(x) \oplus \alpha$.*

For the bisection method it will be important to know that attainment of a value can be checked by converting this question to feasibility. The following proposition explains how this can be done.

Proposition 10.2.5 *$f(x) = \alpha$ for some $x \in S$ if and only if the following nonhomogenous max-linear system has a solution:*

$$A \otimes x \oplus c = B \otimes x \oplus d,$$
$$f(x) \oplus \alpha' = f'(x) \oplus \alpha,$$

where $\alpha' < \alpha$, $f'(x) = f'^T \otimes x$ and $f'_j < f_j$ for every $j \in N$.

Proof The statement follows from Lemma 7.4.1 and Lemma 10.2.4.   □

This result has a useful consequence for programs with integer entries.

Corollary 10.2.6 *If all entries in* MLP^{max} *or* MLP^{min} *are integer then an integer objective function value is attained by a real feasible solution if and only if it is attained by an integer feasible solution.*

Proof It follows immediately from Theorem 10.2.3 and Proposition 10.2.5.   □

For a computational complexity estimate it will be useful to know the computational complexity of the attainment of a value. To do this, for given MLP^{min} or MLP^{max} we denote in this chapter

$$K = \max\left\{|a_{ij}|, |b_{ij}|, |c_i|, |d_j|, |f_j| ; i \in M, j \in N\right\}. \qquad (10.4)$$

Corollary 10.2.7 *If all entries in* MLP^{max} *or* MLP^{min} *and α are integer then the decision problem whether $f(x) = \alpha$ for some $x \in S \cap \mathbb{Z}^n$ can be solved using $O(mn(m+n)K')$ operations where $K' = \max(K+1, |\alpha|)$.*

Proof For α' and f_j' in Proposition 10.2.5 we can take $\alpha - 1$ and $f_j - 1$ respectively. Using Proposition 10.2.1, Theorem 10.2.2 and Proposition 10.2.5 the computational complexity then is

$$O\left((m+1)(n+1)(m+n+2)K'\right) = O\left(mn(m+n)K'\right). \qquad □$$

Before we compile bisection methods for MLP^{min} and MLP^{max} we need to prove a simple property of max-linear programs, which justifies the bisection search.

Proposition 10.2.8 *If $x, y \in S$, $f(x) = \alpha < \beta = f(y)$ then for every $\gamma \in (\alpha, \beta)$ there is a $z \in S$ satisfying $f(z) = \gamma$.*

Proof Let $\lambda = 0$, $\mu = \beta^{-1} \otimes \gamma$, $z = \lambda \otimes x \oplus \beta \otimes y$. Then $\lambda \oplus \mu = 0$, $z \in S$ by Proposition 7.1.1 and by Lemma 10.0.7 we have

$$f(z) = \lambda \otimes f(x) \oplus \mu \otimes f(y) = \alpha \oplus \beta^{-1} \otimes \gamma \otimes \beta = \gamma. \qquad □$$

10.2.2 Bounds and Attainment of Optimal Values

We start by proving criteria for the existence of optimal solutions. For simplicity we denote $\inf_{x \in S} f(x)$ by f^{min}, similarly $\sup_{x \in S} f(x)$ by f^{max}.

First let us consider the lower bound. We may assume without loss of generality that in (10.2) we have $c \geq d$. Let $M^> = \{i \in M; c_i > d_i\}$. For $r \in M^>$ we denote

$$L_r = \min_{k \in N} f_k \otimes c_r \otimes b_{rk}^{-1}$$

and

$$L = \max_{r \in M^>} L_r.$$

Recall that $\max \emptyset = -\infty$ by definition.

Lemma 10.2.9 *If $c \geq d$ then $f(x) \geq L$ for every $x \in S$.*

Proof If $M^> = \emptyset$ then the statement follows trivially since $L = -\infty$. Let $x \in S$ and $r \in M^>$. Then

$$(B \otimes x)_r \geq c_r$$

and so

$$x_k \geq c_r \otimes b_{rk}^{-1}$$

for some $k \in N$. Hence $f(x) \geq f_k \otimes x_k \geq f_k \otimes c_r \otimes b_{rk}^{-1} \geq L_r$ and the theorem statement follows. □

A very simple criterion for the existence of a lower bound is given in the next statement.

Theorem 10.2.10 $f^{\min} = -\infty$ *if and only if $c = d$.*

Proof If $c = d$ then $\alpha \otimes x \in S$ for any $x \in \mathbb{R}^n$ and every $\alpha < 0$ small enough. Hence by letting $\alpha \longrightarrow -\infty$ we have $f(\alpha \otimes x) = \alpha \otimes f(x) \longrightarrow -\infty$.

If $c \neq d$ then without loss of generality $c \geq d$ and the statement now follows by Lemma 10.2.9 since $L > -\infty$. □

Let us now discuss the upper bound. We prove two lemmas before presenting the main result, Theorem 10.2.13.

Lemma 10.2.11 *Let $c \geq d$. If $x \in S$ and $(A \otimes x)_i > c_i$ for all $i \in M$ then $x' = \alpha \otimes x \in S$ and $(A \otimes x')_i = c_i$ for some $i \in M$, where*

$$\alpha = \max_{i \in M} \left(c_i \otimes (A \otimes x)_i^{-1} \right). \tag{10.5}$$

Proof Let $x \in S$. If

$$(A \otimes x)_i > c_i$$

for every $i \in M$ then $A \otimes x = B \otimes x$. For every $\alpha \in \mathbb{R}$ we also have

$$A \otimes (\alpha \otimes x) = B \otimes (\alpha \otimes x).$$

It follows from the choice of α that

$$(A \otimes (\alpha \otimes x))_i = \alpha \otimes (A \otimes x)_i \geq c_i$$

for every $i \in M$, with equality for at least one $i \in M$. Hence $x' \in S$ and the lemma follows. $\square$

Let us denote

$$U = \max_{r \in M} \max_{j \in N} f_j \otimes a_{rj}^{-1} \otimes c_r.$$

Lemma 10.2.12 *If $c \geq d$ then the following hold:*

(a) *If $x \in S$ and $(A \otimes x)_r \leq c_r$ for some $r \in M$ then $f(x) \leq U$.*
(b) *If $A \otimes x = B \otimes x$ has no nontrivial solution then $f(x) \leq U$ for every $x \in S$.*

Proof (a) Since

$$a_{rj} \otimes x_j \leq c_r$$

for all $j \in N$, we have

$$f(x) \leq \max_{j \in N} f_j \otimes a_{rj}^{-1} \otimes c_r \leq U.$$

(b) If $S = \emptyset$ then the statement holds trivially. Let $x \in S$. Then

$$(A \otimes x)_r \leq c_r$$

for some $r \in M$ since otherwise $A \otimes x = B \otimes x$, and the statement now follows from (a). $\square$

Theorem 10.2.13 $f^{\max} = +\infty$ *if and only if $A \otimes x = B \otimes x$ has a nontrivial solution.*

Proof We may assume without loss of generality that $c \geq d$. If $A \otimes x = B \otimes x$ has no solution then the statement follows from Lemma 10.2.12. If it has a solution, say z, $z \neq \varepsilon$, then for all sufficiently large $\alpha \in \mathbb{R}$ we have

$$A \otimes (\alpha \otimes z) = B \otimes (\alpha \otimes z) \geq c \oplus d$$

and hence $\alpha \otimes z \in S$. The statement now follows by letting $\alpha \longrightarrow +\infty$. $\square$

Theorem 10.2.13 provides a criterion for the existence of an upper bound, which is less simple than that for the lower bound, but still enables us to answer this question in pseudopolynomial time.

We can now discuss the question of attainment of $f^{\min}$ and $f^{\max}$. In both cases the answer is affirmative: We will show that the maximal (minimal) value is attained if $S \neq \emptyset$ and $f^{\max} < +\infty$ [$f^{\min} > -\infty$]. Due to continuity of f this will be proved by showing that both for minimization and maximization the set S can be reduced to a compact subset. To achieve this we denote for $j \in N$:

$$h_j = \min \left(\min_{r \in M} a_{rj}^{-1} \otimes c_j, \min_{r \in M} b_{rj}^{-1} \otimes d_j, f_j^{-1} \otimes L \right), \tag{10.6}$$

$$h'_j = \min\left(\min_{r \in M} a_{rj}^{-1} \otimes c_j, \min_{r \in M} b_{rj}^{-1} \otimes d_j\right) \tag{10.7}$$

and $h = (h_1, \ldots, h_n)^T$, $h' = (h'_1, \ldots, h'_n)^T$. Clearly, h' is finite. Note that h is finite if and only if $f^{\min} > -\infty$.

First we show the attainment of $f^{\min}$.

Proposition 10.2.14 *For any $x \in S$ there is an $x' \in S$ such that $x' \geq h$ and $f(x) = f(x')$.*

Proof Let $x \in S$. It is sufficient to set $x' = x \oplus h$ since if $x_j < h_j$, $j \in N$ then x_j is not active on any side of any equation or in the objective function and therefore changing x_j to h_j will not affect validity of any equation or the objective function value. □

Corollary 10.2.15 *If $f^{\min} > -\infty$ and $S \neq \emptyset$ then there is a compact set $\overline{S}$ such that*

$$f^{\min} = \min_{x \in \overline{S}} f(x).$$

Proof Note that h is finite since $f^{\min} > -\infty$. By Proposition 10.2.14 there is an $\tilde{x} \in S, \tilde{x} \geq h$. Then

$$\overline{S} = S \cap \left\{x \in \mathbb{R}^n; h_j \leq x_j \leq f_j^{-1} \otimes f(\tilde{x}), j \in N\right\}$$

is a compact subset of S and $\tilde{x} \in S$. If there was a $y \in S$ such that

$$f(y) < \min_{x \in \overline{S}} f(x) \leq f(\tilde{x})$$

then by Proposition 10.2.14 there is a $y' \geq h$, $y' \in S$, $f(y') = f(y)$. Hence

$$f_j \otimes y'_j \leq f(y') = f(y) \leq f(\tilde{x})$$

for every $j \in N$ and thus $y' \in \overline{S}$, $f(y') < \min_{x \in \overline{S}} f(x)$, a contradiction. □

Now we prove the attainment of $f^{\max}$.

Proposition 10.2.16 *For any $x \in S$ there is an $x' \in S$ such that $x' \geq h'$ and $f(x) \leq f(x')$.*

Proof Let $x \in S$ and $j \in N$. It is sufficient to set $x' = x \oplus h'$, since if $x_j < h'_j$ then x_j is not active on any side of any equation and therefore changing x_j to h'_j does not invalidate any equation. The rest follows from isotonicity of $f(x)$. □

Let

$$\overline{S}' = S \cap \{x \in \mathbb{R}^n; h'_j \leq x_j \leq f_j^{-1} \otimes U, j \in N\}.$$

Corollary 10.2.17 *If $f^{\max} < +\infty$ then*

$$f^{\max} = \max_{x \in \overline{S}'} f(x).$$

Proof The statement follows immediately from Proposition 10.2.16, Theorem 10.2.13 and Lemma 10.2.12. □

The next statement summarizes the desired result:

Corollary 10.2.18 *If $S \neq \emptyset$ and $f^{\min} > -\infty$ [$f^{\max} < +\infty$] then $S^{\min} \neq \emptyset$ [$S^{\max} \neq \emptyset$].*

We conclude this subsection by a technical statement that will be useful in the algorithms.

It follows from Lemma 10.2.9 that $f^{\max} > L$. However this information is not useful if $c = d$, since then $L = -\infty$. Because we will need a lower bound for $f^{\max}$, even when $c = d$, we define $L' = f(h')$ and formulate the following.

Corollary 10.2.19 *If $x \in S$ then $x' = x \oplus h'$ satisfies $f(x') \geq L'$ and thus $f^{\max} \geq L'$.*

10.2.3 The Algorithms

In this subsection we present the minimization and maximization algorithms for the case of real entries; those for integer entries are presented in the next section.

It follows from Proposition 10.2.1 and Theorem 10.2.2 that in pseudopolynomial time either a feasible solution to (10.2) can be found or it can be decided that no such solution exists. Due to Theorems 10.2.10 and 10.2.13 we can also recognize the cases when the objective function is unbounded. We may therefore assume that a feasible solution exists, the objective function is bounded (from below or above depending on whether we wish to minimize or maximize) and hence an optimal solution exists (Corollary 10.2.18). If $x^0 \in S$ is found then using the scaling (if necessary) proposed in Lemma 10.2.11 or Corollary 10.2.19 we find (another) x^0 satisfying $L \leq f(x^0) \leq U$ or $L' \leq f(x^0) \leq U$ (see Lemmas 10.2.9 and 10.2.12). The use of the bisection method applied to either $(L, f(x^0))$ or $(f(x^0), U)$ for finding a minimizer or maximizer of $f(x)$ is then justified by Proposition 10.2.8. The algorithms are based on the fact that (see Proposition 10.2.5) checking the existence of an $x \in S$ satisfying $f(x) = \alpha$ for a given $\alpha \in \mathbb{R}$, can be converted to a feasibility problem. They stop when the interval of uncertainty is shorter than a given precision $\varepsilon > 0$.

Algorithm 10.2.20 MAXLINMIN (max-linear minimization)
Input: $A = (a_{ij})$, $B = (b_{ij}) \in \mathbb{R}^{m \times n}$, $f = (f_1, \ldots, f_n)^T \in \mathbb{R}^n$, $c = (c_1, \ldots, c_m)^T$,
$d = (d_1, \ldots, d_m)^T \in \mathbb{R}^m$, $c \geq d$, $c \neq d$, $\varepsilon > 0$.
Output: $x \in S$ such that $f(x) - f^{\min} \leq \varepsilon$.

1. If $L = f(x)$ for some $x \in S$ then stop ($f^{\min} = L$).
2. Find an $x^0 \in S$. If $(A \otimes x^0)_i > c_i$ for all $i \in M$ then scale x^0 by α defined in (10.5).
3. $L(0) := L, U(0) := f(x^0), r := 0$.
4. $\alpha := \frac{1}{2}(L(r) + U(r))$.
5. Check whether $f(x) = \alpha$ is satisfied by some $x \in S$ and in the positive case find one.
   If yes then $U(r + 1) := \alpha$, $L(r + 1) := L(r)$.
   If not then $U(r + 1) := U(r)$, $L(r + 1) := \alpha$.
6. $r := r + 1$.
7. If $U(r) - L(r) \leq \varepsilon$ then stop else go to 4.

Theorem 10.2.21 *The algorithm MAXLINMIN is correct and the number of iterations before termination is*

$$O\left(\log_2 \frac{U - L}{\varepsilon}\right).$$

Proof Correctness follows from Proposition 10.2.8 and Lemma 10.2.9. Since $c \neq d$ we have at the end of step 2: $f(x^0) \geq L > -\infty$ (Lemma 10.2.9) and $U(0) := f(x^0) \leq U$ by Lemma 10.2.12. Thus the number of iterations is $O(\log_2 \frac{U-L}{\varepsilon})$, since after every iteration the interval of uncertainty is halved. □

The maximization algorithm has many similarities with the minimization algorithm; however, for the proof we need to consider it separately.

Algorithm 10.2.22 MAXLINMAX (max-linear maximization)
Input: $A = (a_{ij})$, $B = (b_{ij}) \in \mathbb{R}^{m \times n}$, $f = (f_1, \ldots, f_n)^T \in \mathbb{R}^n$, $c = (c_1, \ldots, c_m)^T$,
$d = (d_1, \ldots, d_m)^T \in \mathbb{R}^m$, $\varepsilon > 0$.
Output: $x \in S$ such that $f^{\max} - f(x) \leq \varepsilon$ or an indication that $f^{\max} = +\infty$.

1. If $U = f(x)$ for some $x \in S$ then stop ($f^{\max} = U$).
2. Check whether $A \otimes x = B \otimes x$ has a solution. If yes, stop ($f^{\max} = +\infty$).
3. Find an $x^0 \in S$ and set $x^0 := x^0 \oplus h'$ where h' is as defined in (10.7).
4. $L(0) := f(x^0)$, $U(0) := U, r := 0$.
5. $\alpha := \frac{1}{2}(L(r) + U(r))$.
6. Check whether $f(x) = \alpha$ is satisfied by some $x \in S$ and in the positive case find one.
   If yes then $U(r + 1) := U(r)$, $L(r + 1) := \alpha$.
   If not then $U(r + 1) := \alpha$, $L(r + 1) := L(r)$.
7. $r := r + 1$.
8. If $U(r) - L(r) \leq \varepsilon$ then stop else go to 5.

Theorem 10.2.23 *The algorithm MAXLINMAX is correct and the number of iterations before termination is*

$$O\left(\log_2 \frac{U - L'}{\varepsilon}\right).$$

Proof Correctness follows from Proposition 10.2.8 and Lemma 10.2.12. By Lemma 10.2.12 and Corollary 10.2.19 $U \geq f(x^0) \geq L'$ and thus the number of iterations is $O(\log_2 \frac{U-L'}{\varepsilon})$, since after every iteration the interval of uncertainty is halved.   □

10.2.4 The Integer Case

The algorithms of the previous section may immediately be applied to $MLP^{\min}$ or $MLP^{\max}$ when all input data are integer. However, we show that in such a case $f^{\min}$ and $f^{\max}$ are integers and therefore the algorithms find an *exact* solution once the interval of uncertainty is of length one, since then either $L(r)$ or $U(r)$ is the optimal value. Note that L and U are now integers and integrality of $L(r)$ and $U(r)$ can easily be maintained during the run of the algorithms. This implies that the algorithms will find exact optimal solutions in a finite number of steps and we will prove that their computational complexity is pseudopolynomial. The symbol $fr(k)$ will stand for the fractional part of $k \in \mathbb{Z}$, that is, $fr(k) = k - \lfloor k \rfloor$.

Theorem 10.2.24 *If A, B, c, d, f are integer, $S \neq \emptyset$ and $f^{\min} > -\infty$ then $S^{\min} \cap \mathbb{Z}^n \neq \emptyset$ (and therefore $f^{\min} \in \mathbb{Z}$).*

Proof Due to Corollary 10.2.18 it is sufficient to prove that for any $z \in S$ there is a $z^* \in S \cap \mathbb{Z}^n$ such that $f(z^*) \leq f(z)$. Let $z = (z_1, \ldots, z_n)^T \in S$. Without loss of generality, suppose $z \notin \mathbb{Z}^n$ and denote

$$N(z) = \left\{ j \in N; z_j \notin \mathbb{Z} \right\},$$

$$J = \left\{ j \in N(z); fr(z_j) = \min_{k \in N(z)} fr(z_k) \right\}.$$

Let $z'_j = \lfloor z_j \rfloor$ for $j \in J$ and $z'_j = z_j$ otherwise. Then $z' \leq z$, thus $f(z') \leq f(z)$, and $z' \in S$ since the validity of equations is unaffected: if z_j, $j \in J$ was active on one side of an equation then z_k for some $k \in J$ is active on the other side, by minimality there are no terms in this equation between a_{ij} and $a_{ij} + z_j$ and so the new values of both sides are a_{ij}; if z_j, $j \in J$ was not active then the transition $z \longrightarrow z'$ does not affect this equation at all. After at most n repetitions of this operation we get the sequence $z', z'', z''', \ldots,$ whose last term is the wanted $z^* \in \mathbb{Z}^n$.   □

Theorem 10.2.25 *If A, B, c, d, f are integer, $S \neq \emptyset$ and $f^{\max} < +\infty$ then $f^{\max} \in \mathbb{Z}$ (and therefore $S^{\max} \cap \mathbb{Z}^n \neq \emptyset$).*

Proof Suppose $c \geq d$, $f^{\max} \notin \mathbb{Z}$ and let $z = (z_1, \ldots, z_n)^T \in S^{\max}$. For any $x \in \mathbb{R}^n$ denote

$$F(x) = \{ j \in N; \, f_j \otimes x_j = f(x) \}.$$

We take one fixed $j \in F(z)$ (hence $z_j \notin \mathbb{Z}$) and show that it is possible to increase z_j without invalidating any equation, which will be a contradiction.

Due to integrality of all entries it is not possible that equality in an equation is achieved by both integer and noninteger components of z. Hence the increase of z_j only forces the noninteger components of z to increase.

At the same time an equality of the form $(A \otimes z)_i = c_i$ (if any) cannot be attained by noninteger components, thus $a_{ij} \otimes z_j < c_i$ and $b_{ij} \otimes z_j < c_i$ whenever $z_j \notin \mathbb{Z}$ and $(A \otimes z)_i = c_i$, hence there is always scope for an increase of $z_j \notin \mathbb{Z}$. □

Integer modifications of the algorithms are now straightforward since L, L' and U are also integer: we only need to ensure that the algorithms start from an integer vector (see Theorem 10.2.3) and that the integrality of both ends of the intervals of uncertainty is maintained, for instance by taking one of the integer parts of the middle of the interval.

We start with the minimization. Note that

$$L, L', U \in [-3K, 3K], \tag{10.8}$$

where K has been defined by (10.4).

Algorithm 10.2.26 INTEGER MAXLINMIN (integer max-linear minimization)
Input: $A = (a_{ij})$, $B = (b_{ij}) \in \mathbb{Z}^{m \times n}$, $f = (f_1, \ldots, f_n)^T \in \mathbb{Z}^n$, $c = (c_1, \ldots, c_m)^T$, $d = (d_1, \ldots, d_m)^T \in \mathbb{Z}^m$, $c \geq d$, $c \neq d$.
Output: $x \in S^{\min} \cap \mathbb{Z}^n$.

1. If $L = f(x)$ for some $x \in S \cap \mathbb{Z}^n$ then stop ($f^{\min} = L$).
2. Find $x^0 \in S \cap \mathbb{Z}^n$. If $(A \otimes x^0)_i > c_i$ for all $i \in M$ then scale x^0 by α defined in (10.5).
3. $L(0) := L$, $U(0) := f(x^0)$, $r := 0$.
4. $\alpha := \lfloor \frac{1}{2}(L(r) + U(r)) \rfloor$.
5. Check whether $f(x) = \alpha$ is satisfied by some $x \in S \cap \mathbb{Z}^n$ and in the positive case find one.
   If x exists then $U(r+1) := \alpha$, $L(r+1) := L(r)$.
   If it does not then $U(r+1) := U(r)$, $L(r+1) := \alpha$.
6. $r := r + 1$.
7. If $U(r) - L(r) = 1$ then stop ($U(r) = f^{\min}$) else go to 4.

Theorem 10.2.27 *The algorithm INTEGER MAXLINMIN is correct and terminates after using $O(mn(m+n)K \log K)$ operations.*

Proof Correctness follows from the correctness of MAXLINMIN and from Theorem 10.2.24. For computational complexity first note that the number of iterations is $O(\log(U - L)) \leq O(\log 6K) = O(\log K)$. The computationally prevailing part of the algorithm is the checking whether $f(x) = \alpha$ for some $x \in S \cap \mathbb{Z}^n$ when α is given. By Corollary 10.2.7 this can be done using $O(mn(m + n)K')$ operations where $K' = \max(K + 1, |\alpha|)$. Since $\alpha \in [L, U]$, using (10.8) we have $K' = O(K)$. Hence the computational complexity of checking whether $f(x) = \alpha$ for some $x \in S \cap \mathbb{Z}^n$ is $O(mn(m + n)K)$ and the statement follows.   □

Again, for the same reasons as before, we present the maximization algorithm in full.

Algorithm 10.2.28 INTEGER MAXLINMAX (integer max-linear maximization)
Input: $A = (a_{ij})$, $B = (b_{ij}) \in \mathbb{Z}^{m \times n}$, $f = (f_1, \dots, f_n)^T \in \mathbb{Z}^n$, $c = (c_1, \dots, c_m)^T$, $d = (d_1, \dots, d_m)^T \in \mathbb{Z}^m$.
Output: $x \in S^{\max} \cap \mathbb{Z}^n$ or an indication that $f^{\max} = +\infty$.

1. If $U = f(x)$ for some $x \in S \cap \mathbb{Z}^n$ then stop ($f^{\max} = U$).
2. Check whether $A \otimes x = B \otimes x$ has a solution. If yes, stop ($f^{\max} = +\infty$).
3. Find an $x^0 \in S \cap \mathbb{Z}^n$ and set $x^0 := x^0 \oplus h'$ where h' is as defined in (10.7).
4. $L(0) := f(x^0)$, $U(0) := U$, $r := 0$.
5. $\alpha := \lceil \frac{1}{2}(L(r) + U(r)) \rceil$.
6. Check whether $f(x) = \alpha$ is satisfied by some $x \in S \cap \mathbb{Z}^n$ and in the positive case find one.
   If x exists then $U(r + 1) := U(r)$, $L(r + 1) := \alpha$.
   If not then $U(r + 1) := \alpha$, $L(r + 1) := L(r)$.
7. $r := r + 1$.
8. If $U(r) - L(r) = 1$ then stop ($L(r) = f^{\max}$) else go to 5.

Theorem 10.2.29 *The algorithm INTEGER MAXLINMAX is correct and terminates after using $O(mn(m + n)K \log K)$ operations.*

Proof Correctness follows from the correctness of MAXLINMAX and from Theorem 10.2.25. The computational complexity part follows the lines of the proof of Theorem 10.2.27 after replacing L by L'.   □

10.2.5 An Example

Let us consider the max-linear program (minimization) in which

$$f = (3, 1, 4, -2, 0)^T,$$

$$A = \begin{pmatrix} 17 & 12 & 9 & 4 & 9 \\ 9 & 0 & 7 & 9 & 10 \\ 19 & 4 & 3 & 7 & 11 \end{pmatrix},$$

$$B = \begin{pmatrix} 2 & 11 & 8 & 10 & 9 \\ 11 & 0 & 12 & 20 & 3 \\ 2 & 13 & 5 & 16 & 4 \end{pmatrix},$$

$$c = \begin{pmatrix} 12 \\ 15 \\ 13 \end{pmatrix}, \qquad d = \begin{pmatrix} 12 \\ 12 \\ 3 \end{pmatrix}$$

and the starting vector is

$$x^0 = (-6, 0, 3, -5, 2)^T.$$

Clearly, $f(x^0) = 7$, $M^> = \{2, 3\}$ and the lower bound is

$$L = \max_{r \in M^>} \min_{k \in N} f_k \otimes c_r \otimes b_{rk}^{-1}$$

$$= \max (\min (7, 16, 7, -7, 12), \min (14, 1, 12, -5, 9)) = -5.$$

Record of the run of INTEGER MAXLINMIN for this problem:
Iteration 1: Check whether $L = -5$ is attained by $f(x)$ for some $x \in S$ by solving the system

$$\begin{pmatrix} 17 & 12 & 9 & 4 & 9 & 12 \\ 9 & 0 & 7 & 9 & 10 & 15 \\ 19 & 4 & 3 & 7 & 11 & 13 \\ 3 & 1 & 4 & -2 & 0 & -6 \end{pmatrix} \otimes w = \begin{pmatrix} 2 & 11 & 8 & 10 & 9 & 12 \\ 11 & 0 & 12 & 20 & 3 & 12 \\ 2 & 13 & 5 & 16 & 4 & 3 \\ 2 & 0 & 3 & -3 & -1 & -5 \end{pmatrix} \otimes w.$$

There is no solution, hence $L(0) := -5$, $U(0) := 7$, $r := 0$, $\alpha := 1$.
Check whether $f(x) = 1$ is satisfied by some $x \in S$ by solving

$$\begin{pmatrix} 17 & 12 & 9 & 4 & 9 & 12 \\ 9 & 0 & 7 & 9 & 10 & 15 \\ 19 & 4 & 3 & 7 & 11 & 13 \\ 3 & 1 & 4 & -2 & 0 & 0 \end{pmatrix} \otimes w = \begin{pmatrix} 2 & 11 & 8 & 10 & 9 & 12 \\ 11 & 0 & 12 & 20 & 3 & 12 \\ 2 & 13 & 5 & 16 & 4 & 3 \\ 2 & 0 & 3 & -3 & -1 & 1 \end{pmatrix} \otimes w.$$

There is a solution $x = (-6, 0, -3, -5, 1)^T$. Hence $U(1) := 1$, $L(1) := -5$, $r := 1$, $U(1) - L(1) > 1$.
Iteration 2: Check whether $f(x) = -2$ is satisfied by some $x \in S$ by solving

$$\begin{pmatrix} 17 & 12 & 9 & 4 & 9 & 12 \\ 9 & 0 & 7 & 9 & 10 & 15 \\ 19 & 4 & 3 & 7 & 11 & 13 \\ 3 & 1 & 4 & -2 & 0 & -3 \end{pmatrix} \otimes w = \begin{pmatrix} 2 & 11 & 8 & 10 & 9 & 12 \\ 11 & 0 & 12 & 20 & 3 & 12 \\ 2 & 13 & 5 & 16 & 4 & 3 \\ 2 & 0 & 3 & -3 & -1 & -2 \end{pmatrix} \otimes w.$$

There is no solution. Hence $U(2) := 1$, $L(2) := -2$, $r := 2$, $U(2) - L(2) > 1$.

Iteration 3: Check whether $f(x) = 0$ is satisfied by some $x \in S$ by solving

$$\begin{pmatrix} 17 & 12 & 9 & 4 & 9 & 12 \\ 9 & 0 & 7 & 9 & 10 & 15 \\ 19 & 4 & 3 & 7 & 11 & 13 \\ 3 & 1 & 4 & -2 & 0 & -1 \end{pmatrix} \otimes w = \begin{pmatrix} 2 & 11 & 8 & 10 & 9 & 12 \\ 11 & 0 & 12 & 20 & 3 & 12 \\ 2 & 13 & 5 & 16 & 4 & 3 \\ 2 & 0 & 3 & -3 & -1 & 0 \end{pmatrix} \otimes w.$$

There is no solution. Hence $U(3) := 1$, $L(3) := 0$, $U(1) - L(1) = 1$, stop, $f^{\min} = 1$, an optimal solution is $x = (-6, 0, -3, -5, 1)^T$.

10.3 Exercises

Exercise 10.3.1 Prove that the problem of minimizing the function

$$2^{x_1} + \cdots + 2^{x_n}$$

subject to one-sided max-linear constraints $A \otimes x = b$ is *NP*-complete. (Hint: Find a polynomial transformation of the classical minimum set covering problem to this problem with a matrix A over $\{0, -1\}$, $b = 0$)

Exercise 10.3.2 Find a minimizer of the function $\max(x_1, x_2, x_3, x_4, x_5)$ subject to the constraints $A \otimes x \oplus c = B \otimes x \oplus d$, where

$$A = \begin{pmatrix} 49 & 31 & 82 & 38 & 35 \\ 44 & 51 & 79 & 81 & 94 \\ 45 & 51 & 64 & 53 & 88 \end{pmatrix}, \qquad B = \begin{pmatrix} 55 & 21 & 23 & 23 & 44 \\ 62 & 30 & 84 & 17 & 31 \\ 59 & 47 & 19 & 23 & 92 \end{pmatrix},$$

$$c = \begin{pmatrix} 43 \\ 18 \\ 90 \end{pmatrix}, \qquad d = \begin{pmatrix} 98 \\ 44 \\ 11 \end{pmatrix}.$$

$[x^{\min} = (19, 19, 16, 19, -2)^T]$

Exercise 10.3.3 Find a maximizer of the function $\max(x_1, x_2, x_3, x_4, x_5)$ subject to the constraints $A \otimes x \oplus c = B \otimes x \oplus d$, where

$$A = \begin{pmatrix} 95 & 49 & 46 & 44 & 92 \\ 23 & 89 & 2 & 62 & 74 \\ 61 & 76 & 82 & 79 & 18 \end{pmatrix}, \qquad B = \begin{pmatrix} 41 & 41 & 35 & 14 & 60 \\ 94 & 89 & 81 & 20 & 27 \\ 92 & 6 & 1 & 20 & 20 \end{pmatrix},$$

$$c = \begin{pmatrix} 2 \\ 75 \\ 45 \end{pmatrix}, \qquad d = \begin{pmatrix} 93 \\ 47 \\ 42 \end{pmatrix}.$$

$[x^{\max} = (-2, 14, 8, 11, 1)^T]$

Chapter 11
Conclusions and Open Problems

The aim of this book is two-fold: to provide an introductory text to max-algebra and to present results on advanced topics. Chapters 1–5 aim to be a guide through basic max-algebra, and possibly to accompany an undergraduate or postgraduate course. Chapters 6–10 are focused on more advanced topics with emphasis on feasibility and reachability.

In the case of feasibility the most important results are: complete resolution of the eigenvalue-eigenvector problem using $O(n^3)$ algorithms; methods for solving two-sided systems of max-linear equations of pseudopolynomial computational complexity; full characterization of strongly regular matrices and the simple image sets of max-linear mappings; $O(n^3)$ algorithms for three presented types of matrix regularity and a polynomial algorithm for finding all essential coefficients of a characteristic maxpolynomial.

Basic reachability problems are solvable in polynomial time. These include the question of reachability of eigenspaces by matrix orbits (for irreducible matrices) and robustness (for irreducible and reducible matrices). Max-linear programs with two-sided constraints can be solved in pseudopolynomial time for problems with integer entries.

There are a number of problems that seem to be unresolved at the time of printing this book. We list some of them:

OP1: Is it possible to multiply out two $n \times n$ matrices in max-algebra in time better than $O(n^3)$?

OP2: Although strong regularity and Gondran–Minoux regularity can be checked in $O(n^3)$ time, it is still not clear whether it is possible to check the strong linear independence or Gondran–Minoux independence in polynomial time.

OP3: Although the question of the existence of permutations of both parities, optimal for the assignment problem for a matrix, is decidable in $O(n^3)$ time, it is not clear whether the best optimal permutations of both parities can be found in polynomial time.

OP4: Although the two-sided max-linear systems with integer entries are solvable in pseudopolynomial time, it is still not clear whether this problem is polynomially solvable or NP-complete.

P. Butkovič, *Max-linear Systems: Theory and Algorithms,*
Springer Monographs in Mathematics 151,
DOI 10.1007/978-1-84996-299-5_11, © Springer-Verlag London Limited 2010

OP5: Although all essential coefficients of a characteristic maxpolynomial can be found in polynomial time, it is still not clear whether the problem of finding all coefficients is polynomially solvable or *NP*-complete.

OP6: Can the pseudopolynomial algorithms for solving max-linear programs with finite entries be extended to problems with non-finite entries?

OP7: Although it is clear that the greatest corner of a characteristic maxpolynomial is equal to the principal eigenvalue, it is not clear how to interpret the other corners.

OP8: One of the hardest problems in max-algebra seems to be the generalized eigenproblem. Although some progress is presented in Chap. 9, probably no method is available of any kind, exact or approximate (including heuristics), to find at least one generalized eigenvalue for general matrices. In particular, we know that there is at most one generalized eigenvalue if the matrices are symmetric (Theorem 9.1.6), yet there is no clear description of the unique (possible) eigenvalue.

References

1. Ahuja, R. K., Magnanti, T., & Orlin, J. B. (1993). *Network flows: Theory, algorithms and applications*. Englewood Cliffs: Prentice Hall.
2. Akian, M., Bapat, R., & Gaubert, S. (2004). Perturbation of eigenvalues of matrix pencils and optimal assignment problem. *Comptes Rendus de L'Academie Des Sciences Paris, Série I, 339*, 103–108.
3. Akian, M., Bapat, R., & Gaubert, S. (2007). Max-plus algebras. In L. Hogben, R. Brualdi, A. Greenbaum, & R. Mathias (Eds.), *Discrete mathematics and its applications: Vol. 39. Handbook of linear algebra*. Boca Raton: Chapman & Hall/CRC. Chap. 25.
4. Akian, M., Gaubert, S., & Guterman, A. (2009). Linear independence over tropical semirings and beyond. In G. L. Litvinov & S. N. Sergeev (Eds.), *Contemporary mathematics series: Vol. 495. Proceedings of the international conference on tropical and idempotent mathematics* (pp. 1–38). Providence: AMS.
5. Akian, M., Gaubert, S., & Kolokoltsov, V. (2002). Invertibility of functional Galois connections. *Comptes Rendus de L'Academie Des Sciences Paris, 335*, 883–888.
6. Akian, M., Gaubert, S., & Walsh, C. (2005). Discrete max-plus spectral theory. ESI Preprint 1485. In G. L. Litvinov & V. P. Maslov (Eds.), *Contemporary mathematics series: Vol. 377. Idempotent mathematics and mathematical physics* (pp. 53–77). Providence: AMS.
7. Allamigeon, X., Gaubert, S., & Goubault, E. (2009). Computing the extreme points of tropical polyhedra. arXiv:0904.3436.
8. Baccelli, F. L., Cohen, G., Olsder, G.-J., & Quadrat, J.-P. (1992). *Synchronization and linearity*. Chichester: Wiley.
9. Balcer, Y., & Veinott, A. F. (1973). Computing a graph's period quadratically by node condensation. *Discrete Mathematics, 4*, 295–303.
10. Bapat, R. B. (1995). Pattern properties and spectral inequalities in max algebra. *SIAM Journal on Matrix Analysis and Applications, 16*, 964–976.
11. Bapat, R. B., & Raghavan, T. E. S. (1997). *Nonnegative matrices and applications*. Cambridge: Cambridge University Press.
12. Bapat, R. B., Stanford, D., & van den Driessche, P. (1993). *The eigenproblem in max algebra* (DMS-631-IR). University of Victoria, British Columbia.
13. Berman, A., & Plemmons, R. J. (1979). *Nonnegative matrices in the mathematical sciences*. New York: Academic Press.
14. Bezem, M., Nieuwenhuis, R., & Rodríguez-Carbonell, E. (2010). Hard problems in max-algebra, control theory, hypergraphs and other areas. *Information Processing Letters, 110*(4), 133–138.
15. Binding, P. A., & Volkmer, H. (2007). A generalized eigenvalue problem in the max algebra. *Linear Algebra and Its Applications, 422*, 360–371.
16. Braker, J. G. (1993). *Algorithms and applications in timed discrete event systems*. Thesis, University of Delft.

17. Braker, J. G., & Olsder, G. J. (1993). The power algorithm in max algebra. *Linear Algebra and Its Applications*, *182*, 67–89.
18. Brualdi, R. A., & Ryser, H. (1991). *Combinatorial matrix theory*. Cambridge: Cambridge University Press.
19. Burkard, R. E., & Butkovič, P. (2003). Finding all essential terms of a characteristic max-polynomial. *Discrete Applied Mathematics*, *130*, 367–380.
20. Burkard, R. E., & Butkovič, P. (2003). Max algebra and the linear assignment problem. *Mathematical Programming Series B*, *98*, 415–429.
21. Burkard, R. E., & Çela, E. (1999). Linear assignment problems and extensions. In P. M. Pardalos & D.-Z. Du (Eds.), *Handbook of combinatorial optimization. Supplement Volume A* (pp. 75–149). Dordrecht: Kluwer Academic Publishers.
22. Burkard, R. E., Dell'Amico, M., & Martello, S. (2009). *Assignment problems*. Philadelphia: SIAM.
23. Butkovič, P. (1985). Necessary solvability conditions of systems of linear extremal equations. *Discrete Applied Mathematics*, *10*, 19–26.
24. Butkovič, P. (1994). Strong regularity of matrices—a survey of results. *Discrete Applied Mathematics*, *48*, 45–68.
25. Butkovič, P. (1995). Regularity of matrices in min-algebra and its time-complexity. *Discrete Applied Mathematics*, *57*, 121–132.
26. Butkovič, P. (2000). Simple image set of (max, +) linear mappings. *Discrete Applied Mathematics*, *105*, 73–86.
27. Butkovič, P. (2003). Max-algebra: the linear algebra of combinatorics? *Linear Algebra and Its Applications*, *367*, 313–335.
28. Butkovič, P. (2003). On the complexity of computing the coefficients of max-algebraic characteristic polynomial and characteristic equation. *Kybernetika*, *39*, 129–136.
29. Butkovič, P. (2007). A note on the parity assignment problem. *Optimization*, *56*(4), 419–424.
30. Butkovič, P. (2008). Finding a bounded mixed-integer solution to a system of dual inequalities. *Operations Research Letters*, *36*, 623–627.
31. Butkovič, P. (2008). Permuted max-algebraic (tropical) eigenvector problem is NP-complete. *Linear Algebra and Its Applications*, *428*, 1874–1882.
32. Butkovič, P., & Aminu, A. (2008). Introduction to max-linear programming. *IMA Journal of Management Mathematics*, *20*(3), 1–17.
33. Butkovič, P., & Cuninghame-Green, R. A. (1992). An algorithm for the maximum cycle mean of an $n \times n$ bivalent matrix. *Discrete Applied Mathematics*, *35*, 157–162.
34. Butkovič, P., & Cuninghame-Green, R. A. (2007). The eigenproblem for matrix powers in max-algebra. *Linear Algebra and Its Applications*, *421*, 370–381.
35. Butkovič, P., & Hegedüs, G. (1984). An elimination method for finding all solutions of the system of linear equations over an extremal algebra. *Ekonomicko-matematický obzor*, *20*, 203–215.
36. Butkovič, P., & Hevery, F. (1985). A condition for the strong regularity of matrices in the minimax algebra. *Discrete Applied Mathematics*, *11*, 209–222.
37. Butkovič, P., & Lewis, S. (2007). On the job rotation problem. Discrete. *Optimization*, *4*, 163–174.
38. Butkovič, P., & Murfitt, L. (2000). Calculating essential terms of a characteristic maxpolynomial. *Central European Journal of Operations Research*, *8*, 237–246.
39. Butkovič, P., & Plávka, J. (1989). On the dependence of the maximum cycle mean of a matrix on permutations of the rows and columns. *Discrete Applied Mathematics*, *23*, 45–53.
40. Butkovič, P., & Schneider, H. (2005). Applications of max-algebra to diagonal scaling of matrices. *Electronic Journal of Linear Algebra*, *13*, 262–273.
41. Butkovič, P., & Schneider, H. (2007). *On the visualisation scaling of matrices* (Preprint 2007/12). University of Birmingham.
42. Butkovič, P., & Tam, K. P. (2009). On some properties of the image set of a max-linear mapping. In G. L. Litvinov & S. N. Sergeev (Eds.), *Contemporary mathematics series: Vol. 495. Tropical and idempotent mathematics* (pp. 115–126). Providence: AMS.

43. Butkovič, P., Schneider, H., & Sergeev, S. (2007). Generators, extremals and bases of max cones. *Linear Algebra and Its Applications, 421*, 394–406.
44. Butkovič, P., Gaubert, S., & Cuninghame-Green, R. A. (2009). Reducible spectral theory with applications to the robustness of matrices in max-algebra. *SIAM Journal on Matrix Analysis and Applications, 31*(3), 1412–1431.
45. Carré, B. A. (1971). An algebra for network routing problems. *Journal of the Institute of Mathematics and Its Applications, 7*, 273.
46. Cechlárová, K. (2005). Eigenvectors of interval matrices over max-plus algebra. *Discrete Applied Mathematics, 150*(1–3), 2–15.
47. Cechlárová, K., & Cuninghame-Green, R. A. (2003). Soluble approximation of linear systems in max-plus algebra. *Kybernetika (Prague), 39*(2), 137–141. Special issue on max-plus algebras (Prague, 2001).
48. Chen, W., Qi, X., & Deng, S. (1990). The eigenproblem and period analysis of the discrete event systems. *Systems Science and Mathematical Sciences, 3*(3), 243–260.
49. Cochet-Terrasson, J., Cohen, G., Gaubert, S., Mc Gettrick, M., & Quadrat, J.-P. (1998). Numerical computation of spectral elements in max-plus algebra. In *IFAC conference on system structure and control*.
50. Cohen, G., Dubois, D., Quadrat, J.-P., & Viot, M. (1983). *Analyse du comportement périodique de systèmes de production par la thèorie des dioïdes* (Rapports de Recherche No 191). INRIA, Le Chesnay.
51. Cohen, G., Dubois, D., Quadrat, J.-P., & Viot, M. (1985). A linear-system-theoretic view of discrete-event processes and its use for performance evaluation in manufacturing. *IEEE Transactions on Automatic Control, AC-30*(3).
52. Cohen, G., Gaubert, S., & Quadrat, J.-P. (2004). Duality and separation theorems in idempotent semimodules. *Linear Algebra and Its Applications, 379*, 395–422. Tenth conference of the international linear algebra society.
53. Cohen, G., Gaubert, S., & Quadrat, J.-P. (2006). Projection and aggregation in maxplus algebra. In *Systems control found appl. Current trends in nonlinear systems and control* (pp. 443–454). Boston: Birkhäuser.
54. Cohen, G., Gaubert, S., Quadrat, J.-P., & Singer, I. (2005). Max-plus convex sets and functions. In *Contemp. math.: Vol. 377. Idempotent mathematics and mathematical physics* (pp. 105–129). Providence: Amer. Math. Soc.
55. Conforti, M., Di Summa, M., Eisenbrand, F., & Wolsey, L. (2006). *Network formulations of mixed-integer programs* (CORE discussion paper 117). Université Catholique de Louvain, Belgium.
56. Coppersmith, D., & Winograd, S. (1990). Matrix multiplication via arithmetic progressions. *Journal of Symbolic Computation, 9*, 251–280.
57. Cuninghame-Green, R. A. (1960). Process synchronisation in a steelworks—a problem of feasibility. In Banbury & Maitland (Eds.), *Proc. 2nd int. conf. on operational research* (pp. 323–328). London: English University Press.
58. Cuninghame-Green, R. A. (1962). Describing industrial processes with interference and approximating their steady-state behaviour. *Operations Research Quarterly, 13*, 95–100.
59. Cuninghame-Green, R. A. (1976). Projections in minimax algebra. *Mathematical Programming, 10*(1), 111–123.
60. Cuninghame-Green, R. A. (1979). *Lecture notes in economics and math systems: Vol. 166. Minimax algebra*. Berlin: Springer. (Downloadable from http://web.mat.bham.ac.uk/P.Butkovic/).
61. Cuninghame-Green, R. A. (1981). Minimax algebra. *Bulletin—Institute of Mathematics and Its Applications, 17*(4), 66–69.
62. Cuninghame-Green, R. A. (1983). The characteristic maxpolynomial of a matrix. *Journal of Mathematical Analysis and Applications, 95*, 110–116.
63. Cuninghame-Green, R. A. (1991). Minimax algebra and applications. *Fuzzy Sets and Systems, 41*, 251–267.
64. Cuninghame-Green, R. A. (1995). Maxpolynomial equations. *Fuzzy Sets and Systems, 75*(2), 179–187.

65. Cuninghame-Green, R. A. (1995). Minimax algebra and applications. In *Advances in imaging and electron physics* (Vol. 90, pp. 1–121). New York: Academic Press.
66. Cuninghame-Green, R. A., & Butkovič, P. (1995). Extremal eigenproblem for bivalent matrices. *Linear Algebra and Its Applications, 222,* 77–89.
67. Cuninghame-Green, R. A., & Butkovič, P. (1995). Discrete-event dynamic systems: the strictly convex case. *Annals of Operation Research, 57,* 45–63.
68. Cuninghame-Green, R. A., & Butkovič, P. (2003). The equation Ax = By over (max, +). *Theoretical Computer Science, 293,* 3–12.
69. Cuninghame-Green, R. A., & Butkovič, P. (2004). Bases in max-algebra. *Linear Algebra and Its Applications, 389,* 107–120.
70. Cuninghame-Green, R. A., & Butkovič, P. (2008). Generalised eigenproblem in max algebra. IEEE Xplore. *Discrete Event Systems* (pp. 236–241).
71. Cuninghame-Green, R. A., & Meijer, P. F. (1980). An algebra for piecewise-linear minimax problems. *Discrete Applied Mathematics, 2,* 267–294.
72. Dantzig, G. B., Blattner, W., & Rao, M. R. (1967). Finding a cycle in a graph with minimum cost to time ratio with application to a ship routing problem. In P. Rosenstiehl (Ed.), *Theory of graphs* (pp. 77–84). Paris: Dunod.
73. Dell'Amico, M., & Martello, S. (1997). The *k*-cardinality assignment problem. *Discrete Applied Mathematics, 76*(1–3), 103–121. Second International Colloquium on Graphs and Optimization (Leukerbad, 1994).
74. Denardo, E. V. (1977). Periods of connected networks and powers of nonnegative matrices. *Mathematics of Operations Research, 2*(1), 20–24.
75. De Schutter, B. (2000). On the ultimate behavior of the sequence of consecutive powers of a matrix in the max-plus algebra. *Linear Algebra and Its Applications, 307*(1–3), 103–117.
76. Develin, M., & Sturmfels, B. (2004). Tropical convexity. *Documenta Mathematica, 9,* 1–27.
77. Elsner, L., & van den Driessche, P. (1999). On the power method in max-algebra. *Linear Algebra and Its Applications, 302/303,* 17–32.
78. Elsner, L., & van den Driessche, P. (2001). Modifying the power method in max algebra. *Linear Algebra and Its Applications, 332/334,* 3–13. Proceedings of the Eighth Conference of the International Linear Algebra Society (Barcelona, 1999).
79. Elsner, L., & van den Driessche, P. (2008). Bounds for the Perron root using max eigenvalues. *Linear Algebra and Its Applications, 428,* 2000–2005.
80. Engel, G. M., & Schneider, H. (1973). Cyclic and diagonal products on a matrix. *Linear Algebra and Its Applications, 7,* 301–335.
81. Engel, G. M., & Schneider, H. (1975). Diagonal similarity and equivalence for matrices over groups with 0. *Czechoslovak Mathematical Journal, 25,* 389–403.
82. Fiedler, M., & Pták, V. (1967). Diagonally dominant matrices. *Czechoslovak Mathematical Journal, 92,* 420–433.
83. Garey, M. R., & Johnson, D. S. (1979). *Computers and intractability. A guide to the theory of NP-completeness.* San Francisco: W. H. Freeman and Co.
84. Gaubert, S. (1992). *Théorie des systèmes linéaires dans les dioïdes.* Thèse, Ecole des Mines de Paris.
85. Gaubert, S. (2007). Private communication.
86. Gaubert, S., & Butkovič, P. (1999). Sign-nonsingular matrices and matrices with unbalanced determinant in symmetrised semirings. *Linear Algebra and Its Applications, 301,* 195–201.
87. Gaubert, S., & Katz, R. (2006). Max-plus convex geometry. In *Lecture notes in comput. sci.: Vol. 4136. Relations and Kleene algebra in computer science* (pp. 192–206). Berlin: Springer.
88. Gaubert, S., & Katz, R. (2006). Reachability problems for products of matrices in semirings. *International Journal of Algebra and Computation, 16*(3), 603–627.
89. Gaubert, S., & Katz, R. (2007). The Minkowski theorem for max-plus convex sets. *Linear Algebra and Its Applications, 421,* 356–369.
90. Gaubert, S., Butkovič, P., & Cuninghame-Green, R. A. (1998). Minimal (max, +) realization of convex sequences. *SIAM Journal on Control and Optimization, 36*(1), 137–147.

91. Gaubert, S. et al. (1998). *Algèbres max-plus et applications en informatique et automatique.* 26ème école de printemps d'informatique théorique Noirmoutier.
92. Gavalec, M. (2000). Linear matrix period in max-plus algebra. *Linear Algebra and Its Applications, 307,* 167–182.
93. Gavalec, M. (2000). Polynomial algorithm for linear matrix period in max-plus algebra. *Central European Journal of Operations Research, 8,* 247–258.
94. Gavalec, M., & Plávka, J. (2006). Computing an eigenvector of a Monge matrix in max-plus algebra. *Mathematical Methods of Operations Research, 63*(3), 543–551.
95. Giffler, B. (1963). Scheduling general production systems using schedule algebra. *Naval Research Logistics Quarterly, 10,* 237–255.
96. Giffler, B. (1968). Schedule algebra: a progress report. *Naval Research Logistics Quarterly, 15,* 255–280.
97. Gondran, M. (1975). Path algebra and algorithms. In B. Roy (Ed.), *NATO advanced study inst. ser., ser. C: Math. and phys. sci.: Vol. 19. Combinatorial programming: methods and applications* (pp. 137–148). Reidel: Dordrecht. (Proc NATO Advanced Study Inst, Versailles, 1974).
98. Gondran, M., & Minoux, M. (1977). Valeurs propres et vecteur propres dans les dioïdes et leur interprétation en théorie des graphes. *Bulletin de la Direction des Etudes et Recherches Serie C Mathematiques et Informatiques* (2), 25–41.
99. Gondran, M., & Minoux, M. (1978). L'indépendance linéaire dans les dioïdes. *Bulletin de la Direction Etudes et Recherches. EDF, Série C, 1,* 67–90.
100. Gondran, M., & Minoux, M. (1984). Linear algebra of dioïds: a survey of recent results. *Annals of Discrete Mathematics, 19,* 147–164.
101. Gunawardena, J. (Ed.) (1998). *Publications of the INI Cambridge: Vol. 11. Idempotency.* Cambridge: Cambridge University Press.
102. Heidergott, B., Olsder, G. J., & van der Woude, J. (2005). *Max plus at work: modeling and analysis of synchronized systems. A course on max-plus algebra.* Princeton: Princeton University Press.
103. Helbig, S. (1988). Caratheodory's and Krein-Milman's theorems in fully ordered groups. *Commentationes Mathematicae Universitatis Carolinae, 29,* 157–167.
104. Itenberg, I., Mikhalkin, I., & Shustin, E. (2009). *Oberwolfach seminars: Vol. 35. Tropical algebraic geometry* (2nd ed.). Berlin: Springer.
105. Joswig, M. (2005). Tropical halfspaces. In J. E. Goodman, J. Pach, & E. Welzl (Eds.), *MSRI publications: Vol. 52. Combinatorial and computational geometry* (pp. 409–431). Cambridge: Cambridge University Press.
106. Karp, R. M. (1978). A characterization of the minimum cycle mean in a digraph. *Discrete Mathematics, 23,* 309–311.
107. Katz, R., Schneider, H., & Sergeev, S. (2010). *Commuting matrices in max-algebra* (Preprint 2010/03). University of Birmingham, School of Mathematics.
108. Kolokoltsov, V. N., & Maslov, V. P. (1997). *Idempotent analysis and its applications.* Dordrecht: Kluwer Academic Publishers.
109. Lawler, E. (1976). *Combinatorial optimization—networks and matroids.* New York: Holt, Rinehart and Winston.
110. Litvinov, G. L., & Maslov, V. P. (2005). Idempotent analysis and mathematical physics. In *Contemporary mathematics: Vol. 377. International workshop,* February 3–10, 2003, Erwin Schrödinger International Institute for Mathematical Physics, Vienna, Austria. Providence: American Mathematical Society.
111. MacLane, S., & Birkhoff, G. (1979). *Algebra.* New York: Macmillan.
112. McEneaney, W. (2005). *Systems & control: foundations & applications. Max-plus methods for nonlinear control and estimation.* Boston: Birkhäuser.
113. Merlet, G. (2010). Semigroup of matrices acting on the max-plus projective space. *Linear Algebra and Its Applications, 432,* 1923–1935.
114. Molnárová, M., & Pribiš, J. (2000). Matrix period in max-algebra. *Discrete Applied Mathematics, 103*(1–3), 167–175.

115. Munro, I. (1971). Efficient determination of the transitive closure of a directed graph. *Information Processing Letters*, *1*, 56–58.
116. Murfitt, L. (2000). *Discrete-event dynamic systems in max-algebra*. Thesis, University of Birmingham.
117. Nachtigall, K. (1997). Powers of matrices over an extremal algebra with applications to periodic graphs. *Mathematical Methods of Operations Research*, *46*, 87–102.
118. Nussbaum, R. D. (1991). Convergence of iterates of a nonlinear operator arising in statistical mechanics. *Nonlinearity*, *4*, 1223–1240.
119. Olsder, G. J., & Roos, C. (1988). Cramér and Cayley-Hamilton in the max algebra. *Linear Algebra and Its Applications*, *101*, 87–108.
120. Papadimitriou, C. H., & Steiglitz, K. (1998). *Combinatorial optimization—algorithms and complexity*. Mineola: Dover Publications.
121. Plávka, J. (1996). Static maximum cycle mean problem of a trivalent matrix. *Optimization*, *37*(2), 171–176.
122. Plávka, J. (2001). On eigenproblem for circulant matrices in max-algebra. *Optimization*, *50*(5–6), 477–483.
123. Plus, M. (1990). Linear systems in (max,+) algebra. In *Proceedings of 29th conference on decision and control*, Honolulu.
124. Robertson, N., Seymour, P. D., & Thomas, R. (1999). Permanents, Pfaffian orientations and even directed circuits. *Annals of Mathematics*, *150*(2), 929–975.
125. Rockafellar, R. T. (1970). *Convex analysis*. Princeton: Princeton University Press.
126. Rosen, K. H. et al. (2000). *Handbook of discrete and combinatorial mathematics*. New York: CRC Press.
127. Rothblum, U. G., Schneider, H., & Schneider, M. H. (1994). Scaling matrices to prescribed row and column maxima. *SIAM Journal on Matrix Analysis and Applications*, *15*, 1–14.
128. Schneider, H. (1988). The influence of the marked reduced graph of a nonnegative matrix on the Jordan form and on related properties: a survey. *Linear Algebra and Its Applications*, *84*, 161–189.
129. Schneider, H., & Schneider, M. H. (1991). Max-balancing weighted directed graphs and matrix scaling. *Mathematics of Operations Research*, *16*(1), 208–222.
130. Semančíková, B. (2006). Orbits in max-min algebra. *Linear Algebra and Its Applications*, *414*, 38–63.
131. Sergeev, S. (2007). Max-plus definite matrix closures and their eigenspaces. *Linear Algebra and Its Applications*, *421*, 182–201.
132. Sergeev, S. (2008). Private communication.
133. Sergeev, S. (2009). Max-algebraic powers of irreducible matrices in the periodic regime: an application of cyclic classes. *Linear Algebra and Its Applications*, *431*, 1325–1339.
134. Sergeev, S. (2009). *Cyclic classes and attraction cones in max algebra* (Preprint 2009/05). University of Birmingham, School of Mathematics. E-print arXiv:0903.3960.
135. Sergeev, S. Spectrum of two-sided eigenproblem in max algebra: every system of intervals is realizable. *Discrete Applied Mathematics*, to appear.
136. Sergeev, S. (2009). Multiorder Kleene stars and cyclic projectors in the geometry of max cones. In G. L. Litvinov & S. N. Sergeev (Eds.), *Contemporary mathematics: Vol. 495. Tropical and idempotent mathematics* (pp. 317–342). Providence: American Mathematical Society.
137. Sergeev, S. (2010). Max-algebraic attraction cones of nonnegative irreducible matrices. *Linear Algebra and Its Applications*, submitted.
138. Sergeev, S., & Schneider, H. (2010). *CSR expansions of matrix powers in max algebra* (Preprint 2010/02). University of Birmingham, School of Mathematics. E-print arXiv:0912.2534.
139. Sergeev, S., Schneider, H., & Butkovič, P. (2009). On visualization scaling, subeigenvectors and Kleene stars in max algebra. *Linear Algebra and Its Applications*, *431*, 2395–2406.
140. Straubing, H. (1983). A combinatorial proof of the Cayley-Hamilton theorem. *Discrete Mathematics*, *43*, 273–279.

141. Sturmfels, B., Santos, F., & Develin, M. (2005). Discrete and computational geometry. In J. E. Goodman, J. Pach, & E. Welzl (Eds.), *Mathematical sciences research institute publications: Vol. 52. On the tropical rank of a matrix* (pp. 213–242). Cambridge: Cambridge University Press.

142. Tarjan, R. E. (1972). Depth-first search and linear graph algorithms. *SIAM Journal on Computing, 1*(2), 146–160.

143. Thomassen, C. (1986). Sign-nonsingular matrices and even cycles in directed graphs. *Linear Algebra and Its Applications, 75*, 27–41.

144. Vorobyov, N. N. (1967). Extremal algebra of positive matrices. *Elektronische Informationsverarbeitung und Kybernetik, 3*, 39–71 (in Russian).

145. Vorobyov, N. N. (1970). Extremal algebra of nonnegative matrices. *Elektronische Informationsverarbeitung und Kybernetik, 6*, 303–311 (in Russian).

146. Wagneur, E. (1988). Finitely generated moduloïds: The existence and unicity problem for bases. In J. L. Lions & A. Bensoussan (Eds.), *Lecture notes in control and inform. sci.: Vol. 111. Analysis and optimization of systems*, Antibes, 1988 (pp. 966–976). Berlin: Springer.

147. Wagneur, E. (1991). Moduloïds and pseudomodules: 1. Dimension theory. *Discrete Mathematics, 98*, 57–73.

148. Walkup, E. A., & Boriello, G. (1988). A general linear max-plus solution technique. In J. Gunawardena (Ed.), *Idempotency* (pp. 406–415). Cambridge.

149. Zimmermann, K. (1976). *Extremální algebra*. Výzkumná publikace Ekonomicko-matematické laboratoře při Ekonomickém ústavě ČSAV 46 Praha (in Czech).

150. Zimmermann, U. (1981). *Annals of discrete mathematics: Vol. 10. Linear and combinatorial optimization in ordered algebraic structures*. Amsterdam: North-Holland.

Index

P. Butkovič, *Max-linear Systems: Theory and Algorithms*,
Springer Monographs in Mathematics 151,
DOI 10.1007/978-1-84996-299-5, © Springer-Verlag London Limited 2010